Study Guide to Accompany

Biology

THIRD EDITION

SOLOMON • BERG • MARTIN • VILLEE

by

Ronald S. DANIEL
California State Polytechnic University, Pomona

Sharon Callaway DANIEL
Orange Coast College, Costa Mesa

Ronald L. TAYLOR
Orange County Health Care Agency

SAUNDERS COLLEGE PUBLISHING
Harcourt Brace College Publishers

Fort Worth • Philadelphia • Boston • New York • Chicago • Orlando
San Francisco • Atlanta • Dallas • London • Toronto • Austin • San Antonio

Ronald S. Daniel/Sharon Callaway Daniel/ Ronald L. Taylor: Study Guide to accompany BIOLOGY, 3/E, by Solomon/Berg/Martin/Villee

ISBN 0-03-072803-7

5 021 987654

PREFACE

□

This study guide provides a means to learn the content of biology, and to an extent, it also acquaints the beginning student with some of the experiential aspects of biology. It is designed for use with the third edition of the textbook *Biology* by Solomon, Berg, Martin, and Villee.

While it is of utmost importance that students learn the subject matter of biology, it is also important that they appreciate the ways that biologists think and perceive the world about them. Accordingly, we believe that students should be engaged in the processes of biology while learning its content. This study guide offers learning aids and study strategies that we have found effective in achieving these goals. The approach we have taken has also been influenced by feedback from our students about pedagogical methods they find valuable.

The variety of exercises we have included accommodates the different cognitive learning styles among students. Consequently, study guide chapters contain the following sections: *Chapter Outline and Concept Review*, *Key Terms*, *Self Test*, and *Visual Foundations*. Each section has a specific purpose, and the whole is designed to provide a well-rounded learning experience.

Chapter Outline and Concept Review: This section outlines the major headings of each textbook chapter. Students can quickly see the principal topics and concepts presented in the textbook. Major concepts from each chapter are presented in the form of fill-in questions, or "Concept Questions." Answers for the blanks within "Concept Questions" are in the *Answers to Concept Questions, Definitions, and Self Tests* section in the back of the study guide.

Key Terms: This section provides some of the prefixes and suffixes (word roots) that are used in the textbook. Word roots are used to construct terms that match the definitions provided. Each *Key Terms* section also lists important terms that appear in that particular chapter. Answers for the exercises are provided in the *Answers* section of the study guide.

Self Test: The questions in this section will help the student develop study skills, and they provide a means to re-examine, integrate, and analyze textbook material. The *Self Test* also helps the student evaluate his or her mastery of the subject matter, as well as identify material that needs additional study. The questions provide a fairly comprehensive sampling of the material presented in each chapter. Answers to these questions are also provided in the *Answers* section.

Visual Foundations: Selected visual elements from the textbook are presented in this section. The exercises are designed to involve the student directly in the learning process, specifically as pertains to visual and mechanical learning styles. Coloring illustrations reveals gaps in one's understanding of graphic material.

ACKNOWLEDGEMENTS

We thank our students who helped us learn what does and does not work. In particular, we are indebted to Stefan Fornalski and George J. Gosztola of Orange Coast College who worked through the study guide and informed us of anything that was ambiguous or unclear. Our special thanks go to our colleague Kenneth S. Unsworth for his assistance in revising the *Key Terms* section for this edition.

We appreciate the contributions of our students and colleagues who checked the clarity and accuracy of our work, and we thank all of the instructors of biology whose creative ideas and dedication inspired us to prepare this study guide.

Ronald S. Daniel, PhD
Professor of Biological Sciences
California State Polytechnic University, Pomona

Sharon Callaway Daniel, MS
Professor of Biological Sciences
Orange Coast College, Costa Mesa, California

Ronald L. Taylor, PhD
Program Manager, Orange County Health Care Agency

CONTENTS

❑

Preface   iii
Table of Contents   v
To the Student   vii

Part I: The Organization of Life   1
   1 A View of Life   3
   2 Atoms and Molecules: The Chemical Basis of Life   11
   3 The Chemistry of Life: Organic Compounds   19
   4 Cellular Organization   27
   5 Biological Membranes   35

Part II: Energy Transfer Through Living Systems   41
   6 The Energy of Life   43
   7 Energy-Releasing Pathways and Biosynthesis   49
   8 Photosynthesis: Capturing Energy   55

Part III: The Continuity of Life: Genetics   61
   9 Chromosomes, Mitosis, and Meiosis   63
   10 The Basic Principles of Heredity   69
   11 DNA: The Carrier of Genetic Information   77
   12 RNA and Protein Synthesis: Expression of Genetic Information   83
   13 Gene Regulation: The Control of Gene Expression   91
   14 Genetic Engineering   97
   15 Human Genetics   103
   16 Genes and Development   109

Part IV: Evolution   115
   17 Evolution: Mechanism and Evidence   117
   18 Population Genetics   123
   19 Speciation and Macroevolution   129
   20 The Origin and Evolutionary History of Life   135
   21 The Evolution of Primates   141

Contents

Part V: The Diversity of Life 147

22 The Classification of Organisms 149
23 Viruses and Kingdom Prokaryotae 155
24 The Protist Kingdom 165
25 Kingdom Fungi 173
26 The Plant Kingdom: Seedless Plants 181
27 The Plant Kingdom: Seed Plants 189
28 The Animal Kingdom: Animals Without a Coelom 197
29 The Animal Kingdom: The Coelomate Protostomes 205
30 The Animal Kingdom: The Deuterostomes 213

Part VI: Structures and Life Processes in Plants 221

31 Plant Structure, Growth, and Differentiation 223
32 Leaf Structure and Function 233
33 Stems and Plant Transport 239
34 Roots and Mineral Nutrition 247
35 Reproduction in Flowering Plants 253
36 Plant Hormones and Responses 259

Part VII: Structures and Life Processes in Animals 263

37 The Animal Body: Tissues, Organs, and Organ Systems 265
38 Protection, Support, and Movement: Skin, Skeleton, and Muscle 273
39 Neural Control: Neurons 283
40 Neural Regulation: Nervous Systems 291
41 Sense Organs 301
42 Internal Transport 311
43 Internal Defense: Immunity 325
44 Gas Exchange 335
45 Processing Food and Nutrition 343
46 Osmoregulation and Disposal of Metabolic Wastes 355
47 Endocrine Regulation 363
48 Reproduction 371
49 Development 381
50 Animal Behavior 389

Part VIII: Ecology 395

51 Ecology and the Geography of Life 397
52 Population Ecology 405
53 Community Ecology 411
54 Ecosystems and the Ecosphere 417
55 Humans in the Environment 425

Answers to Concept Questions, Definitions, and Self Tests 431

TO THE STUDENT

□

We prepared this Study Guide to help you learn biology. We have included the ways of studying and learning that our most successful students find helpful. If you study the textbook, use this study guide regularly, and follow the guidelines provided here, you will learn more about biology and, in the process, you are likely to earn a better grade. It is our sincere desire to provide you with enjoyable experiences during your excursion into the world of life.

Study guide chapters contain the following sections: *Chapter Outline and Concept Review*, *Key Terms*, *Self Test*, and *Visual Foundations*. The ways that you can use these sections are described below.

CHAPTER OUTLINE AND CONCEPT REVIEW

This section contains an outline and "Concept Questions" that are placed in context throughout the chapter outline. Concept Questions contain blanks for you to fill in. Once completed, they summarize most of the major concepts presented in each chapter.

> ■ 1 *Concept Questions* appear like this throughout the *Chapter Outline and Concept Review* sections.

We suggest that you pay particular attention to these concepts. Be sure that you understand them. Examine the textbook material to be sure that you have correctly answered all of the questions (blank spaces) within these concept statements. Once you have filled in the blanks, check your work in the *Answers to Concept Questions, Definitions, and Self Tests* section at the back of the study guide.

KEY TERMS

The language of biology is fundamental to an understanding of the subject. Many terms used in biology have common word roots. If you know word roots, you can frequently figure out the meaning of a new term. The *Key Terms* section provides you with a list of some of the prefixes and suffixes (word roots) that are used in that chapter.

Learn the meanings of these prefixes and suffixes, then use them to construct terms in the blank spaces next to the definitions provided. Once you have filled in the blanks for all of the definitions, check your work in the *Answers* section at the back of the study guide.

Make sure that you know the terms listed under the heading "Other Terms You Should Know." Some words have shades of meaning; that is, there may be a subtle difference in their use in different contexts. Some key terms are presented more than once in the study guide. When you learn a term, be sure that you know its meaning in the context of the material presented in that portion of the textbook.

Learning the language of biology also means that you can communicate in its language, and effective communication in any language means that you can read, write, and speak the language. Use the glossary in your textbook to learn how to pronounce new terms. Say the words out loud. Listen to them carefully. You will discover that pronouncing words aloud will not only teach you verbal communication, it will also help you remember them.

SELF TEST

The questions in this section were selected for two purposes. First, and perhaps most importantly, they help you develop study skills. They provide a means to re-examine, integrate, and analyze selected textbook material. To answer many of them, you must understand how material presented in one part of a chapter relates to material presented in other parts of the chapter. The *Self Test* also provides a means to evaluate your mastery of the subject matter, and to identify material that needs additional study.

The questions do not cover everything presented in the textbook. Although they provide a fairly comprehensive sampling of each chapter, the questions may not cover all of the material that your instructor expects you to know.

Once you have studied the textbook material and your notes, close your textbook, put away your notes, and try answering the questions in the *Self Test* section. Check your answers in the *Answers* section when you have completed a *Self Test*. Don't be discouraged if you miss some answers. Take each question that you missed one at a time, go back to the textbook and your notes, and determine what you missed and why. Do not merely copy down the correct answers — that defeats the purpose of the questions. If you dig the answers out of the textbook, you will find that your understanding of the material will improve. Some points that you thought you already knew will become clearer, and you will develop a better understanding of the relationships between different aspects of biology.

VISUAL FOUNDATIONS

Selected visual elements from the textbook are presented in this section. Coloring the parts of biological illustrations causes one to focus on the subject more carefully.

Coloring also helps to clearly delineate the parts and reveals the relationship of parts to one another and to the whole.

Begin by coloring the open boxes (❑). Colored boxes will provide you with a quick reference guide to the illustration when you come back to study it later. Close your textbook while you are doing this section. After completing the *Visual Foundations* section, use your textbook to check your work. Make notes about anything you missed for future reference. Come back to these figures periodically for review. In order to be consistent, items to color are all listed in the singular form, even when there may be more than one of them to color. For example, we may ask you to color the "mitochondrion," even though there are several mitochondria shown in the illustration.

The figures in your textbook and in this study guide have been carefully selected to illustrate principles and help you learn the material. Be sure that you understand the figures and their legends.

OTHER STUDY SUGGESTIONS

When possible, read material in the textbook *before* it is covered in class. Don't be concerned if you don't understand all of the material at first, and don't try to memorize every detail in this first reading. Reading before class will contribute greatly to your understanding of the material as it is presented in class, and it will help you take better notes. As soon as possible after the lecture, recopy all of your class notes. You should do this while the classroom presentation is still fresh in your mind. If you are prompt and disciplined about it, you will be able to recall additional details and include them in your notes. Use material in this study guide and your textbook to fill in the gaps.

What is the best way for you to study? That's something you have to decide for yourself. Reading all of the chapters assigned by your instructor is basic, but it is not all there is to studying.

A technique that many successful students have found beneficial is to organize a study group of 4-6 students. Meet periodically, especially before examinations, and go over the material together. Bring along your textbooks, study guides, and expanded notes. Have each member of the group ask every other member several questions about the material you are reviewing. Discuss the answers as a group until the principles and details are clear to everyone. Use the combined notes of all members to fill in any gaps in your notes. Consider the reasons for correct and incorrect answers to questions in the Concept Questions and *Self Test* section. This study technique can do wonders. There is probably no better way to find out how well you know the material than to verbalize it in front of a critical audience. Just ask any biology instructor!

If you follow the guidelines presented here and carefully complete the work provided in this study guide, you will make giant strides toward understanding the principles of biology. You will lay a foundation for a more concentrated study of those areas of the field that you find most interesting. If you decide to specialize in some area of biology, you will find the general background you obtained in this course to be invaluable.

PART I

The Organization of Life

CHAPTER 1

❑

A View of Life

O U T L I N E

Characteristics of living things
Transmission of information
Evolution: A unifying concept
Energy flow through living systems
The scientific method

Living things are organized entities that share a common set of characteristics that distinguishes them from nonliving things. These characteristics include being composed of one or more cells, growth and development, metabolism, homeostasis, movement, responsiveness, reproduction, and adaptation to environmental change. Organisms transmit information from one generation to the next by means of DNA. They transmit information internally by means of chemical and electrical signals, and from one organism to another by releasing chemicals into the environment, by visual displays, and/or by sound production. Perhaps the most important thing living things do is evolve. Evolution is responsible for the diversity of life on the planet. Living things are organized at a variety of levels, each more complex than the previous, from the chemical level up through the ecosystem level. To study the millions of organisms that have evolved on our planet, scientists have developed a binomial and hierarchical system of nomenclature. In this system, all organisms are assigned to one of five kingdoms -- Prokaryotae, Protista, Fungi, Plantae, or Animalia. Living things require a continuous input of energy to maintain the chemical transactions and cellular organization essential to their existence. They are categorized either as producers, consumers, or decomposers, depending on how they obtain and process energy. Living things are studied by a system of observation, hypothesis, experiment, more observation, and revised hypothesis known as the scientific method.

CHAPTER OUTLINE AND CONCEPT REVIEW (Fill in the blanks)

I. INTRODUCTION
II. LIFE CAN BE DEFINED IN TERMS OF THE CHARACTERISTICS OF LIVING THINGS

■ **1** Nine of the essential characteristics of life include _____
_____
_____.

A. Living things are composed of cells

■ **2** The _____ asserts that all living things are composed of cells.

B. Living things grow and develop

■ **3** Living things grow by increasing _____ of cells.

3

■ 4 _____ is a term that encompasses all the changes that occur during the life of an organism.

C. Metabolism includes the chemical processes essential to growth, repair, and reproduction

■ 5 Metabolism can be described as the sum of all the _____ that take place in the organism.

■ 6 _____ is the tendency of organisms to maintain a constant internal environment.

D. Movement is a basic property of cells
E. Living things respond to stimuli
F. Living things reproduce

■ 7 In general, living things reproduce in one of two ways: _____ _____.

G. Populations evolve and become adapted to the environment

■ 8 The process by which living things become adapted to their environments, or change, is called _____.

III. INFORMATION MUST BE TRANSMITTED WITHIN INDIVIDUALS, BETWEEN INDIVIDUALS, AND FROM ONE GENERATION TO THE NEXT

A. DNA transmits information from one generation to the next

■ 9 _____ are the units of hereditary material.

B. Information is transmitted by many types of molecules and by nervous systems

■ 10 Information in living systems is carried by two major categories of chemical messengers, _____ , and by electrical impulses.

IV. EVOLUTION IS THE PRIMARY UNIFYING CONCEPT OF BIOLOGY

A. Natural selection is an important mechanism by which evolution proceeds

■ 11 Charles Darwin's famous book, "_____ _____," published in 1859, had and still has a profound influence on the biological sciences, as well as impacting on the thinking of society in general.

■ 12 Darwin's theory of natural selection was based on the observations that individuals within species (a)_____ significantly, and more individuals are produced than can possibly (b)_____, and individuals with the most advantageous characteristics are the most likely to (b)_____ and pass those characteristics on to their offspring.

B. Populations evolve as a result of selective pressures from changes in the environment

■ 13 Natural selection favors organisms with traits that enable them to cope with _____. These organisms are most likely to survive and produce offspring.

■ 14 Well adapted organisms are the products of _____.

C. Biological organization reflects the course of evolution

■ 15 The basic unit of organization in an organism is the _____.

■ 16 The basic unit of ecological organization is the _____.

■ 17 All of the communities of organisms on Earth are collectively referred to as the _____.

D. Millions of kinds of organisms have evolved on our planet

■ 18 The science of classifying and naming organisms is _____.

■ 19 The scientific name of an organism consists of the _____ names for that organism.

■ 20 The broadest of the taxonomic groups is the _____.

■ 21 The five kingdoms are: (a)_____, consisting of the unicellular bacteria; (b)_____, unicellular organisms <u>with</u> distinct nuclei such as protozoa and algae; (c)_____, heterotrophic decomposers that include molds and yeasts; (d)_____, complex multicellular producers that can make food by photosynthesizing; and (e)_____, complex multicellular organisms that must obtain food from a "ready made" source.

V. LIFE DEPENDS ON A CONTINUOUS INPUT OF ENERGY

A. Energy flows through individual cells and organisms

■ 22 _____ is the process by which molecular energy is released for the activities of life.

B. Energy flows through ecosystems

■ 23 A self-sufficient ecosystem contains three major categories of organisms: _____. It must also have a continuous input of energy.

VI. BIOLOGY IS STUDIED USING THE SCIENTIFIC METHOD

A. Science is based on systematic thought processes

■ 24 The _____ is a system of observation, hypothesis, experiment, more observation, and revised hypothesis.

■ 25 _____ reasoning are two categories of systematic, scientific thought processes.

B. Predictions are tested by observation and experiment

■ 26 A _____ is an educated guess. It is based on presumptive data and, once formulated, it is tested by more observations and carefully controlled experiments.

C. A well-supported hypothesis may become a theory

■ 27 A hypothesis becomes a theory only when it is supported by a large body of _____.

KEY TERMS

Frequently Used Prefixes and Suffixes: Use combinations of these prefixes and suffixes to generate terms for the definitions below.

Prefixes	The Meaning		Suffixes	The Meaning
a-	without, not, lacking		-logy	"study of"
auto-	self, same		-stasis	equilibrium caused by opposing forces
bio-	life		-troph	nutrition, growth, "eat"
eco-	home		-zoa	animal
hetero-	different, other			
homeo-	similar, "constant"			
proto-	first, earliest form of			

Prefix	Suffix	Definitions
_____	_____	1. The study of life.
_____	_____	2. The study of groups of organisms interacting with one another and their nonliving environment.
_____	-sexual	3. Reproduction without sex.
_____	_____	4. Maintaining a constant internal environment.
_____	-system	5. A community along with its nonliving environment.
_____	-sphere	6. The planet earth including all living things.
_____	_____	7. An organism that produces its own food.
_____	_____	8. An organism that is dependent upon other organisms for food, energy, and oxygen.
_____	_____	9. Animal-like protists.

Other Terms You Should Know

adaptation
amoeboid motion
animal
atoms
bacterium (-ia)
binomial system of nomenclature
cell
cellular level
cellular respiration
chemical level
cilium (-ia)
class
community (-ties)
consumer
control group
controlled experiment
cuticle

decomposer
deductive reasoning
development
division
DNA
ecosphere
enzyme
evolve
experimental group
family (-lies)
flagellum (-la)
fungus (-gi)
gametangium (-ia)
genes
genus (-nera)
growth
homeostatic mechanism
hormones

hypothesis (-ses)
inductive leap
inductive reasoning
metabolism
molecule
mutation
natural selection
order
organ
organ system
organelle
organism
phylum (-la)
placebo
plant
plasma membrane
population
producer

proteins
Protista
scientific principle or law
sexual reproduction
species

specific epithet
sponge
stimulus (-li)
stomatum (-ta)
taxonomist

taxonomy
theory (-ries)
theory of organic evolution
tissue

SELF TEST

Multiple Choice: Some questions may have more than one correct answer.

1. The characteristic(s) common to all living things include(s)
   a. reproduction.
   b. autotrophic synthesis.
   c. heterotrophic synthesis.
   d. growth.
   e. adaptation.

2. Maintenance of internal physiological equilibrium in a changing environment by an organism is known as
   a. response.
   b. adaptation.
   c. homeostasis.
   d. metabolism.
   e. cyclosis.

3. The Prokaryotae
   a. lack a nuclear membrane.
   b. include bacteria.
   c. contain membrane-bound organelles.
   d. are simple plants.
   e. comprise one kingdom.

4. A snake is a member of the kingdom/phylum
   a. Prokaryotae/vertebrate.
   b. Fungi/ascomycete.
   c. Protista/Aves.
   d. Animalia/chordate.
   e. Plantae/angiosperm.

5. Considering the rigors of the scientific method, in which order would the following occur?
   a. hypothesis, data from experiments, law, theory
   b. data from experiments, hypothesis, law, theory
   c. hypothesis, data from experiments, law, theory
   d. hypothesis, data from experiments, theory, law
   e. hypothesis, theory, data from experiments, law

6. To comply with scientific method, a hypothesis must
   a. be true.
   b. withstand rigorous investigations.
   c. be popular and fundable.
   d. be testable.
   e. eventually reach the status of a principle.

The Federal Center for Disease Control tested 600 female prostitutes who regularly use condoms. They found that none of the women had the AIDS virus. Use this study and data to answer questions 7-9.

7. To determine the effectiveness of using condoms in preventing AIDS among women, this study was lacking in not having
   a. enough subjects.
   b. male subjects.
   c. controls.
   d. subjects who do not use condoms during sex.
   e. women who are not prostitutes.

8. In terms of determining a relationship between use of condoms by women and the kind of sexual behavior that might lead to contraction of AIDS, this experiment
   a. is useless.
   b. is well-designed.
   c. should be repeated with more subjects.
   d. could be called a theory.
   e. should include male subjects.

9. The experiment
   a. proves that condoms are effective in preventing AIDS.
   b. suggests that condoms may be effective in preventing AIDS.
   c. indicates that AIDS is not transmitted through sexual activity.
   d. probably means that none of the partners of the female prostitutes had AIDS.
   e. shows that women are less prone to contract AIDS than men.

10. Organisms that can use sunlight energy to synthesize complex food molecules are known as
    a. heterotrophs.   d. consumers.
    b. cyanobacteria.   e. producers.
    c. autotrophs.

11. The kingdom(s) containing organisms that have a true nucleus and are unicellular is/are
    a. Prokaryotae.   d. Plantae.
    b. Protista.   e. Animalia.
    c. Fungi.

12. The kingdom(s) containing organisms that are both single-celled and have a "nucleoid" area that is not separated from the cytoplasm by a membrane is/are
    a. Prokaryotae.   d. Plantae.
    b. Protista.   e. Animalia.
    c. Fungi.

13. The kingdom(s) containing organisms that have a spinal cord is/are
    a. Prokaryotae.   d. Plantae.
    b. Protista.   e. Animalia.
    c. Fungi.

VISUAL FOUNDATIONS

Questions 14-17 refer to the following illustration.

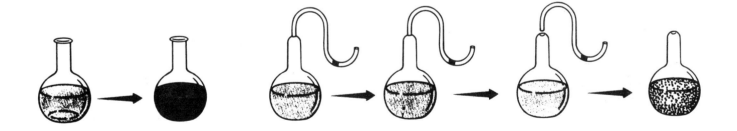

14. Why is the experiment represented in this illustration an excellent example of experimental design?
    a. It proved that spontaneous generation can occur.
    b. It is a controlled experiment.
    c. It was the first time S-shaped flasks were used.
    d. It proved that bacteria existed.
    e. None of the above are true.

15.  If Pasteur had used flasks shaped like the one illustrated here instead of the rounded flasks shown in the illustration above, would his results have been different? _____ If so, why?

    _____

16. If Pasteur had used T-shaped necks (as illustrated here) instead of S-shaped necks, would his results have been different? _____ If so, why? _____

17. How might this experiment have been affected if Pasteur had chosen to shake the broth in the flasks with S-shaped necks, potentially splashing broth into the neck of the flask?
    a. Bacteria would have grown in all the flasks with S-shaped necks.
    b. The results would have been the same.
    c. The effectiveness of the control would have been diminished.
    d. Bacteria might have grown in some of the flasks with S-shaped necks.
    e. The results might not have validated Pasteur's hypothesis.

Atoms and Molecules:
The Chemical Basis of Life

O U T L I N E

Chemical elements
Atoms
Subatomic particles
Molecules and compounds
Chemical bonds
Molecular mass
Oxidation and reduction
Inorganic compounds
Water
Acids and bases

Living things, like everything else on our planet, are made of atoms and molecules. Therefore, to understand life is first to understand the interaction and activity of atoms and molecules. Living things share remarkably similar chemical compositions. Although the matter of the universe is composed of 92 chemical elements, about 98% of an organism's mass is made up of just six elements -- oxygen, carbon, hydrogen, nitrogen, calcium, and phosphorus. The smallest portion of an element that retains its chemical properties is the atom. Atoms, in turn, consist of subatomic particles, the most important of which are protons, neutrons, and electrons. The number and arrangement of electrons determine the chemical behavior of atoms. Molecules and compounds are formed by the joining of atoms. The forces that hold atoms together are called chemical bonds. The unique properties of water are determined by its chemical bonds. Water is the molecule most important to the evolution of life and to the maintenance of life as we know it. Oxidation and reduction are chemical reactions that play critical roles in life's processes. Acids, bases, and salts also play important roles.

CHAPTER OUTLINE AND CONCEPT REVIEW (Fill in the blanks)

I. INTRODUCTION

II. MATTER IS COMPOSED OF CHEMICAL ELEMENTS

■ 1 An element is a substance that cannot be separated into simpler parts by

_____.

■ **2** Of the (a)_____(#?) elements, (b)_____ is the heaviest, and (c)_____ is the lightest. About 98% of an organism's weight is made up of the six elements (d)_____
_____.

III. ATOMS ARE THE BASIC PARTICLES OF ELEMENTS
IV. ATOMS CONSIST OF SUBATOMIC PARTICLES

■ **3** The three main parts of an atom are the (a)_____
_____. The (b)_____ are in the nucleus, and the (c)_____ occupy energy levels and orbitals around the nucleus.

■ **4** (a)_____ have one unit of positive charge, (b)_____ are neutral, and (c)_____ have a negative charge.

A. Protons and neutrons make up the atomic nucleus

■ **5** The number of protons in an atom is called the _____.

■ **6** The total number of protons and neutrons in an atom is referred to as the
_____.

B. Isotopes differ in number of neutrons

■ **7** The (a)_____ of isotopes vary, but the (b)_____ remains constant.

C. Electrons are located in energy levels outside the nucleus

■ **8** A neutral atom has (a)_____ net electrical charge because it has the same number of (b)_____.

■ **9** Electrons with the same energy comprise an _____.

V. ATOMS FORM MOLECULES AND COMPOUNDS

■ **10** A chemical compound consists of _____
_____ that are combined in a fixed ratio.

A. Chemical formulas describe chemical compounds

■ **11** A (a)_____ formula shows types, numbers and arrangement of atoms in a compound, whereas a (b)_____ formula uses chemical symbols to describe the chemical composition of a compound.

B. Chemical equations describe chemical reactions

■ **12** (a)_____ are substances that participate in the reaction, (b)_____ are substances formed by the reaction

■ **13** Arrows are used to indicate the direction of a reaction; reactants are placed to the (a)_____ of arrows, products are placed to the (b)_____ of arrows

VI. CHEMICAL BONDS HOLD ATOMS TOGETHER

■ **14** The electrons in the outermost energy level of an atom are called _____ electrons.

■ **15** A _____ forms when two or more atoms are joined by chemical bonds.

■ 16 Energy that is available to do work is potential energy, while energy that is actually doing work is referred to as kinetic energy. Chemical bonds represent _____ chemical energy.

A. Electrons are shared in covalent bonds

B. A molecule has a characteristic size and shape

■ 17 The _____ of molecules largely determines their functions.

C. Covalent bonds can be nonpolar or polar

■ 18 Covalent bonds are (a)_____ if the electrons are shared equally between the two atoms; and they are (b)_____ if one atom has a greater affinity for electrons than the other.

D. Atoms gain or lose electrons to form ionic bonds

■ 19 Ionic compounds are comprised of positively charged ions called (a)_____, and negatively charged ions called (b)_____.

E. Hydrogen bonds are weak attractions involving partially charged hydrogen atoms

■ 20 Weak hydrogen bonds form between a hydrogen atom in one molecule and a relatively _____ atom in the same or another molecule.

F. Van der Waals forces and hydrophobic attractions are other interactions between atoms

■ 21 Single Van der Waals "bonds" are very weak, but they become significant in large numbers, and when the shape of the _____ involved allow close contact between atoms.

■ 22 Hydrophobic interactions occur between groups of _____ molecules

VII. COMPOUNDS HAVE MOLECULAR MASS

■ 23 The molecular mass of a compound is the sum of the _____
_____.

VIII. OXIDATION INVOLVES THE LOSS OF ELECTRONS; REDUCTION INVOLVES THE GAIN OF ELECTRONS

■ 24 When iron rusts, it loses electrons and is said to be (a)_____, while at the same time oxygen gains electrons and is (b)_____. Reactions involving oxidation <u>and</u> reduction are called (c)_____ reactions.

IX. INORGANIC COMPOUNDS ARE RELATIVELY SIMPLE COMPOUNDS WITHOUT BACKBONES

■ 25 Organic compounds are relatively complex and they always contain the element _____.

X. WATER IS ESSENTIAL TO LIFE

A. Water molecules are polar

■ 26 Each water molecule can form up to four _____ bonds with adjacent water molecules.

B. Water is an excellent solvent

C. Water exhibits both cohesive and adhesive forces

■ 27 Cohesion of water molecules is due to _____ between the molecules.

D. Water helps maintain a stable temperature

■ 28 Water has a high _____, which helps organisms maintain a relatively constant internal temperature.

E. Density of water is maximum at 4° C

■ 29 Ice is less dense than liquid water at 4° C due to the fact that it is _____ that join water molecules. This is why ice floats.

F. Water molecules dissociate slightly

■ 30 Water has a _____ (slight or strong?) tendency to form ions.

XI. ACIDS YIELD HYDROGEN IONS; BASES ARE PROTON ACCEPTORS

■ 31 A base is a proton acceptor that generally dissociates in solution to yield (a)_____ _____. An acid is a proton donor that dissociates in solution to yield (b)_____.

■ 32 The pH scale covers the range (a)_____. Neutrality is at pH (b)_____. A solution with a pH above 7 is a(an) (c)_____. A solution with a pH below 7 is a(an) (d)_____.

A. Salts form from acids and bases

■ 33 When the (a)_____ of an acid is replaced by some other cation, a salt forms. Salts provide (b)_____ ions that are essential for fluid balance, nerve and muscle function, and many other body functions.

B. Buffers minimize pH change

■ 34 (a)_____ are comprised of a weak acid and a salt of that acid, or a weak base and a salt of that base. They resist changes in the pH of a solution when (b)_____ are added.

KEY TERMS

Frequently Used Prefixes and Suffixes: Use combinations of these prefixes and suffixes to generate terms for the definitions below.

Prefixes	The Meaning	Suffixes	The Meaning
equi-	equal	-hedron	face
hydr(o)-	water	-phobic	fearing
neutro-	neutral		
non-	not		
radio-	beam, ray		
tetra-	four		

Prefix	Suffix	Definitions
_____	-n	1. An uncharged subatomic particle.
_____	-nuclide	2. An isotope that emits high-energy radiation when it decays.
_____	-librium	3. The condition of a chemical reaction when the forward and reverse reaction rates are equal.
_____ _____		4. The shape of a molecule when it appears as a three-dimensional pyramid.
_____	-ation	5. The process whereby ionic compounds combine chemically with water and dissolve.
_____ _____		6. Not attracted to water; insoluble in water.
_____	-electrolyte	7. A substance that does not form ions when dissolved in water, and therefore does not conduct an electric current.

Other Terms You Should Know

acid
adhesive surface tension
anion
atom
atomic mass unit (AMU)
atomic mass
atomic nucleus (-ei)
atomic number
Avogadro's number
base
buffer
calorie
capillary action
cation
chemical bond
chemical compound
chemical equation
chemical formula (-ae)
chemical symbol
cohesive
covalent bond
covalent compound
dalton
dissociate

double bond
electrolyte
electron configuration
electron
electron shell
electronegativity
element
energy level
heat of vaporization
hybridized
hydrogen bond
hydrophobic interaction
inorganic compound
ion
ionic bond
ionic compound
ionize
isotope
mole
molecular biology
molecular mass
molecule
neutral
nonpolar covalent bond

orbital
organic compound
oxidation
pH
polar covalent bond
product
proton
quantum jump
quantum (-ta)
radioisotope
reactant
redox reaction
reduction
salt
single bond
solvent
specific heat
structural formula (-ae)
trace element
triple bond
valence electron
van der Waals forces

SELF TEST

Multiple Choice: Some questions may have more than one correct answer.

Use the following chemical equation to answer questions 1-6.

$$2 H_2 + O_2 \longrightarrow 2 H_2O \text{ (water)}$$

1. How many oxygen atoms are involved in this reaction?
   a. one   d. four
   b. two   e. none
   c. three

2. The lightest reactant has a mass number of
   a. one.   d. eight.
   b. two.   e. ten.
   c. larger than the product.

3. The reactant(s) in the formula is/are
   a. hydrogen.   d. water.
   b. oxygen.   e. the arrow.
   c. H_2O.

4. The product has a molecular mass of
   a. one.   d. sixteen.
   b. twelve.   e. eighteen.
   c. larger than either reactant alone.

5. The product(s) is/are
   a. hydrogen.   d. water.
   b. oxygen.   e. the arrow.
   c. H_2O.

6. The direction in which this reaction tends to occur is
   a. right to left.   d. left to right.
   b. from products to reactants.   e. from reactants to products.
   c. from hydrogen and oxygen to water.

7. Electrolytes result from ionization of
   a. salts.   d. bases.
   b. acids.   e. alcohols.
   c. hydrogen.

8. Adhesion might be described as
   a. attraction of atoms of different substances.   d. a form of chemical bonding.
   b. attraction of atoms within one substance.   e. capillary action.
   c. arrested cohesion.

9. A solution is made acidic by
   a. hydroxyl ions.   d. protons.
   b. water.   e. electrons.
   c. hydrogen ions.

10. A quantum is the amount of energy required to
    a. create an isotope.
    b. form a covalent bond.
    c. form a hydrogen bond.
    d. move an electron to a higher energy level.
    e. add a neutron to a stable atom.

11. The symbol $_6C$ stands for
    a. calcium with six electrons.
    b. carbon with six protons.
    c. carbon with three protons and three neutrons.
    d. calcium with three protons and three electrons.
    e. an isotope of carbon.

12. The outer electron energy level of an atom
    a. may contain two or more orbitals.
    b. determines the element.
    c. contains the same number of electrons as protons.
    d. often contains at least one neutron.
    e. determines chemical properties.

13. If the number of neutrons in an atom is changed, it will also change
    a. atomic mass.
    b. atomic number.
    c. the element.
    d. the number of electrons.
    e. the elements chemical properties.

14. H–O–H is/are
    a. a structural formula.
    b. water.
    c. a chemical formula.
    d. a molecule.
    e. a structure with the molecular weight 18.

15. When two atoms combine by sharing one pair of their electrons, the bond is a
    a. divalent, single.
    b. covalent, single.
    c. divalent, double.
    d. covalent, double.
    e. covalent, polar.

16. Oxidation has occurred when
    a. an atom gains an electron.
    b. a molecule loses an electron.
    c. hydrogen bonds break.
    d. an atom loses an electron.
    e. an ion gains an electron.

17. A substance containing equal amounts of hydrogen and hydroxyl ions would be
    a. acidic, pH below 7.
    b. alkaline, pH below 7.
    c. acidic, pH above 7.
    d. alkaline, pH above 7.
    e. neutral, pH 7.

18. Cations are
    a. negative atoms.
    b. positive atoms.
    c. isotopes.
    d. incomplete products.
    e. negative molecules.

19. Which of the following is/are among the most common elements found in organisms?
    a. carbon
    b. oxygen
    c. hydrogen
    d. nitrogen
    e. calcium

20. A form of an element that is progressing toward a more stable state by emitting radiation is called a(n)
    a. Dalton atom
    b. radioisotope
    c. enerton
    d. quantum unitron
    e. radionuclide

VISUAL FOUNDATIONS

Color the parts of the illustration below as indicated.

RED   ☐   electron

YELLOW   ☐   neutron

BLUE   ☐   proton

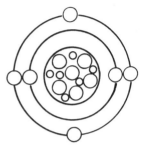 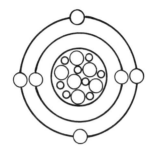

CHAPTER 3

□

The Chemistry of Life: Organic Compounds

OUTLINE

The carbon atom
Isomers
Functional groups
Polymers
Carbohydrates
Lipids
Proteins
Nucleic acids

Most of the chemical compounds present in organisms are organic compounds. Carbon is the central component of organic compounds, probably because it forms bonds with a greater number of different elements than any other type of atom. The major groups of organic compounds are the carbohydrates, lipids, proteins, and nucleic acids. Each group is characterized by the presence of one or more specific groups of atoms that convey specific properties to the molecules. Many organic molecules are polymers, that is, chains of similar units joined in a row. Carbohydrates serve as energy sources for cells, and as structural components of fungi, plants, and some animals. Carbohydrates are categorized by their degree of complexity into monosaccharides, disaccharides, and polysaccharides. Lipids function as biological fuels, as structural components of cell membranes, and as hormones. Among the biologically important lipids are the neutral fats, phospholipids, steroids, carotenoids, and waxes. Proteins serve as structural components of cells and tissues and as the facilitators of thousands of different chemical reactions. Proteins are made up of amino acid subunits and exhibit four levels of structural organization, each more complex than the previous. The structure of a protein helps determine its biological activity. DNA and RNA are nucleic acids that transmit hereditary information and determine what proteins a cell manufactures. They are made up of subunits called nucleotides. Some single and double nucleotides, such as ATP, cyclic AMP, and NAD+, are central to cell function.

CHAPTER OUTLINE AND CONCEPT REVIEW (Fill in the blanks)

I. INTRODUCTION

■ 1 _____
        are the four major "classes" of macromolecules in living systems.

19

II. THE CARBON ATOM

■ **2** Carbon atoms can form _____ (#?) bonds with a large variety of other atoms.

■ **3** Carbon-containing molecules have a three-dimensional structure due to the _____ nature of their bond angles.

III. ISOMERS HAVE THE SAME MOLECULAR FORMULA

■ **4** Isomers have the same molecular formula but different _____.

IV. FUNCTIONAL GROUPS FORM BONDS WITH OTHER MOLECULES

■ **5** Each class of organic compound is characterized by specific _____ groups with distinctive properties.

■ **6** Polarity in a functional group is caused by _____ at opposite ends of a bond.

V. MANY BIOLOGICAL MOLECULES ARE POLYMERS

■ **7** Macromolecules produced by linking monomers together are called _____.

VI. CARBOHYDRATES INCLUDE SUGARS, STARCHES, AND CELLULOSE

■ **8** Carbohydrates contain the elements _____ _____ in a ratio of about 1:2:1.

A. Monosaccharides are simple sugars

■ **9** The simple sugar _____ is an important and nearly ubiquitous fuel molecule in living cells.

B. Disaccharides consist of two monosaccharide units

■ **10** Disaccharides form when two monosaccharides are bonded by a _____ linkage.

C. Polysaccharides are large polymers

■ **11** Long chains of _____ link together to form a polysaccharide.

D. Some modified and complex carbohydrates are important biological molecules

■ **12** A carbohydrate combined with a protein forms (a)_____; carbohydrates combined with lipids are called (b)_____.

VII. LIPIDS ARE FATS OR FATLIKE SUBSTANCES

■ **13** Lipids are composed of the elements _____.

A. Neutral fats are composed of glycerol and fatty acids

■ **14** A fat consists of one (a)_____ molecule combined with one to three (b)_____ molecules.

B. Phospholipids are components of cell membranes

■ **15** A phospholipid molecule assumes a distinctive configuration in water because of its _____ property.

C. Carotenoids are plant pigments

D. Steroids contain four rings of carbon atoms

VIII. PROTEINS ARE MACROMOLECULES FORMED FROM AMINO ACIDS

A. Amino acids contain a carboxyl and an amino group

■ **16** Proteins are composed of the elements _____ _____ .

■ **17** All amino acids contain (a)_____ groups but vary in the (b)_____ .

B. Polypeptide chains are built from amino acids

■ **18** Proteins are large, complex molecules made of (a)_____ molecules joined by (b)_____ bonds.

C. Proteins have four levels of organization

■ **19** The levels of organization distinguishable in protein molecules are the primary, and the _____ .

D. Protein structure determines function

■ **20** A mutation that results in a change in (a)_____ sequence or a change in (b)_____ structure can disrupt the biological activity of a protein.

IX. DNA AND RNA ARE NUCLEIC ACIDS

A. Nucleic acids consist of nucleotide subunits

■ **21** Each nucleotide subunit in a nucleic acid is composed of a (a)_____ _____ nitrogenous base, a five-carbon (b)_____ , and an inorganic (c)_____ group.

B. Some single and double nucleotides are central to cell function

■ **22** The energy for life functions is supplied mainly by the nucleotide _____ .

KEY TERMS

Frequently Used Prefixes and Suffixes: Use combinations of these prefixes and suffixes to generate terms for the definitions below.

Prefixes	The Meaning	Suffixes	The Meaning
amphi-	two, both, both sides	-lysis	breaking down, decomposition
amylo-	starch	-mer	part
di-	two, twice, double	-philic	loving, friendly, lover
hex-	six	-plast	formed, molded, "body"
hydr(o)-	water	-saccharide	sugar
iso-	equal, "same"		
macro-	large, long, great, excessive		
mono-	alone, single, one		
poly-	much, many		
tri-	three		

Prefix	Suffix	Definitions
_____	_____	1. A compound that has the same molecular formula as another compound but a different structure and different properties.
_____	-molecule	2. A large molecule consisting of thousands of atoms.
_____	_____	3. A single organic compound that links with similar compounds in the formation of a polymer.
_____	_____	4. The degradation of a compound by the addition of water.
_____	_____	5. A simple sugar.
_____	-ose	6. A sugar that consists of six carbons.
_____	_____	7. Two monosaccharides covalently bonded to one another.
_____	_____	8. A macromolecule consisting of repeating units of simple sugars.
_____	_____	9. A starch-forming granule.
_____	-glyceride	10. A compound formed of one fatty acid and one glycerol molecule.
_____	-glyceride	11. A compound formed of two fatty acids and one glycerol molecule.
_____	-glyceride	12. A compound formed of three fatty acids and one glycerol molecule.
_____	_____	13. Having a strong affinity for water; water soluble.
_____	-pathic	14. Having one hydrophilic end and one hydrophobic end.
_____	-peptide	15. A compound consisting of two amino acids.
_____	-peptide	16. A compound consisting of a long chain of amino acids.

Other Terms You Should Know

alpha carbon	essential amino acid	NAD+
alpha helix (-ices)	ester bond	neutral fat
amino acid	fatty acid	nucleoside
ATP	fibrous protein	nucleotide
beta pleated sheet	functional group	organic compound
carbohydrate	geometric isomer	peptide bond
carotenoid	globular protein	phosphodiester linkage
cellulose	glucose	phospholipid
chitin	glycerol	plastid
condensation	glycine	polymer
conformation	glycogen	polyunsaturated fatty acid
cyclase	glycolipid	primary structure
dehydration synthesis (-ses)	glycoprotein	protein
denaturation	glycosidic linkage	purine
diacylglycerol	hydrocarbon	pyrimidine
DNA	hydrophobic	quaternary structure
domain	isoprene unit	RNA
enantiomer	lipid	saturated fatty acid
enzyme	monoacylglycerol	secondary structure

sickle cell anemia
side chain
starch
steroid

stick formula
structural isomer
sugar
tertiary structure

triacylglycerol
unsaturated fatty acid

SELF TEST

Multiple Choice: Some questions may have more than one correct answer.

1. Hydrolysis results in
   a. smaller molecules.
   b. larger molecules.
   c. reduced ambient water.
   d. increase in polymers.
   e. increase in monomers.

2. Glycogen is
   a. a molecule in animals.
   b. a molecule in plants.
   c. found only in the liver and muscle of mammals.
   d. composed of simple carbohydrate subunits.
   e. more soluble than starch.

3. Proteins contain
   a. glycerol.
   b. polypeptides.
   c. peptide bonds.
   d. amino acids.
   e. nucleotides.

4. Nucleotides
   a. are polypeptide polymers.
   b. are polymers of simple carbohydrates.
   c. contain a base, phosphate, and sugar.
   d. comprise complex proteins.
   e. store genetic information.

5. A monomer is to a polymer as a
   a. chain is to a link.
   b. link is to a chain.
   c. monosaccharide is to a polysaccharide.
   d. necklace is to a bead.
   e. bead is a necklace.

6. Polymers are
   a. made from monomers.
   b. formed by dehydration synthesis.
   c. found in monomers.
   d. macromolecules.
   e. smaller than monomers.

7. $C_6H_{12}O_6$ is the formula for
   a. glucose.
   b. a type of monosaccharide.
   c. cellulose.
   d. fructose.
   e. a type of disaccharide.

8. Phospholipids
   a. are found in cell membranes.
   b. form a cubelike structure in water.
   c. contain glycerol, fatty acids, and phosphorus.
   d. are polar.
   e. are polypeptide polymers.

9. Compared to unsaturated fatty acids, saturated fatty acids
   a. contain more hydrogens.
   b. contain more double bonds.
   c. are more like oils.
   d. are more prone to be solids.
   e. contain more carbohydrates.

10. The main structural components of cells and tissues are
    a. carbon-based molecules.
    b. carbohydrates, lipids, and DNA.
    c. basically strings of covalently bonded carbons.
    d. organic compounds.
    e. DNA, RNA, and ATP.

11. An isomer can be described as one of a group of molecules with the same atomic composition but different
    a. elements.
    b. compounds.
    c. arrangement of atoms.
    d. arrangement of covalent bonds.
    e. placement of groups.

12. Proteins
    a. are a major source of energy.
    b. contain carbon, hydrogen, and oxygen.
    c. can combine with carbohydrates.
    d. are found in both plants and animals.
    e. are polypeptide polymers.

13. The complex carbohydrate storage molecule in plants is
    a. glycogen.
    b. starch.
    c. usually in an extracellular position.
    d. composed of simple carbohydrate subunits.
    e. more soluble than the animal storage molecule.

VISUAL FOUNDATIONS

Color the parts of the illustration below as indicated.

RED ☐ hydrophobic end
BLUE ☐ hydrophilic end

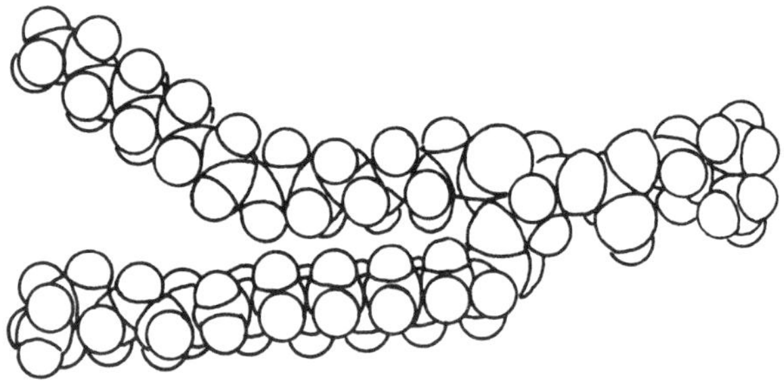

Questions 14-16 refer to the illustration above.

14. The illustration represents
    a. a polypeptide.
    b. cellulose.
    c. glycogen.
    d. lecithin.
    e. starch.

15. The hydrophobic end of this molecule is composed of
    a. amino acids.
    b. fatty acids.
    c. glucose.
    d. phosphate groups.
    e. steroids.

16. This kind of molecule contributes to the formation of
    a. the lipid bilayer.
    b. cell membranes.
    c. RNA.
    d. cellulose.
    e. complex proteins.

CHAPTER 4

□

Cellular Organization

O U T L I N E

The cell theory
Characteristics of cells
Limits on cell size
How cells are studied
Eukaryotic and prokaryotic cells
Parts of a eukaryotic cell

The cell is the basic unit of living organisms. Because all cells come from preexisting cells, they have similar needs and therefore share many fundamental features. Most cells are microscopically small because of limitations on the movement of materials across the plasma membrane. Cells are studied by a combination of methods, including microscopy and cell fractionation. Unlike prokaryotic cells, eukaryotic cells are complex, containing a variety of membranous organelles. Although each organelle has its own particular structure and function, all of the organelles of a cell work together in an integrated fashion. Most cells are surrounded by an extracellular matrix that is secreted by the cells themselves.

CHAPTER OUTLINE AND CONCEPT REVIEW (Fill in the blanks)

I. INTRODUCTION

■ 1 The cell is the smallest unit capable of carrying on _____.

II. THE CELL IS THE SMALLEST UNIT OF LIFE

■ 2 All (a)_____ are made up of cells, and all cells arise from
(b)_____.

III. BECAUSE OF THEIR COMMON ORIGINS AND COMMON NEEDS, CELLS SHARE MANY FUNDAMENTAL FEATURES

■ 3 *All* cells are enclosed by a _____.

IV. LIMITATIONS ON TRANSPORT ACROSS THE PLASMA MEMBRANE IMPOSE LIMITS ON CELL SIZE

■ 4 In general, the smaller the cell, the larger the (a)_____,
and the faster the (b)_____.

V. CELLS ARE STUDIED BY A COMBINATION OF METHODS

■ 5 Cells are studied by using important techniques and tools; for example, powerful
(a)_____ microscopes are used to resolve ultrastructure, and cell
components are separated by a method known as (b)_____.

VI. EUKARYOTIC CELLS ARE COMPLEX AND CONTAIN MEMBRANOUS ORGANELLES; PROKARYOTIC CELLS ARE SIMPLER

■ 6 _____ cells are relatively lacking in complexity and their genetic material is not enclosed by membranes.

■ 7 _____ cells are relatively complex and possess both membrane-bound organelles and a "true" nucleus.

VII. THE MANY PARTS OF A EUKARYOTIC CELL INTERACT IN AN INTEGRATED FASHION

■ 8 Unlike animal cells, plant cells have rigid (a)_____ of cellulose, storage organelles called (b)_____, and exceptionally large (c)_____ that transport and store materials in the cytoplasm.

A. The cell nucleus is the primary command center of the cell

■ 9 The genetic material in the eukaryotic nucleus consists of hereditary units called _____ that contain codes for producing proteins.

■ 10 Chromosomes are composed of a complex of DNA and protein called _____.

B. The internal membrane system is made up of membranous organelles that interact by means of vesicles

■ 11 The _____ is a complex of intracytoplasmic membranes and cisternae that compartmentalize the cytoplasm.

■ 12 RER is studded with _____ that are involved in protein synthesis.

■ 13 Some proteins constructed on RER are transported by _____ for secretion to the outside or insertion in other membranes.

■ 14 The _____ processes, sorts, and modifies proteins. It consists of stacks of platelike membranes and vesicles, some of which are filled with cellular products.

■ 15 _____ are small sacs containing digestive enzymes that can break down (lyse) complex molecules, foreign substances, and "dead" organelles.

C. Mitochondria and chloroplasts are energy-converting organelles

■ 16 The chemical reactions that convert food energy to ATP (cellular respiration) take place in organelles called _____.

■ 17 _____ are organelles that contain green pigments that trap light energy for photosynthesis.

■ 18 _____ membranes contain chlorophyll that traps sunlight energy and converts it to chemical energy in ATP.

■ 19 The matrix within chloroplasts, where carbohydrates are synthesized, is called the (a)_____, while the ATP that supplies energy to drive the process is synthesized on membranes called (b)_____

D. The cytoskeleton is a dynamic network of protein fibers

■ 20 The cytoskeleton is a network of protein filaments that are responsible for both _____ of cells.

■ 21 _____ are both made up of globular protein subunits that constantly polymerize and depolymerize.

■ 22 _____ appear to be the organizing centers for microtubule formation in animal cells.

■ 23 Cell motility is accomplished by two types of movable, whiplike structures that extend from the cell surface called _____.

■ 24 The rapid association and disassociation of (a) _____ microfilaments and (b)_____ filaments results in a sliding motion that generates force and movement.

E. Most cells are surrounded by an extracellular matrix

■ 25 Plant cell walls are composed primarily of _____ and smaller quantities of other polysaccharides.

■ 26 In some animal cells glycoproteins and glycolipids form a _____ coating.

KEY TERMS

Frequently Used Prefixes and Suffixes: Use combinations of these prefixes and suffixes to generate terms for the definitions below.

Prefixes	The Meaning		Suffixes	The Meaning
chloro-	green		-plast(id)	formed, molded, "body"
chromo-	color		-some	body
cyto-	cell			
leuco-	white (without color)			
lyso-	loosening, decomposition			
micro-	small			
myo-	muscle			
pro-	"before"			

Prefix	Suffix	Definitions
_____	-phyll	1. A green pigment that traps light for photosynthesis.
_____	_____	2. Organelles containing pigments that give fruits and flowers their characteristic colors.
_____	-plasm	3. Cell contents exclusive of the nucleus.
_____	-skeleton	4. A complex network of protein filaments within the cell.
glyoxy-	_____	5. A microbody containing enzymes used to convert stored fats in plant seeds to sugars.

_____ _____   6. An organelle that is not pigmented and is found primarily in roots and tubers, where it is used to store starch.

_____ _____   7. An organelle containing digestive enzymes.

_____ -filaments   8. Small, solid filaments, 7 nm in diameter, that make up part of the cytoskeleton of eukaryotic cells.

_____ -tubules   9. Small, hollow filaments, 25 nm in diameter, that make up part of the cytoskeleton of eukaryotic cells.

_____ -villi   10. Small, finger-like projections from cell surfaces that increase surface area.

_____ -sin   11. A muscle protein which, together with actin, is responsible for muscle contraction.

peroxi- _____   12. An organelle containing enzymes that split hydrogen peroxide, rendering it harmless.

_____ _____   13. Precursor organelles.

_____ -karyotes   14. Single-celled organisms, including the bacteria and cyanobacteria; organisms that evolved before organisms with nuclei.

_____ -sol   15. Cell contents exclusive of the nucleus and organelles; the fluid component of the cytoplasm.

Other Terms You Should Know

9 + 2 arrangement
9 x 3 structure
actin
actin filament
antibody (-dies)
basal body
carotenoid
cell center
cell fractionation
cell theory (-ries)
cell wall
cellular respiration
cellulose
centrifuge
centriole
chlorophyll a and b
chloroplast
chromatin
chromosome
cilium (-ia)
confocal fluorescence
   microscope
crista (-ae)
DNA
dynein
electron microscope
endomembrane system

endoplasmic reticulum
enzyme
eukaryote
flagellum (-la)
fluorescence microscope
gene
glycocalyx (-yces)
glycoprotein
Golgi complex
granum (-na)
intermediate filament
intermembrane space
internal membrane system
kinesin
lamin
light microscope
lumen
magnification
matrix (-ices)
messenger RNA
microtubule-organizing
   center
mitochondrion (-ria)
Nomarski differential
   interference microscope
nuclear pores
nucleoid

nucleolus (-li)
nucleoplasm
nucleus
organelle
outer membrane
phase contrast microscope
plasma membrane
plastid
primary cell wall
protoplasm
replication
resolving power
ribosome
RNA
rough ER
scanning electron microscope
secondary cell wall
smooth ER
spindle
stress fiber
stroma
thylakoid
thylakoid space
transcription
translation
transmission electron
   microscope

transport vesicle   ultrastructure   vesicle
tubulin   vacuole

SELF TEST

Multiple Choice: Some questions may have more than one correct answer.

1. The membranes that partition the cytoplasm of eukaryotic cells (endomembrane system) include
   a. Golgi complex.   d. transport vesicles.
   b. lysosomes.   e. plasma membrane.
   c. endoplasmic reticulum.

2. The Golgi complex functions to
   a. modify proteins.   d. produce polysaccharides.
   b. process proteins.   e. add carbohydrates to proteins.
   c. form glycoproteins.

3. The "cytoskeleton" of eukaryotic cells
   a. changes constantly.   d. extends into the nucleus.
   b. includes microfilaments.   e. includes protein.
   c. includes some DNA.

4. Which of the following is in the nucleolus, but not normally found in the rest of chromatin?
   a. DNA   d. RNA
   b. protein   e. ribosomes
   c. chromosomes

5. The part(s) of a mitochondrion that are rich in enzymes is/are the
   a. cristae.   d. intermembrane space.
   b. outer membrane.   e. inner membrane.
   c. matrix.

6. Actin, myosin, and tubulin are
   a. proteins.   d. components of filaments.
   b. in chromatin.   e. components of the plasma membrane.
   c. constructed on ribosomes.

7. Lysosomes
   a. contain digestive enzymes.   d. break down complex molecules.
   b. possess a membrane.   e. break down organelles.
   c. contain nucleic acids.

8. Cell membrane functions include
   a. energy transduction.   d. sorting genetic material.
   b. selective permeability.   e. concentration of reactants.
   c. isolation of different chemical reactions.

9. Cells are small at least in part because as size increases the surface-to-volume ratio
   a. doubles.   d. decreases.
   b. decreases to half.   e. reduces efficiency of cell activities.
   c. increases.

10. The cell theory states that
    a. new cells come from preexisting cells.   d. living things are composed of cells.
    b. all cells are descended from ancient cells.   e. cells contain genetic material.
    c. cells divide.

11. Chloroplasts and mitochondria both
    a. are found in plant cells.   d. are found in animal cells.
    b. have two membranes.   e. contain a matrix.
    c. contain DNA.

12. The high resolution attained by the electron microscope is attributed to its use of
    a. fixed specimens.   d. grid patterns.
    b. electromagnetic lenses.   e. electrons instead of light.
    c. short wavelengths.

13. Which of the following cells contain plastids?
    a. animal   d. some prokaryotic
    b. plant   e. all cells
    c. some eukaryotic

VISUAL FOUNDATIONS

Color the parts of the illustration below as indicated.

RED   ❑ plasma membrane
GREEN ❑ nucleoid region
YELLOW ❑ cell wall

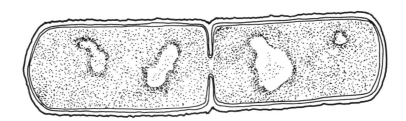

Questions 14 and 15 pertain to the illustration above.

14. Is this a prokaryotic cell or a eukaryotic cell? _____

15. Which of the colored parts is/are unique to this kind of cell? _____

Color the parts of the illustration below as indicated.

RED	❑	plasma membrane
GREEN	❑	nucleus
YELLOW	❑	cell wall
BLUE	❑	prominent vacuole
ORANGE	❑	chloroplast
BROWN	❑	mitochondrion
TAN	❑	internal membrane system

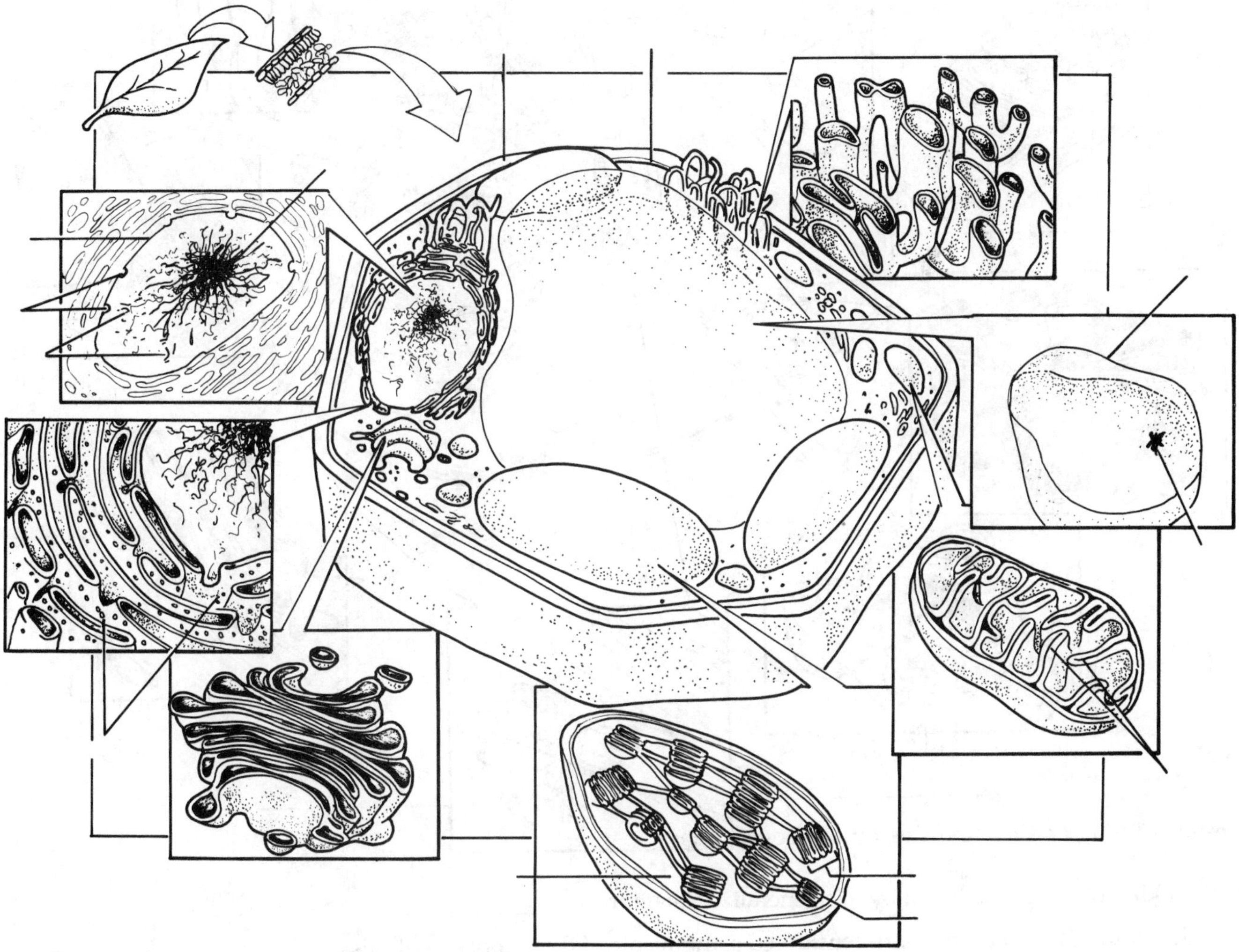

Questions 16–18 pertain to the illustration above.

16. What kind of cell is illustrated by this generalized diagram? _____

17. Which of the colored parts is/are considered components of the cytoplasm of this cell?

_____

18. Which of the colored parts is/are characteristic of this kind of cell and not generally associated with other types of cells? _____

Color the parts of the illustration below as indicated.

RED   ❑   plasma membrane

GREEN   ❑   nucleus

YELLOW   ❑   centriole

BLUE   ❑   lysosome

ORANGE   ❑   mitochondrion

BROWN   ❑   internal membrane system

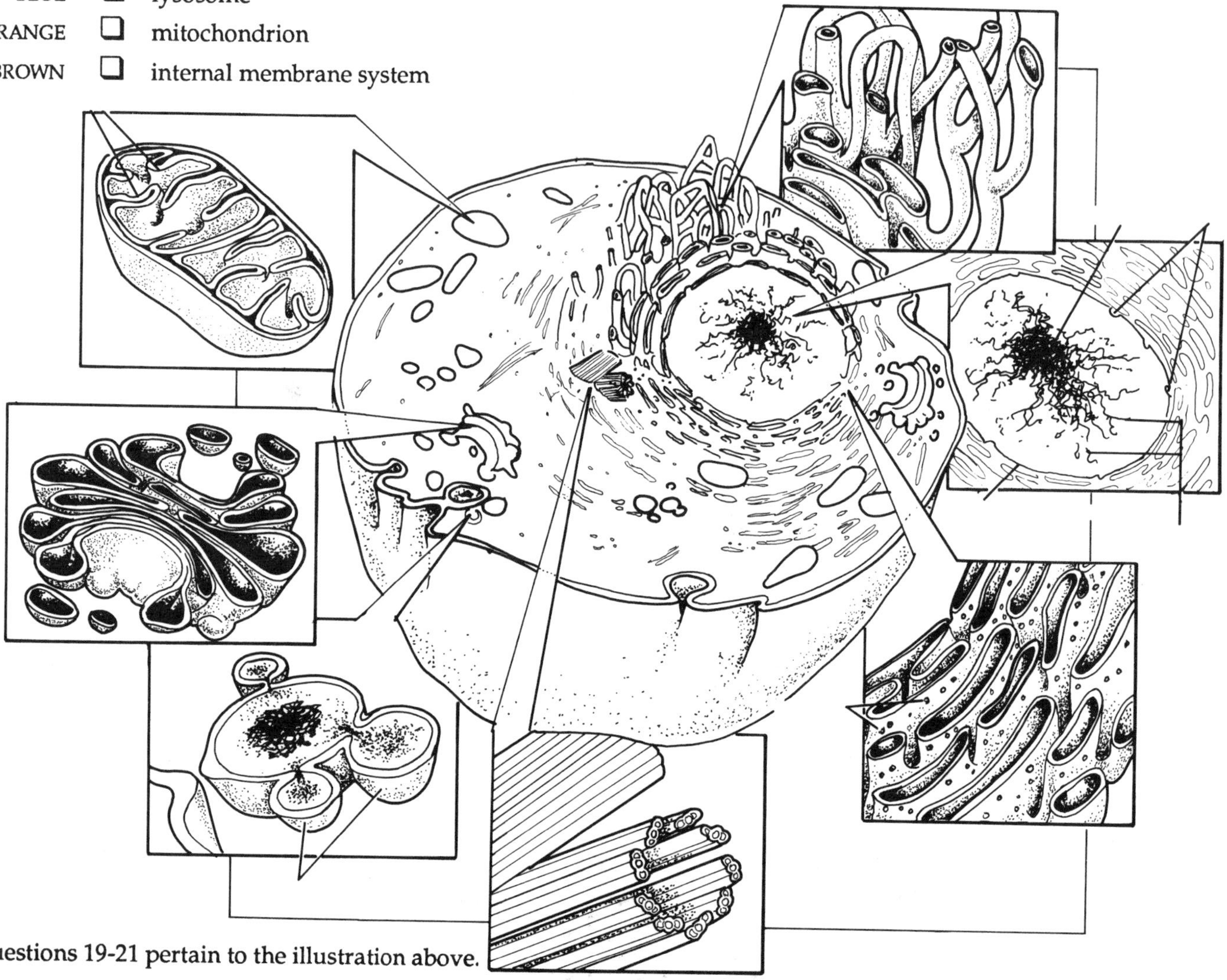

Questions 19-21 pertain to the illustration above.

19. What kind of cell is illustrated by this generalized diagram? _____

20. Which of the colored parts is/are considered components of the cytoplasm of this cell?

_____

21. Which of the colored parts is/are characteristic of this kind of cell and not generally associated with other types of cells? _____

CHAPTER 5

□

Biological Membranes

OUTLINE

Structure of biological membranes
Transport across membranes
Specialized contacts between cells

Biological membranes enclose cells, separating the interior world of the cell from the exterior. In eukaryotes, they also enclose a variety of internal structures and form extensive internal membrane systems. Membranes function to provide work surfaces for many chemical reactions, to regulate movement of materials in and out of the cell, and to transmit signals and information between the environment and the interior of the cell. Membranes are phospholipid bilayers with both internal and peripheral proteins associated with them. Membranes have the ability to seal themselves, round up and form closed vesicles, and fuse with other membranes. Molecules, depending to a large extent on their size and electrical charge, pass through membranes in a variety of ways. In multicellular organisms, cell membranes have specialized structures associated with them that allow neighboring cells to form strong connections with each other or to establish rapid communications between adjacent cells.

CHAPTER OUTLINE AND CONCEPT REVIEW (Fill in the blanks)

I. INTRODUCTION

■ 1 All cells are physically separated from the external environment by a _____.

II. BIOLOGICAL MEMBRANES ARE LIPID BILAYERS WITH ASSOCIATED PROTEINS

■ 2 The theory of membrane structure known as the (a)_____ model holds that membranes consist of a dynamic, fluid (b)_____ and embedded globular (c)_____.

■ 3 The physical properties of biological membranes are due primarily to the characteristics of their _____.

A. Phospholipids associate as bilayers in water because each molecule is roughly cylindrical and has hydrophobic and hydrophilic regions

■ 4 The cylindrical shape and strongly _____ character of phospholipid molecules are features responsible for the formation of the bilayer.

■ 5 The headgroups of phospholipid molecules are said to be (a)_____ because they readily associate with water. On the other hand, the (b)_____ ends turn away from water and associate with each other.

B. Biological membranes behave like two-dimensional fluids

■ 6 The crystal-like properties of many phospholipid bilayers is due to the orderly arrangement of (a)_____ on the outside and (b)_____ on the inside.

■ 7 The two-dimensional fluid property of the lipid bilayer is due to the constant motion of _____.

C. Biological membranes tend to form closed vesicles and to fuse with other membranes
D. Integral proteins are partially or fully embedded in the lipid bilayer; peripheral proteins are bound to the membrane surface

■ 8 _____ have hydrophobic stretches of amino acids that contact fatty acid chains of the lipid bilayer.

■ 9 _____ have hydrophobic amino acids that are buried inside the molecule away from water.

E. Membranes are asymmetric; each side of the bilayer has its own set of proteins

■ 10 The differences in characteristics of the outer and inner surfaces of membranes is the result of asymmetrical distribution of constituent _____.

F. Functions of membrane proteins include reception of signals, binding of specific molecules, transport across the membrane, and catalyzing certain reactions

■ 11 A variety of membrane functions is made possible by the diversity of _____ molecules in membranes.

III. LIPID BILAYERS ALLOW ONLY CERTAIN MOLECULES TO CROSS FREELY; ALL OTHERS REQUIRE SPECIAL TRANSPORT SYSTEMS

■ 12 Biological membranes are said to be _____ because they allow some substances to pass through while "blocking" others.

A. Diffusion, through random motion, results in a net movement of particles from a region of high concentration to a region of low concentration

■ 13 Diffusion rate is affected by temperature and the _____ _____ of the moving molecules.

■ 14 (a)_____ is diffusion of a solute through a selectively permeable membrane. Diffusion of a solvent through a selectively permeable membrane is called (b)_____.

■ 15 The _____ of a solution is determined by the amount of dissolved substances in the solution.

B. Carrier-mediated transport requires special integral membrane proteins to move small molecules and ions

■ 16 The two forms of carrier-mediated transport are _____ _____.

■ 17 _____ is the process by which cells expend energy in order to move ions and molecules against a concentration gradient.

C. Exocytosis and endocytosis are ways of transporting large particles by means of vesicles or vacuoles

IV. SPECIALIZED CONTACTS (JUNCTIONS) FORM BETWEEN SOME CELLS OF MULTICELLULAR ORGANISMS

■ 18 Three types of junctions, the _____
_____, are specialized structures associated with the plasma membrane of animal cells.

A. Desmosomes are points of attachment between some animal cells
B. Tight junctions seal off intercellular spaces between some animal cells
C. Gap junctions contain pores that permit transfer of small molecules and ions between some animal cells
D. Plasmodesmata are cytoplasmic channels that allow movement of molecules and ions between plant cells.

■ 19 Plasmodesmata in plant cells are functionally equivalent to _____ in animal cells.

KEY TERMS

Frequently Used Prefixes and Suffixes: Use combinations of these prefixes and suffixes to generate terms for the definitions below.

Prefixes	The Meaning	Suffixes	The Meaning
endo-	within	-cyto(sis)	cell
exo-	outside, outer, external		
hyper-	over		
hypo-	under		
iso-	equal, "same"		
phago-	eat, devour		
pino-	drink		

Prefix	Suffix	Definitions
_____	_____	1. A process whereby materials are taken into the cell.
_____	_____	2. The process whereby waste or secretion products are ejected from a cell by fusion of a vesicle with the plasma membrane.
_____	-tonic (-osmotic)	3. Having an osmotic pressure or solute concentration that is greater than a standard solution.
_____	-tonic (-osmotic)	4. Having an osmotic pressure or solute concentration that is less than a standard solution.
_____	-tonic (-osmotic)	5. Having an osmotic pressure or solute concentration that is the same as a standard solution.
_____	_____	6. The engulfing of microorganisms, foreign matter, and other cells by a cell.
_____	_____	7. A type of endocytosis whereby fluid is engulfed by vesicles originating at the cell surface.

Other Terms You Should Know

active transport
amphipathic molecule
carrier proteins
carrier-mediated transport
coated pit
coated vesicle
concentration gradient
cotransport
desmosome
desmotubule
dialysis
diffusion
endosome
enzyme

facilitated diffusion
fluid mosaic model
gap junction
globular
impermeable
integral membrane protein
membrane fusion
membrane transport
microvillus (-li)
osmosis
osmotic pressure
peripheral protein
permeable
physiologic saline

plasma membrane
plasmodesma (-ata)
plasmolysis
receptor protein
receptor-mediated
endocytosis
selectively permeable
sodium-potassium pump
solute
solvent
tight junction
transmembane protein
turgor pressure

SELF TEST

Multiple Choice: Some questions may have more than one correct answer.

Red blood cells are placed in three beakers containing the following solutions: beaker A, distilled water; beaker B, isotonic solution; beaker C, 5% salt solution. Use this information to answer questions 1-3.

1. The cells in beaker A
   a. shriveled.
   b. swelled.
   c. plasmolyzed.
   d. were unaffected.
   e. probably eventually burst.

2. The cells in beaker B
   a. shriveled.
   b. swelled.
   c. plasmolyzed.
   d. were unaffected.
   e. probably eventually burst.

3. The cells in beaker C
   a. shriveled.
   b. swelled.
   c. plasmolyzed.
   d. were unaffected.
   e. probably eventually burst.

4. A molecule is called amphipathic when it
   a. prevents free passage of substances.
   b. is in a membrane.
   c. has hydrophobic and hydrophilic regions.
   d. is a lipid.
   e. is embedded in a bilipid layer.

5. If membranes were not fluid and dynamic, which of the following might still occur normally?
   a. active transport
   b. facilitated diffusion
   c. simple diffusion
   d. endo- and exocytosis
   e. osmosis

6. Active transport
   a. moves molecules against a gradient.
   b. does not occur in prokaryotes.
   c. requires use of ATP.
   d. occurs in animal cells, not in plant cells.
   e. moves substances both into and out of cells.

7. Facilitated diffusion
   a. moves molecules against a gradient.
   b. does not occur in prokaryotes.
   c. requires use of ATP.
   d. involves protein channels.
   e. moves substances both into and out of cells.

8. Cell membranes are
   a. asymmetrical.
   b. 11–12 nm thick.
   c. partly nonpolar, hydrophobic fatty acid chains.
   d. composed mostly of phospholipids and nucleic acids.
   e. all similar in basic structure.

9. Cell walls
   a. are large plasma membranes.
   b. contain carbohydrates.
   c. are thicker than membranes.
   d. are fluid.
   e. contain desmosomes.

10. Diffusion rate depends on
    a. the flow of water.
    b. concentration gradient.
    c. energy from the cell.
    d. the plasma membrane.
    e. kinetic energy.

11. Plasma membranes of eukaryotic cells have a large amount of
    a. sterols.
    b. phospholipid.
    c. rigidity.
    d. fluidity.
    e. protein.

12. Endocytosis may include
    a. secretion vacuoles.
    b. pinocytosis.
    c. phagocytosis.
    d. combination of inbound particles with proteins.
    e. receptor-mediation.

13. Plasmodesmata
    a. are channels in cytoplasm.
    b. connect plant cells.
    c. plant cell structures equivalent to animal cell desmosomes.
    d. are the same as several plasmodesma.
    e. connect ER of adjacent cells.

14. Membrane fusion enables
    a. diversity of proteins.
    b. exocytosis.
    c. endocytosis.
    d. fusion of vesicles and plasma membrane.
    e. formation of vesicles.

Visual Foundations on next page ☞

VISUAL FOUNDATIONS

Color the parts of the illustration below as indicated.

RED ❑ transmembrane protein
GREEN ❑ hydrophilic heads
YELLOW ❑ hydrophobic tails
BLUE ❑ glycolipid
ORANGE ❑ glycoprotein
BROWN ❑ cholesterol
TAN ❑ cell exterior
PINK ❑ cytoplasm
VIOLET ❑ carbohydrate chains

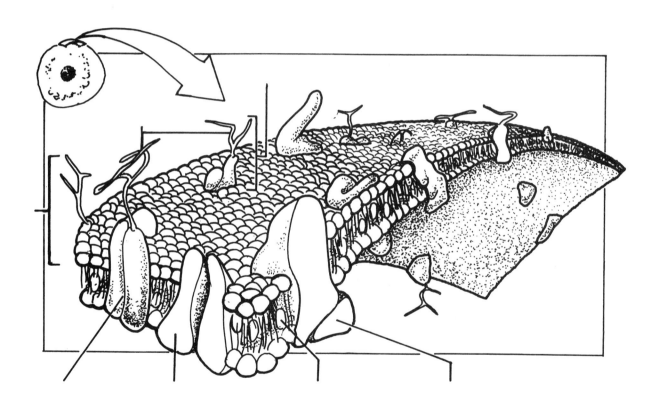

PART II

Energy Transfer Through Living Systems

The Energy of Life

O U T L I N E

Energy for biological work
The laws of thermodynamics
Energy transformation and metabolic reactions
Adenosine triphosphate (ATP) is the energy
currency of the cell
Enzymes are chemical regulators

Life depends on a continuous input of energy. The myriad chemical reactions of cells that enable them to grow, move, maintain and repair themselves, reproduce, respond to stimuli, etc. involve energy transformations. These transformations are governed by the laws of thermodynamics that explain why organisms cannot produce energy but must continuously capture it from somewhere else, and why in every energy transaction, some energy is dissipated as heat. Some chemical reactions occur spontaneously, releasing free energy which is then available to perform work. Other reactions are not spontaneous, requiring an input of free energy before they can occur. The energy released from spontaneous reactions is stored in the chemical bonds of adenosine triphosphate (ATP), and the energy consumed in energy-requiring reactions is taken from the chemical bonds of ATP. In general, for each energy-requiring reaction occurring in a cell there is an energy-releasing reaction coupled to it, and ATP is what links them. Chemical reactions in organisms are regulated by enzymes, protein catalysts that affect the speed of a chemical reaction without being consumed in the process. Enzymes lower the amount of energy needed to activate reactions. They are highly specific and work by forming temporary chemical compounds with their substrates. Whereas some enzymes consist solely of protein, others consist of a protein and an additional chemical component. Enzymes usually work in teams, with the product of one enzyme-controlled reaction serving as the substrate for the next. Cells regulate enzyme activity by controlling the amount of enzyme produced and the shape of the enzyme. Enzymes work best at specific temperatures and pH levels, and most can be inhibited by certain chemical substances.

CHAPTER OUTLINE AND CONCEPT REVIEW (Fill in the blanks)

I. INTRODUCTION

■ 1 Life depends on a continuous supply of outside energy. Fortunately, (a)_____ capture outside sources of energy and incorporate it in chemical bonds (food). Some of that chemical energy is then transferred to the consumers, and finally to the (b)_____ who eventually consume them all.

II. BIOLOGICAL WORK REQUIRES ENERGY

■ **2** Energy is the capacity to do _____.

■ **3** Forms of energy include (list six forms) _____

_____.

A. Energy can be described as potential or kinetic

■ **4** Energy is in one of two forms: (a)_____ is "stored energy" and (b) _____ is "energy of motion."

B. Heat is a form of energy that can be conveniently measured

III. TWO LAWS OF THERMODYNAMICS GOVERN ENERGY TRANSFORMATIONS

■ **5** The study of energy and its transformations is called _____.

A. The quantity of energy in the universe does not change

■ **6** The first law of thermodynamics states that energy can be neither (a)_____ _____, however, it can be (b)_____ and changed in form.

B. The entropy of the universe is increasing

■ **7** The term entropy refers to the _____ in the universe.

■ **8** In every energy conversion or transfer some energy is dissipated as _____.

IV. METABOLIC REACTIONS INVOLVE ENERGY TRANSFORMATIONS

■ **9** Metabolism is all of the _____ that occur in cells.

A. Enthalpy is the total heat content of a system

■ **10** The total potential energy of a chemical reactions, or enthalpy, equals the _____ _____ of the reactants and products.

B. Free energy is energy available to do work
C. Chemical reactions may be exothermic or endothermic

■ **11** Chemical reactions may release or absorb heat energy. (a)_____ reactions release heat; (b)_____ reactions absorb heat.

D. Chemical reactions are reversible
E. Reversible reactions reach a state of equilibrium

■ **12** When a chemical reaction is at equilibrium, the difference in free energy between reactants and products is _____.

F. Spontaneous reactions do not require outside energy

■ **13** _____ are spontaneous and they release energy that can perform work.

G. Exergonic and endergonic reactions can be coupled in living systems

■ **14** Exergonic reactions (a)_____[release or require input of?] free energy; endergonic reactions (b)_____[release or require input of?] free energy. Endergonic and exergonic reactions are coupled in organisms.

V. ADENOSINE TRIPHOSPHATE (ATP) IS THE ENERGY CURRENCY OF THE CELL

A. The ATP molecule has three main parts

■ 15 The three main parts of the ATP molecule are _____, _____, and
_____.

B. The bonds between the phosphate groups are unstable

■ 16 Phosphate bonds in ATP are broken by the process known as _____.

C. ATP links exergonic and endergonic reactions

■ 17 _____ enzymes link exergonic degradation of ATP to
endergonic cell activities.

D. ATP cannot be stockpiled

VI. ENZYMES ARE CHEMICAL REGULATORS

■ 18 Enzymes _____ the speed of a chemical reaction without being altered or
consumed.

A. Enzymes lower the activation energy necessary to initiate a chemical reaction

■ 19 Enzymes lower activation energy by forming an unstable intermediate called the
_____.

B. Most enzyme names end in *-ase*
C. Enzymes are very efficient catalysts
D. Enzymes are specific
E. Enzymes work by forming enzyme-substrate complexes

■ 20 The _____ model and the _____ model are
designed to explain the nature of enzyme-substrate binding.

■ 21 Enzymes bind specifically to the _____ of substrates.

F. Many enzymes require cofactors

■ 22 An organic cofactor is called a _____.

G. Enzymes usually work in teams

■ 23 When enzymes work in teams, the (a)_____ from one enzyme-substrate
reaction becomes the (b)_____ for the next enzyme-substrate reaction.

H. The cell regulates enzymatic activity

■ 24 A cell regulates enzymatic activity by influencing the _____ of
the enzyme.

I. Enzymes are most effective at optimal conditions

■ 25 Factors that affect enzyme activity include (list four) _____
_____.

KEY TERMS

Frequently Used Prefixes and Suffixes: Use combinations of these prefixes and suffixes to generate terms for the definitions below.

Prefixes	The Meaning	Suffixes	The Meaning
allo-	other, "another"	-calor(ie)	heat
end(o)-	within	-ergonic	work, "energy"
ex(o)-	outside, outer, external	-steric	"space"
kilo-	thousand	-thermic	heat, hot

Prefix	Suffix	Definitions
_____	_____	1. The amount of heat required to raise the temperature of 1000 grams (1 kg) of water from 14.5^0 C to 15.5^0 C.
_____	_____	2. A process in which heat is transferred to the surroundings.
_____	_____	3. A process in which heat is absorbed from the surroundings.
_____	_____	4. A spontaneous reaction that releases free energy and can therefore perform work.
_____	_____	5. A reaction that requires an input of free energy from the surroundings.
_____	_____	6. Refers to a receptor site on some region of an enzyme molecule other than the active site.

Other Terms You Should Know

activation energy
active site
ADP
allosteric enzyme
allosteric site
apoenzyme
ATP
Calorie (capital C)
catalyst
coenzyme
cofactor
competitive inhibition
coupled reaction

dynamic equilibrium (-ia)
energy
enthalpy
entropy
enzyme
equilibrium (-ia)
equilibrium constant (K)
feedback inhibition
free energy
induced fit model of enzyme action
irreversible inhibitor
kinetic energy

lock and key model of enzyme action
matter
metabolism
noncompetitive inhibition
phosphorylation
potential energy
pyrophosphate group
regulator
reversible inhibitor
spontaneous reaction
substrate
thermodynamics

SELF TEST

Multiple Choice: Some questions may have more than one correct answer.

1. Enzyme activity may be affected by
   a. cofactors.
   b. temperature.
   c. pH.
   d. substrate concentration.
   e. genes.

2. Enzymes
   a. are lipoproteins.
   b. lower activation energy.
   c. speed up biological chemical reactions.
   d. become products after complexing with substrates.
   e. are regulated by genes.

3. The reaction $HCl + NaOH \longrightarrow H_2O + NaCl + heat$
   a. is exothermic.
   b. is endothermic.
   c. requires input of energy.
   d. tends to occur in absence of intervention.
   e. produces products chemically different from reactants.

Use the following formulae to answer questions 4-6:

$$A + B \longrightarrow C + D$$

$$K = \frac{[C] \times [D]}{[A] \times [B]}$$

4. Which of the following mean(s) "concentration of?"
   a. $A + B \longrightarrow$
   b. $\longrightarrow C + D$
   c. brackets
   d. K
   e. $[C] \times [D]$

5. The reaction depicted probably has
   a. a small K value.
   b. a large K value.
   c. substances C and D are different than A and B.
   d. a K value of 10^{-7} or less.
   e. a tendency to move toward equilibrium.

6. The formula for the equilibrium constant indicates that
   a. reactants are divided by products.
   b. products are divided by reactants.
   c. products are multiplied by products.
   d. reactants are multiplied by reactants.
   e. reactants are multiplied by products.

Use the following formula to answer question 7:

$$ATP + H_2O \longrightarrow ADP + P \qquad \Delta G = -7.3 \text{ kcal/mole}$$

7. This reaction
   a. hydrolyzes ATP.
   b. loses free energy.
   c. produces adenosine triphosphate.
   d. is endergonic.
   e. is exergonic.

8. A kilocalorie (kcal) is
   a. a measure of heat energy.
   b. equal to a Calorie.
   c. a way to measure energy generally.
   d. the temperature of water.
   e. an essential nutrient.

9. Compliance with the second law of thermodynamics presumes that
   a. disorder is increasing.
   b. order in a system requires input of energy.
   c. entropy will decrease as order in organisms increases.
   d. all energy will eventually be useless to life.
   e. heat dissipates in all systems.

10. Kinetic energy is
    a. doing work.
    b. stored energy.
    c. in chemical bonds.
    d. energy in motion.
    e. energy of position or state.

11. In the formula $\Delta G = \Delta H - T\Delta S$
    a. free energy decreases as entropy decreases.
    b. entropy and free energy are inversely related.
    c. change in enthalpy is greater than change in free energy.
    d. temperature increase decreases free energy.
    e. increasing enthalpy increases free energy.

VISUAL FOUNDATIONS

Color the parts of the illustration below as indicated.

RED   ☐   active sites
GREEN   ☐   substrates
YELLOW ☐   enzyme
BLUE   ☐   enzyme-substrate complex
ORANGE ☐   products

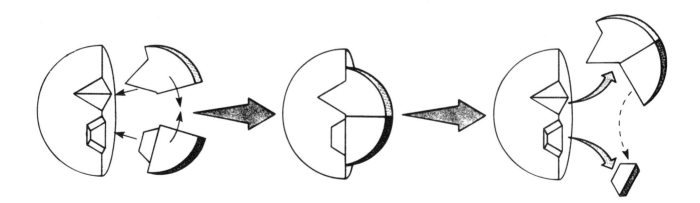

CHAPTER 7

Energy-Releasing Pathways and Biosynthesis

O U T L I N E

Aerobic and anaerobic catabolism
Oxidation-reduction reactions
Stages in aerobic respiration
Energy yield in aerobic respiration of glucose
Energy from nutrients other than glucose
Regulation of aerobic respiration
Anaerobic pathways
Biosynthesis processes

Cells break down nutrients one step at a time. In the process energy is released from chemical bonds in a controlled fashion and transferred to ATP where it is available for cellular work. There are three major pathways by which cells extract energy from nutrients: aerobic respiration, anaerobic respiration, and fermentation. Aerobic respiration involves a series of reactions in which hydrogen is transferred from glucose to oxygen, resulting in the formation of water. Many organisms use nutrients besides glucose, or in addition to it, as a source of energy. These nutrients might be fatty acids or amino acids that become transformed into one of the metabolic intermediates of aerobic respiration. In anaerobic respiration, fuel molecules are broken down in the absence of oxygen, and an inorganic compound such as nitrate or sulfate serves as the final hydrogen (electron) acceptor. Fermentation is a type of anaerobic respiration in which the final electron acceptor is an organic compound derived from the initial nutrient. There is a net gain of only 2 ATPs per glucose molecule in fermentation, compared with 36-38 ATPs produced per glucose molecule by aerobic respiration. Two common types of fermentation are alcohol fermentation and lactate fermentation. Yeast cells carry on alcohol fermentation, in which ethyl alcohol and carbon dioxide are the final products. Lactate fermentation occurs in some fungi and bacteria, and in some animal cells in the absence of sufficient oxygen. In addition to breaking down nutrients, cells also synthesize an array of complex molecules, such as proteins, nucleic acids, lipids, polysaccharides, etc. These biosynthetic reactions are catalyzed by enzymes and require ATP to drive them. At any given time, a cell is in a dynamic state with some molecules being degraded while others are being synthesized.

CHAPTER OUTLINE AND CONCEPT REVIEW (Fill in the blanks)

I. INTRODUCTION

■ 1 _____ is the endergonic aspect of metabolism involving the synthesis of complex molecules.

■ 2 _____ is the exergonic aspect of metabolism involving breakdown of complex molecules.

■ 3 Living cells transfer energy from food to the molecule _____ _____ for later use.

II. CATABOLISM MAY BE AEROBIC OR ANAEROBIC

■ 4 Most cells that live in environments where oxygen is plentiful use the catabolic process called _____ to extract free energy from nutrients.

■ 5 _____ is a means of extracting energy that produces inorganic end products and does not require oxygen.

■ 6 _____ is a means of extracting energy that produces organic end products and does not require oxygen.

■ 7 In aerobic respiration, a fuel molecule is oxidized, yielding the byproducts
(a)_____ with the release of the essential
(b)_____ that is required for life's activities.

III. OXIDATION-REDUCTION REACTIONS OCCUR IN METABOLISM

■ 8 Energy is converted to useful forms (processed) largely as the result of the flow of _____ through cells.

■ 9 (a)_____ molecules release electrons and give up energy, whereas
(b)_____ molecules accept electrons and receive energy.

A. Hydrogens and electrons are transferred in redox reactions

■ 10 _____ are two hydrogen acceptor molecules. They are energized and reduced as hydrogen atoms from oxidized organic compounds are transferred to them.

B. Aerobic respiration is a redox process

■ 11 The redox process in aerobic respiration involves the transfer of hydrogen from glucose to oxygen. In this process, glucose is said to be (a)_____ and oxygen is (b)_____.

IV. THERE ARE FOUR STAGES IN AEROBIC RESPIRATION

■ 12 Three principal types of reactions that occur in aerobic respiration are the _____ and _____ and the "make ready" reactions.

A. In glycolysis, glucose is converted to pyruvate

■ 13 Glycolysis reactions take place in the _____ of the cell.

■ 14 In the first phase of glycolysis, two ATP molecules are consumed and glucose is split into two _____ molecules.

■ **15** In the second phase of glycolysis, each of the molecules resulting from splitting of glucose is oxidized and transformed into a _____ molecule.

■ **16** Glycolysis <u>nets</u> _____ (#?) ATPs.

B. Each pyruvate is converted to acetyl CoA

■ **17** When oxygen is present, pyruvate is converted to acetyl CoA in mitochondria, (a)_____ is reduced, and (b)_____ is released.

C. The citric acid cycle oxidizes acetyl CoA

■ **18** The eight step citric acid cycle completes the oxidation of glucose. For each acetyl group that enters the citric acid cycle, (a)_____ (#?) NAD^+ are reduced to NADH, (b)_____ (#?) molecules of CO_2 are produced, and (c)_____ (#?) hydrogen atoms are removed.

■ **19** _____(#?) acetyl CoAs are completely degraded with two turns of the citric acid cycle.

D. The electron transport system is coupled to chemiosmosis

■ **20** The hydrogens removed during glycolysis, acetyl CoA formation, and the citric acid cycle are first transferred to the primary hydrogen acceptors (a)_____, then they are processed through (b)_____.

■ **21** The (a)_____ consists of a chain of electron acceptors embedded in the inner membrane of mitochondria. (b)_____ is the final acceptor in the chain.

■ **22** The electron transport system provides energy to pump protons into the (a)_____ of the mitochondrion, creating an energy gradient. As protons move down the gradient, the energy released is used by the enzyme (b)_____ to produce ATP.

V. THE AEROBIC RESPIRATION OF EACH GLUCOSE MOLECULE YIELDS A MAXIMUM OF 36 TO 38 ATPs

VI. NUTRIENTS OTHER THAN GLUCOSE ALSO PROVIDE ENERGY

■ **23** Human beings and many other animals usually obtain most of their energy by oxidizing _____. Amino acids are also used.

■ **24** Amino groups are metabolized by a process called _____.

■ **25** The _____ components of neutral fats can be use as fuel.

■ **26** Fatty acids are oxidized and split into acetyl groups by the process of _____ _____. The acetyl molecules enter the citric acid cycle.

VII. CELLS REGULATE AEROBIC RESPIRATION

VIII. ANAEROBIC PATHWAYS ARE LESS EFFICIENT THAN AEROBIC RESPIRATION

■ **27** Fermentation is an anaerobic process in which the final acceptor of electrons from NADH is an _____.

A. Two common types of fermentation are alcohol fermentation and lactate fermentation

■ 28 When hydrogens from NAD are transferred to acetaldehyde, _____ is formed.

■ 29 When hydrogens from NAD are transferred to pyruvate, _____ is formed.

B. Efficiency of anaerobic versus aerobic metabolism

■ 30 Fermentation yields a net gain of only (a)_____(#?) ATPs per glucose molecule, compared with about (b)_____(#?) ATPs per glucose molecule in aerobic respiration.

IX. METABOLISM ALSO INCLUDES BIOSYNTHETIC PROCESSES

KEY TERMS

Frequently Used Prefixes and Suffixes: Use combinations of these prefixes and suffixes to generate terms for the definitions below.

Prefixes	The Meaning	Suffixes	The Meaning
aero-	air	-be (bios)	life
an-	without, not, lacking	-lysis	breaking down, decomposition
de-	indicates removal, separation		
glyco-	sweet, "sugar"		

Prefix	Suffix	Definitions
_____	_____	1. An organism that requires air or free oxygen to live.
_____	-aerobe	2. An organism that does not require air or free oxygen to live.
_____	-hydrogenation	3. A reaction in which hydrogens are removed from the substrate.
_____	-carboxylation	4. A reaction in which a carboxyl group is removed from a substrate.
_____	-amination	5. A reaction in which an amine group is removed from a substrate.
_____	_____	6. A sequence of reactions that breaks down a molecule of glucose (a sugar) to two molecules of pyruvate.

Other Terms You Should Know

acetyl CoA
aerobic respiration
alcohol fermentation
anabolic
anabolism
anaerobic respiration
ATP synthetase
beta oxidation
catabolic reaction
catabolism

cellular respiration
chemiosmotic model
citric acid cycle
cytochrome
electron transport chain
ethyl alcohol
facultative anaerobe
FAD
fermentation
Krebs cycle

lactate
lactic acid fermentation
"make-ready" reaction
muscle fatigue
NAD
NADH
oxaloacetate
oxidation

oxidative phosphorylation
phosphorylation
redox reaction

reduction
respiratory assembly
strict aerobe

tricarboxylic acid cycle

SELF TEST

Multiple Choice: Some questions may have more than one correct answer.

1. Chemiosmosis involves
   a. flow of protons down an electrical gradient.
   b. flow of protons down a concentration gradient.
   c. pumping of protons into the mitochondrial matrix.
   d. proton channels composed of electrons.
   e. ultimate production of ATP.

2. Glycolysis
   a. generates two ATPs.
   b. is more efficient that aerobic respiration.
   c. takes place in the cristae of mitochondria.
   d. produces two pyruvates.
   e. reduces glucose to H_2O and CO_2.

3. Production of acetyl CoA from pyruvate
   a. is anabolic.
   b. takes place in mitochondria.
   c. takes place in cytoplasm.
   d. takes place in endoplasmic reticulum.
   e. uses derivatives of vitamin B.

4. During oxidative phosphorylation
   a. ATP converts to ADP.
   b. ATP forms ADP.
   c. three ATPs are produced when two electrons pass from NADH to oxygen.
   d. ATP is coupled to electron flow.
   e. energy is required to make ATPs.

5. Complete aerobic metabolism of one mole of glucose
   a. yields more than 30 ATPs.
   b. is more than 50% efficient.
   c. yields more ATP than complete metabolism of fatty acid.
   d. involves glycolysis.
   e. releases about 686 kcal.

6. ATP synthetase
   a. is a cytochrome.
   b. converts ATP to ADP.
   c. forms channels across the inner mitochondrial membrane.
   d. resides in cristae of mitochondria.
   e. couples protons to electrons to form water.

7. In the citric acid cycle
   a. glucose is catabolized to H_2O and CO_2.
   b. NAD+ is a hydrogen acceptor.
   c. FAD is a hydrogen acceptor.
   d. decarboxylation occurs.
   e. oxaloacetate is consumed.

8. In the electron transport system
   a. water is the final electron acceptor.
   b. cytochromes carry electrons.
   c. the final electron acceptor has a positive redox potential.
   d. glucose is a common carrier molecule.
   e. electrons gain energy with each transfer.

9. Cells may obtain energy from large molecules by means of
   a. catabolism.
   b. aerobic respiration.
   c. anaerobic respiration.
   d. fermentation.
   e. aerobic respiration.

VISUAL FOUNDATIONS

Color the parts of the illustration below as indicated.

RED   ☐ ATP synthetase
GREEN   ☐ first complex with FMN
YELLOW   ☐ cytochrome b-c_1 complex
BLUE   ☐ third complex with cytochromes a and a_3
ORANGE   ☐ coenzyme Q
BROWN   ☐ phospholipid bilayer
TAN   ☐ matrix
PINK   ☐ intermembrane space
VIOLET   ☐ cytochrome C

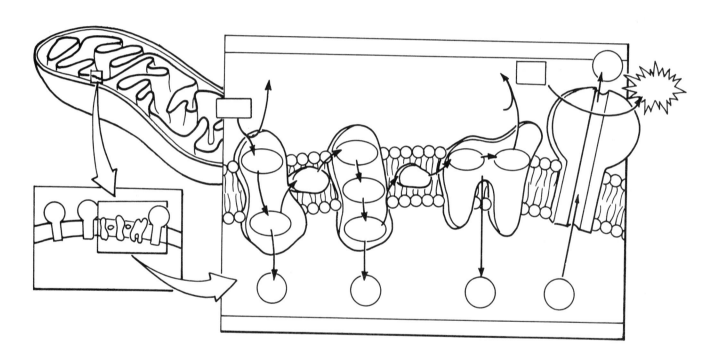

CHAPTER 8

□

Photosynthesis: Capturing Energy

OUTLINE

Chloroplasts
Converting light energy into chemical energy
Light-dependent reactions
Light-independent reactions

The existence and survival of all living things depends on chemosynthesis and photosynthesis, processes by which energy is captured from the environment and converted into the chemical bond energy of carbohydrates. This energy is what fuels the metabolic reactions that sustain all life. Photosynthesis, by far the most prevalent and important process of the two, captures solar energy and uses it to manufacture organic compounds from carbon dioxide and water. Photosynthesis occurs in thylakoid membranes. These membranes exist in photosynthetic prokaryotes as extensions of the plasma membrane, and in eukaryotes as organized structures within chloroplasts. Photosynthesis consists of both light dependent and light independent reactions. In the light dependent reactions, light energy is trapped by chlorophyll and temporarily stored in the chemical bonds of ATP and NADPH. Some of this energy is used to split water into hydrogen and oxygen. The oxygen is released to the environment. In the light independent reactions, additional energy from the ATP and NADPH manufactured earlier is used to make carbohydrate from carbon dioxide and the hydrogen that resulted from the splitting of water. There are two different pathways by which carbon dioxide is assimilated into plants -- the C_3 and C_4 pathways. C_3 is the most prevalent. On bright, hot, dry days when plant cells close their stomata and carbon dioxide cannot get into the leaves, C_3 plants consume oxygen and produce carbon dioxide and water, a process called photorespiration. Photosynthetic efficiency is reduced by photorespiration because it removes some of the intermediates that are used in the C_3 cycle.

CHAPTER OUTLINE AND CONCEPT REVIEW (Fill in the blanks)

I. INTRODUCTION

■ 1 Using the basic raw materials (a) _____,
photosynthetic autotrophs convert (b) _____ energy to stored
(c)_____ energy.

■ 2 _____ are organisms that cannot make their own food.

II. IN EUKARYOTES, PHOTOSYNTHESIS TAKES PLACE IN CHLOROPLASTS

■ 3 The (a)_____ is the fluid-filled region within the inner membrane of
chloroplasts that contains most of the enzymes for photosynthesis. Most
chloroplasts are located in the (b)_____ cells of leaves.

■ 4 _____ are flat, disk-shaped membranes in chloroplasts that are arranged in stacks called grana. Chlorophyll is located in these membranes.

III. IN PHOTOSYNTHESIS, PLANTS CONVERT LIGHT ENERGY INTO THE CHEMICAL ENERGY OF SUGAR MOLECULES

■ 5 (a)_____ traps sunlight energy where it is converted to chemical energy in the form of the two high energy molecules, (b)_____. The energy from these two compounds is used to synthesize carbohydrates.

■ 6 Photosynthesizers obtain hydrogens by splitting (a)_____, which are ultimately combined with (b)_____ to produce carbohydrates.

■ 7 In photosynthesis, the electrons from hydrogen are transferred to (a)_____, and then through a series of (b)_____ _____. Light energy boosts the energy level of electrons as they are transferred from one molecule to another.

A. In light-dependent reactions, light energy is used to make the high-energy compounds, ATP and NADPH

■ 8 When light energy is first absorbed, it is converted to electrical energy in the form of electrons flowing from chlorophyll. Some of this energy is converted to chemical energy during the chemiosmotic synthesis of (a)_____, and some of it is used to split water, a process known as (b)_____.

■ 9 Electrical energy from the flow of chlorophyll electrons is again converted to chemical energy as hydrogens from water are accepted by (a)_____ to form (b)_____.

B. Carbohydrates are produced during the light-independent reactions

■ 10 Light-dependent reactions provide useful chemical energy for synthesis of photosynthetic products. Light-independent reactions transfer the energy from (a)_____ to the bonds in (b)_____ molecules.

IV. THE LIGHT-DEPENDENT REACTIONS CAPTURE ENERGY

A. Light exhibits properties of both waves and particles

■ 11 The lowest energy state for an atom is called the (a)_____. When a molecule absorbs light energy, its electrons are raised (boosted) to higher energy states, and their atoms are said to be (b)_____.

■ 12 (a)_____ occurs when energized electrons return to the ground state, emitting their excess energy in the form of visible light. In photosynthesis, energized electrons leave atoms and pass to an (b)_____ molecule.

B. Chlorophyll is a pigment that absorbs light

■ 13 Chlorophyll absorbs light mainly in the _____ portions of the visible spectrum.

■ 14 Of the several types of chlorophyll in plants, (a)_____ is the bright green form that initiates the light-dependent reactions, and the yellowish-green form, (b)_____, is an accessory pigment. Other yellow and orange accessory pigments in plant cells are (c)_____.

C. Absorption spectra and action spectra indicate that chlorophyll is the main light-gathering pigment of photosynthesis

■ 15 An (a)_____ is a graph that illustrates the relative absorption of different wavelengths of light by a given pigment. It is obtained with an instrument called a (b)_____.

■ 16 The (a)_____ of photosynthesis is a measurement of the <u>actual</u> effectiveness of different wavelengths of light in affecting photosynthesis. It may be greater than can be accounted for by the absorption of chlorophyll alone, the difference accounted for by (b)_____ that transfer energy absorbed from the green wavelengths to chlorophyll.

D. Photosystems I and II are light-harvesting units of chlorophyll and accessory pigment molecules

■ 17 _____ is the process whereby energized electrons provide energy to add a phosphate molecule to ADP.

E. Noncyclic photophosphorylation produces ATP and NADPH

■ 18 Both photosystems are utilized during noncyclic photophosphorylation, during which electrons flow in one direction, yielding one (a)_____ molecule and two (b)_____ molecules.

F. Cyclic photophosphorylation produces ATP

■ 19 Cyclic photophosphorylation takes place in photosystem I, where excited electrons escape from _____ at the reaction center, pass along an electron transport chain, and ultimately return to their point of origin.

G. The transport of protons across thylakoid membranes is coupled with the production of ATP

■ 20 Energy spent as electrons cycle along the transport system is used to pump protons across the (a)_____, where their accumulation in the thylakoid lumen creates a strong (b)_____. When "gates" in the thylakoid membrane "open," protons flow out, providing sufficient energy of motion to generate ATP. This phenomenon is known as (c)_____.

■ 21 Protons leak out of the thylakoid lumen through channels in the ATP synthetase molecule called the _____ complex.

V. THE LIGHT-INDEPENDENT REACTIONS FIX CARBON

■ 22 ATP and NADPH generated in the light-dependent reactions are used to reduce carbon dioxide to a carbohydrate, a process known as _____.

A. Most plants use the Calvin (C_3) cycle to fix carbon

■ 23 The light-dependent reactions form the Calvin cycle, six turns of which produce one six carbon (a)_____ molecule. The cycle begins with the combination of one CO_2 molecule and one five carbon sugar, (b)_____ _____, to form a six carbon molecule. This molecule instantly splits into two three carbon molecules called (c)_____,

which then are phosphorylated to form (d)_____
_____.

■ 24 To produce one six carbon carbohydrate, the light-independent reactions utilize six
molecules of (a)_____, hydrogen obtained from
(b)_____, and energy from (c)_____.

B. Many plants with a tropical origin fix carbon using the C_4 pathway

■ 25 C_4 plants initially fix CO_2 into the four carbon molecule _____.

C. Photorespiration reduces photosynthetic efficiency

■ 26 Photorespiration occurs mainly during bright, hot days when plant stomata are
closed. It reduces photosynthetic efficiency because CO_2 cannot enter the system,
causing the enzyme (a)_____ to bind RuBP to
(b)_____ instead of CO_2.

KEY TERMS

Frequently Used Prefixes and Suffixes: Use combinations of these prefixes and suffixes to generate
terms for the definitions below.

Prefixes	The Meaning		Suffixes	The Meaning
meso-	middle		-lysis	breaking down, decomposition
photo-	light		-plast	formed, molded, "body"
chloro-	green			

Prefix	Suffix	Definitions
_____	-phyll	1. Tissue in the middle of a leaf specialized for photosynthesis.
_____	-synthesis	2. The conversion of solar (light) energy into stored chemical energy by plants, blue-green algae, and certain bacteria.
_____	_____	3. The breakdown (splitting) of water under the influence of light energy trapped by chlorophyll.
_____	-phosphorylation	4. Phosphorylation that uses light as a source of energy.
_____	_____	5. A membranous organelle containing green photosynthetic pigments.
_____	-phyll	6. A green photosynthetic pigment.

Other Terms You Should Know

absorption spectrum	chemosynthetic autotroph	oxaloacetate
action spectrum	fluorescence	P680
bundle sheath cells	grana	P700
C_3 plant	ground state	PGAL
C_4 plant	Hatch-Slack pathway	photon
Calvin cycle	heterotroph	photorespiration
carotenoid	noncyclic	photosynthetic autotroph
CF_0–CF_1 complex	photophosphorylation	photosystem

reaction center   stroma   wavelength
ribulose phosphate   thylakoid
stomata   thylakoid space

SELF TEST

Multiple Choice: Some questions may have more than one correct answer.

1. The more advanced aspects of photosynthesis, generally found only in higher plants, include
   a. photosystem I.   d. noncyclic phosphorylation.
   b. photosystem II.   e. use of oxygen as a hydrogen acceptor.
   c. cyclic phosphorylation.

2. Chlorophyll
   a. contains magnesium in a porphyrin ring.   d. is the only pigment found in most plants.
   b. dissolves in water.   e. is found mostly in the stroma.
   c. is green because it absorbs green portions of the light spectrum.

3. C_4 fixation
   a. replaces C_3 fixation.   d. takes place in bundle sheath cells.
   b. produces PGA.   e. supplements C_3 fixation.
   c. produces oxaloacetate.

4. The reactions of photosystem II
   a. include splitting water with light photons.   d. use $NADP^+$ as a final electron acceptor.
   b. produce H_2O.   e. produce O_2.
   c. assist in producing an electrochemical gradient across the thylakoid membrane.

5. The ATP synthetase enzyme
   a. regulates flow of protons.   d. comprises the CF_0–CF_1 complex.
   b. functions in chemiosmosis.   e. removes phosphate from ATP to form free ADP.
   c. forms a channel for passage of protons to the interior of thylakoids.

6. Photosynthesis involves
   a. enzymes in the stroma.   d. chlorophyll.
   b. enzymes in thylakoids.   e. carotenoids.
   c. a CF_0–CF_1 complex.

7. Most producers are
   a. chemosynthetic heterotrophs. d. photosynthetic heterotrophs.
   b. chemosynthetic autotrophs.   e. photosynthetic consumers.
   c. photosynthetic autotrophs.

8. Both photosystems I and II
   a. are activated by light.   d. occur in higher plants.
   b. form NADPH from $NADP^+$.   e. occur in photosynthetic bacteria.
   c. involve an electron transport system.

9. In photosynthesis, electrons in atoms
   a. are excited.   d. are pushed to higher energy levels by photons of light.
   b. produce fluorescence.   e. release absorbed energy as another wavelength of light.
   c. are accepted by a reducing agent when they escape.

10. Some of the requirements for the light-dependent reactions of photosynthesis include
   a. water.   d. NADP.
   b. photons of light.   e. oxygen.
   c. carbohydrates.

11. The light-independent reactions of photosynthesis
   a. generate PGAL.   d. require ATP.
   b. generate oxygen.   e. require NADPH.
   c. take place in the stroma.

12. Reactions occurring during electron flow in respiration and photosynthesis are
   a. both endergonic.   d. endergonic and exergonic respectively.
   b. both exergonic.   e. neither endergonic nor exergonic.
   c. exergonic and endergonic respectively.

13. Thylakoids
   a. comprise grana.   d. are located in the stroma.
   b. are continuous with the plasma membrane.   e. carry out the light-independent reactions.
   c. possess the basic fluid mosaic membrane structure.

14. Chloroplasts in prokaryotic cells
   a. possess a CF_0–CF_1 complex.   d. contain chlorophyll.
   b. have double membranes.   e. carry out photosynthesis II.
   c. do not exist.

VISUAL FOUNDATIONS

Color the parts of the illustration below as indicated.

RED   ❑   stoma
GREEN   ❑   mesophyll
YELLOW   ❑   spongy mesophyll
BLUE   ❑   vein
ORANGE   ❑   bundle sheath cell
BROWN   ❑   upper epidermis
TAN   ❑   lower epidermis

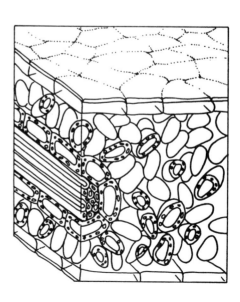

PART III

The Continuity of Life: Genetics

CHAPTER 9

□

Chromosomes, Mitosis, and Meiosis

OUTLINE

Eukaryotic chromosomes
The cell cycle: Mitosis and cytokinesis
Cell division and reproduction
Meiosis

Genetic information is transferred from parent to offspring. In prokaryotes, the information is contained in a single circle of DNA. In eukaryotes, it is carried in the chromosomes contained within the cell nucleus. Chromosomes are made up of DNA, protein, and RNA. The DNA is organized into informational units, or genes, that determine the characteristics of the organism. Genes control the structure of all the proteins of the organism, including the enzymes. Each species is unique due to the information specified by its genes. Genes are passed from parent cell to daughter cell by mitosis, a process that ensures that each new nucleus receives the same number and types of chromosomes as were present in the original nucleus. Interphase is the time period between cell divisions when cell growth, DNA replication, and synthesis of chromosomal and other proteins occur. There are two basic types of reproduction -- asexual and sexual. In asexual reproduction, a single parent cell usually splits, buds, or fragments into two or more individuals. Genetic material identical to that in the parent is distributed to the new individuals. In sexual reproduction, sex cells, or gametes, are produced by meiosis, a special type of cell division that halves the number of chromosomes in the resulting cells. When two gametes fuse, the resulting cell contains the same number of chromosomes as the parent cells. In animals, gametes are the direct products of meiosis. In plants and some other organisms, the products of meiosis are spores that undergo mitosis one or more times before some of the descendents develop into gametes. Unlike asexual reproduction, the new individual in sexual reproduction is not genetically identical to the cells of the parents because the genetic information from both parents is shuffled during meiosis. Various groups of sexually reproducing eukaryotes differ with respect to the roles of mitosis and meiosis in their life cycles.

CHAPTER OUTLINE AND CONCEPT REVIEW (Fill in the blanks)

I. INTRODUCTION

II. EUKARYOTIC CHROMOSOMES ARE MADE UP OF DNA, PROTEIN, AND RNA

■ 1 DNA, protein, and RNA form a complex, the ____chromatin____, that make up chromosomes.

A. DNA is organized into informational units called genes

- 2 Genes code for specific (a)_RNA_ molecules, which in turn code for specific (b)_polypeptides_ .

B. Chromosomes of different species differ in number and informational content

III. THE EUKARYOTIC CELL CYCLE IS AN ALTERNATING SEQUENCE OF CELL GROWTH AND DIVISION

- 3 The period from the beginning of one cell division to the beginning of the next cell division is the _cell cycle_ .

A. Interphase is the time between divisions and includes the duplication of chromosomes

- 4 Interphase is divided into the G_1 phase, or (a)_Gap I_ phase, the S phase, or (b)_Synthesis_ phase, and the G_2 phase, or (c)_Gap II_ phase.

B. Identical chromosomes are distributed into different nuclei during mitosis

- 5 Chromatin threads begin to condense during the _prophase_ stage of mitosis.

- 6 Chromosomes assemble in the equatorial plane of the cell during the _metaphase_ stage of mitosis.

- 7 Sister chromatids separate and the newly formed chromosomes move toward the poles during the _anaphase_ stage of mitosis.

- 8 Nuclei reform in the daughter cells during the _telophase_ stage of mitosis.

C. Two separate daughter cells are formed by cytokinesis

D. Mitosis typically produces two cells genetically identical to the parent cell

E. The cell cycle is controlled by an internal genetic program interacting with external signals

IV. VARIOUS MODES OF REPRODUCTION REQUIRE DIFFERENT TYPES OF CELL DIVISION

- 9 Offspring inherit traits that are virtually identical to those of their single parent when reproduction is _asexual_ .

- 10 In (a)_sexual_ reproduction, offspring receive genetic information from two parents. Haploid (n) gametes from the parents fuse to form a single (b)_zygote diploid_ (2n) cell called the (c)_zygote_ .

- 11 Homologous chromosomes are members of a pair of chromosomes that are similar in _size & structure, centromere_ . They carry genes affecting the same traits.

- 12 One member of each (a)_homologous_ pair is contributed by each parent to the zygote, thereby restoring (b)_diploid #_ in the offspring.

V. DIPLOID CELLS GIVE RISE TO HAPLOID CELLS DURING MEIOSIS

A. Haploid cells produced by meiosis have unique combinations of genes

- 13 Meiosis is a special kind of cell division that produces haploid (a)_daughter cells_ from diploid (b)_cells_ .

B. Meiosis involves segregation of homologous chromosomes into different daughter cells

■ 14 Homologous chromosomes exchange genetic material (crossing over) during the first meiotic (a) _prophase_ , providing more (b)_variation_ among gametes and offspring.

one member of each homologous pair

■ 15 The haploid condition is established as the members of each pair of homologous chromosomes separate during the first meiotic _anaphase_ .

■ 16 The two chromatids of each chromosome separate during the second meiotic _anaphase_ .

VI. MITOSIS AND MEIOSIS OCCUPY DIFFERENT POSITIONS IN VARIOUS KINDS OF LIFE CYCLES

■ 17 Some simple eukaryotes are haploid, producing gametes by (a)_mitosis_. The diploid zygote is restored to haploidy by (b)_meiosis_ .

■ 18 In animals, meiosis reduces diploid _somatic_ cells to haploid _gametes_ .

■ 19 Alternation of generations in plants and some algae involves a spore-forming diploid generation, called the (a)_sporophyte_ , alternating with a gamete-producing haploid generation, called the (b)_gametophyte_ .

KEY TERMS

Frequently Used Prefixes and Suffixes: Use combinations of these prefixes and suffixes to generate terms for the definitions below.

Prefixes	The Meaning		Suffixes	The Meaning
centro-	center		-gen(esis)	production of
chromo-	color		-mere	part
dipl-	double, in pairs		-phyte	plant
gameto-	sex cells, eggs and sperm		-some	body
hapl-	single			
inter-	between			
oo-	egg			
spermato-	seed, "sperm"			
sporo-	spore			

Prefix	Suffix	Definitions
chromo	_some_	1. A dark staining body within the cell nucleus containing genetic information.
inter	-phase	2. The stage in the life cycle of a cell that occurs between successive cell divisions.
hapl	-oid	3. An adjective pertaining to a single set of chromosomes.
dipl	-oid	4. An adjective pertaining to a double set of chromosomes.
centro	_mere_	5. The constricted part or region of a chromosome (often near the center) to which a spindle fiber is attached.

gameo to ~~melo~~ _genesis_ ~~sis~~   6. The process by which gametes (sex cells) are produced.

spermato _genesis_   7. The process whereby sperm are produced.

_oo_ _genesis_   8. The process whereby eggs are produced.

gameo _tophyte_   9. The stage in the life cycle of a plant that produces gametes by mitosis.

sporo _phyte_   10. The stage in the life cycle of a plant that produces spores by meiosis.

Other Terms You Should Know

alternation of generations	gene	oogenesis
anaphase	generation time, (T)	ovum (-va)
anaphase I	genetic recombination	paternal chromosome
anaphase II	genetics	paternal homologue
asexual reproduction	germ line	pericentriolar material
aster	haploid	polar body
bivalent	heredity	polar microtubule
cell cycle	homologous chromosomes	polyploid
cell plate	interkinesis	pre-MPF
centromere	interphase	prophase
chiasma (chiasmata)	karyotype	prophase II
chromatin	kinetochore	S phase
chromosome	kinetochore microtubule	sexual reproduction
clone	maternal chromosome	sister chromatid
colchicine	maternal homologue	sporophyte generation
crossing over	meiosis	synapsis
cyclin	meiosis I	synaptonemal complex
cytokinesis	meiosis II	synthesis phase
cytokinin	metaphase	telophase
diploid	metaphase I	telophase I
flagellum (-la)	metaphase II	telophase II
G_1 phase	mitosis	tetrad
G_2 phase	mitotic spindle	transcription
gamete	MPF	zygote
gametophyte generation	MTOC	

SELF TEST

Multiple Choice: Some questions may have more than one correct answer.

1. In mitosis, cells with 16 chromosomes produce daughter cells with
   a. 32 chromosomes.
   b. 8 chromosomes.
   c. 16 chromosomes.
   d. 8 pairs of chromosomes.
   e. 4 pairs of chromosomes.

2. Gametogenesis typically involves
   a. 2n to 2n.
   b. 2n to n.
   c. reduction division.
   d. mitosis.
   e. meiosis.

3. Eukaryotic chromosomes
   a. contain DNA, RNA, protein.   d. are uncoiled in interphase.
   b. possess asters.   e. align in the equator during prophase.
   c. are distinctly visible in interphase.

4. Typical gametes include
   a. somatic cells.   d. ova.
   b. 2n cells.   e. cells resulting from oogenesis.
   c. sperm.

5. In meiosis, cells with 16 chromosomes produce daughter cells with
   a. 32 chromosomes.   d. 8 pairs of chromosomes.
   b. 8 chromosomes.   e. 4 pairs of chromosomes.
   c. 16 chromosomes.

Use this list to answer questions 6-15 about mitosis:

   a. interphase   e. telophase   h. G_2 phase
   b. prophase   f. T phase   i. S phase
   c. metaphase   g. G_1 phase   j. M phase
   d. anaphase

6. Nuclear membranes break down. b

7. Chromosomes begin to condense by coiling. b

8. Active synthesis and growth. g, a, h

9. The time between mitosis and start of the synthesis phase. g

10. Centromeres divide. d

11. The four stages of mitosis collectively. j

12. Chromosomes are lined up in a central plane. c

13. The time between the synthesis phase and prophase. h

14. DNA replicates. i, a

15. Condensed chromosomes uncoil. e

Use this list to answer questions 16-25 about meiosis:

   a. interphase   d. anaphase I   g. metaphase II
   b. prophase I   e. telophase I   h. anaphase II
   c. metaphase I   f. prophase II   i. telophase II

16. Tetrads form. b

17. Nuclear membranes break down. f, b

18. Members of tetrad separate. d

19. Single chromosomes align at equatorial plane. g

20. Chromatids separate. h

21. Pairs of chromosomes align at equatorial plane. _C_
22. Crossing over takes place. _b_
23. Homologous chromosomes synapse. _b_
24. Reduction of chromosome number from 2n to n. _d_
25. Diploid to haploid. _d_

VISUAL FOUNDATIONS

Color the parts of the illustration below as indicated.

RED	☐	mitosis
GREEN	☐	interphase
YELLOW	☐	cytokinesis
BLUE	☐	S phase
ORANGE	☐	G_2 phase
BROWN	☐	G_1 phase
TAN	☐	generation time

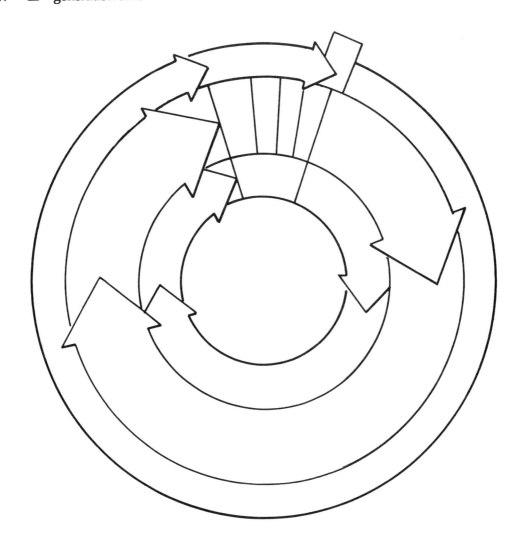

The Basic Principles of Heredity

O U T L I N E

Gregor Mendel
Loci and alleles
A monohybrid cross
The laws of probability
A dihybrid cross
Genetic maps
Sex determination
Genotype and phenotype
Polygenes
Selection, inbreeding, and outbreeding

Experiments conducted in the nineteenth century by Gregor Mendel led to the discovery of the three major principles of heredity: dominance, segregation, and independent assortment. Genes that occupy corresponding loci on homologous chromosomes govern variations of the same characteristic. Suchs genes, called alleles, exist in pairs in diploid organisms. A monohybrid cross is a cross between two individuals that carry different alleles for a single gene locus. Similarly, a dihybrid cross is a cross between two individuals that carry different alleles at each of two gene loci. The results of monohybrid and dihybrid crosses illustrate the basic principles of genetics. The laws of probability are used to predict the results of a cross between two individuals. The order of genes on a chromosome is determined by calculating the frequency that chromatids exchange segments of chromosomal material during meiosis. The sex of most animal species is determined by special chromosomes called sex chromosomes. The relationship between a single pair of alleles at a gene locus and the characteristic it controls may be simple, or it may be complex. For example, dominance may be complete or incomplete; three or more alleles may exist for a given locus; most genes have many different effects; and the presence of a particular allele of one gene pair may determine whether certain alleles of another gene pair are expressed. Also, many characteristics are not inherited through alleles at a single gene locus, but instead result from multiple independent pairs of genes having similar and additive effects. Selection, inbreeding, and outbreeding are used commercially to develop improved strains of plants and animals.

CHAPTER OUTLINE AND CONCEPT REVIEW (Fill in the blanks)

I. INTRODUCTION
II. GREGOR MENDEL FIRST FORMULATED THE PRINCIPLES OF INHERITANCE

A. The principle of dominance states that one gene can mask the expression of another in a hybrid

- ■ 1 A (a)_____ gene may mask the expression of a (b)_____ gene, yielding phenotypes different from the genotype.

B. The principle of segregation states that the genes of a pair separate when gametes are formed

- ■ 2 During meiosis, members of paired genes at each locus _____ so that each gamete contains only one gene from each locus.

- ■ 3 During meiosis, each pair of genes separates independently of _____ located in other homologous chromosomes.

III. ALLELES OCCUPY CORRESPONDING LOCI ON HOMOLOGOUS CHROMOSOMES

- ■ 4 The site of a gene in a chromosome is called its _____.

IV. A MONOHYBRID CROSS INVOLVES INDIVIDUALS WITH DIFFERENT ALLELES FOR A GIVEN GENE LOCUS

- ■ 5 A monohybrid cross begins with homozygous parents who are known as the (a)_____ generation. They differ from one another with respect to their alleles at only one (b)_____.

- ■ 6 Offspring from the P generation cross are heterozygous; they are called the F_1 or (a)_____ generation. Offspring from the F_1 cross are called the (b)_____ generation.

A. Heterozygotes carry two different alleles for a locus; homozygotes carry identical alleles

B. The phenotype of an individual does not always reveal its genotype

C. The Punnett Square predicts the ratios of genotypes and phenotypes of the offspring of a cross

- ■ 7 A monohybrid test cross is a cross between _____ _____ individuals.

D. Monohybrid test cross is used to detect heterozygosity

V. THE LAWS OF PROBABILITY ARE USED TO PREDICT THE LIKELIHOOD OF GENETIC EVENTS

- ■ 8 The probability of a specific event occurring is determined by dividing the number of (a)_____ that occurred by the total number of events. Probability can range from (b)_____ (impossible) to (c)_____ (certain).

A. The product law predicts probabilities of independent events

B. The sum law predicts the probabilities of mutually exclusive events

- ■ 9 The probability that either one or the other of two mutually exclusive events will occur is the _____.

C. The laws of probability can be applied to a variety of calculations

- ■10 The probability of two independent events occurring together is the _____ of the probabilities of each occurring separately.

VI. A DIHYBRID CROSS INVOLVES INDIVIDUALS THAT HAVE DIFFERENT ALLELES AT TWO LOCI

■ 11 A dihybrid cross is a cross between homozygous parents (P generation) that differ with respect to their alleles at _____.

A. The principle of independent assortment states that the members of different gene pairs segregate and assort into gametes independently
B. The mechanics of meiosis are the basis for independent assortment

■ 12 Genes in the same chromosome are said to be _____ and do not assort independently.

VII. THE LINEAR ORDER OF LINKED GENES ON A CHROMOSOME CAN BE "MAPPED" BY CALCULATING THE FREQUENCY OF CROSSING OVER

■ 13 Linked genes are recombined when chromatids exchange genetic material, a process known as _____, that occurs during meiotic prophase.

■ 14 A chromosome can be genetically mapped by determining the frequency of _____ among genes.

VIII. SEX IS COMMONLY DETERMINED BY SPECIAL SEX CHROMOSOMES

■ 15 The sex or gender of many animals is determined by the X and Y sex chromosomes. The other chromosomes in a given organism's genome are called _____.

A. The Y chromosome determines male sex in mammals

■ 16 When a Y-bearing sperm fertilizes an ovum, the result is a(n) (a)_____, and fertilization by an X-bearing sperm produces a(n) (b)_____.

B. X-linked genes have unusual inheritance patterns
C. Sex-influenced genes are autosomal, but their expression is affected by the individual's sex
D. Dosage compensation equalizes the expression of X-linked genes in males and females

■ 17 The effect of X-linked genes is made equivalent in males and females by dose compensation, which is accomplished by a (a)_____ X-chromosome in the male or (b)_____ of one X-chromosome in the female.

IX. THE RELATIONSHIP BETWEEN GENOTYPE AND PHENOTYPE IS NOT ALWAYS STRAIGHTFORWARD

A. Dominance is not always complete
B. Multiple alleles for a locus may exist in a population

■ 18 Multiple alleles are (a)_____(#?) different alleles that can occupy the same (b)_____.

C. A single gene may affect multiple aspects of the phenotype; alleles of different loci may interact to produce a phenotype

■ 19 _____ refers to the many different effects that can often result from a given gene.

■ 20 _____ is when one allele of a gene pair determines whether alleles of other gene pairs are expressed.

X. POLYGENES ACT ADDITIVELY TO PRODUCE A PHENOTYPE

■ 21 It is called _____ when two or more independent pairs of genes have similar and additive effects on a phenotype.

XI. SELECTION, INBREEDING, AND OUTBREEDING ARE USED TO DEVELOP IMPROVED STRAINS

■ 22 Inbreeding increases the _____ of recessive genes.

■ 23 Outbreeding, the mating of unrelated individuals, increases (a)_____. (b)_____ is the improvement of offspring as a result of outbreeding.

KEY TERMS

Frequently Used Prefixes and Suffixes: Use combinations of these prefixes and suffixes to generate terms for the definitions below.

Prefixes	The Meaning
di-	two, twice, double
hemi-	half
hetero-	different, other
homo-	same
mon(o)-	alone, single, one

Prefix	Suffix	Definitions
_____	-oecious	1. Having both male and female parts in one individual.
_____	-zygous	2. Having the same (identical) members of a gene pair.
_____	-zygous	3. Having dissimilar (different) members of a gene pair.
_____	-hybrid	4. Pertaining to the mating of individuals differing in two specific pairs of genes.
_____	-hybrid	5. Pertaining to the mating of individuals differing in one pair of genes.
_____	-oecious	6. Having the male and female parts in two separate and distinct organisms.
_____	-zygous	7. Having only half (one) of a given pair of genes.

Other Terms You Should Know

albinism	dihybrid cross	genotype
allele	dominant	hermaphroditic
anther	dosage compensation	heterozygote advantage
autosome	epistasis	hybrid
Barr body	F_1 generation	hybrid vigor
blending inheritance	F_2 generation	inbreeding
codominance	filial generation	incomplete dominance
controlled pollination	gene	linkage group
crossing over	genetic recombination	locus (-ci)

map unit	pleiotropy	sex-influenced
monohybrid cross	polygenic inheritance	sex-linked
monohybrid test cross	principle of dominance	stigma (-ata)
multiple allele	principle of independent	sum law
normal distribution curve	assortment	test cross
outbreeding	principle of segregation	true breeding
overdominance	product law	variegation
P generation	Punnett square	X chromosome
parental generation	recessive	X-linked
parental type	recombinant gamete	Y chromosome
phenotype	recombinant type	

SELF TEST

Multiple Choice: Some questions may have more than one correct answer.

R and r are genes for flower color. Homozygous dominant and heterozygous genotypes both have red flowers; the homozygous recessive genotype has white flowers. T and t are genes that control plant height. Homozygous dominant plants are tall, heterozygous plants are medium height, and homozygous recessive plants are short. Genes for flower color and height are on different chromosomes. Use these data and the following list to answer questions 1-9. Construct Punnett Squares as needed.

a. RRTT	f. rrtt	k. 1:2:1:2:4:2:1:2:1	p. pink, medium
b. RrTt	g. 1:1	l. red, tall	q. pink, short
c. Rrtt	h. 1:2:1	m. red, medium	r. white, tall
d. rrTT	i. 9:3:3:1	n. red, short	s. white, medium
e. rrTt	j. 1:2:1:1:2:1	o. pink, tall	t. white, short

Plants with the genotypes Rrtt and rrTT are mated. Their offspring are then mated to produce another generation. Questions 1-5 pertain to this last generation.

1. What are the genotypic and phenotypic ratios among offspring?

2. What are all of the phenotypes among offspring?

3. Which of the genotypes listed above are found among offspring?

4. What are the genotypes found among the offspring that are not in the list?

5. What would the phenotypic ratio among offspring be if both flower color and height genes had exhibited incomplete dominance?

Questions 6-9 pertain to the following cross: Rrtt x rrTT.

6. What are the genotypes and phenotypes of the parents?

7. What are the genotypes and phenotypes of the offspring?

8. What is the genotypic ratio among offspring?

9. What is the phenotypic ratio among offspring?

10. Pleiotropy means that a pair of genes
    a. exhibits incomplete dominance.
    b. affects expression of other genes.
    c. has the same effect as another pair of genes.
    d. has multiple effects.
    e. is sex-influenced.

11. Given the following information about crossing over, determine the relative positions of three loci (X,Y,Z) on one chromosome: X and Y = 8%, Y and Z = 5%, X and Z = 3 map units.
    a. X, Y, Z
    b. X, Z, Y
    c. Z, Y, X
    d. Y, Z, X
    e. Z, Y, X

12. Polygenic inheritance means that a pair of genes
    a. exhibits incomplete dominance.
    b. affects expression of other genes.
    c. has the same effect as another pair of genes.
    d. has multiple effects.
    e. is sex-influenced.

13. Epistasis means that a pair of genes
    a. exhibits incomplete dominance.
    b. affects expression of other genes.
    c. has the same effect as another pair of genes.
    d. has multiple effects.
    e. is sex-influenced.

14. Which of the following would be true if hairy toes happened to be a recessive X-linked trait?
    a. All men would have hairy toes.
    b. No women would have hairy toes.
    c. Parents with hairy toes could have a child without hairy toes.
    d. More women than men would have hairy toes.
    e. More men than women would have hairy toes.

15. Phenotypic expression
    a. always involves only one pair of genes.
    b. may involve many pairs of genes.
    c. is partly a function of environmental influences.
    d. refers specifically to the composition of genes.
    e. refers to the appearance of a trait.

16. Linked genes are
    a. inseparable.
    b. generally on the same chromosome.
    c. in separate homologous chromosomes.
    d. in different chromatids.
    e. always separated during crossing over.

17. Which of the following is/are true?
    a. XX is usually female.
    b. XY is usually male.
    c. birds and butterflies do not have sex chromosomes.
    d. hermaphrodites are XX or XY.
    e. XXY is usually male.

18. Which of the following is/are consistent with Mendel's principle of dominance?
    a. All F_1 offspring express the dominant trait.
    b. All F_2 offspring express the dominant trait.
    c. Both organisms in the P generation are homozygous.
    d. Both P generation parents are true breeding.
    e. Only one P generation parent is true breeding.

19. Which of the following applies if two pure-breeding P generation plants are used to ultimately produce an F_2 generation with flower colors in a ratio of 1 red: 2 pink: 1 white? (R is a gene for red; r = white)
    a. All F_1 plants were Rr.
    b. F_1 plants were red and white.
    c. At least one P generation parent was pink.
    d. F_2 genotypic ratio is 3:1.
    e. F_2 genotypic ratio is 1:2:1.

20. Mendel's "factors"
    a. are haploid in gametes.
    b. interact in gametes.
    c. segregate into separate gametes.
    d. are genes.
    e. affect flower color, but not seed color.

VISUAL FOUNDATIONS

Color the parts of the illustration below as indicated.

WHITE	☐	white flowers, white alleles
RED	☐	red flowers, red alleles
GREEN	☐	circle parents
BLUE	☐	circle F_1 generation
ORANGE	☐	circle F_2 generation
PINK	☐	pink flowers

Use an "R" to label red alleles and an "r" to label white alleles.

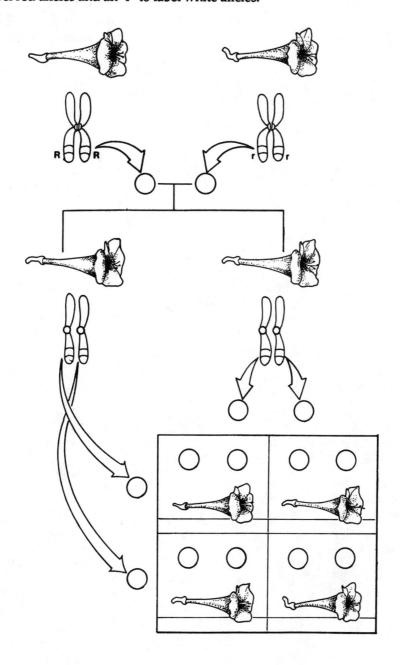

Questions 21–23 pertain to the illustration above.

21. Name the phenomenon illustrated by this figure. _____

22. What color are the flowers of individuals in the F_1 generation? _____

23. What is the phenotypic ratio in the F_2 generation? _____

CHAPTER 11
◻

DNA: The Carrier of Genetic Information

OUTLINE

Genes Code for Proteins
Nucleic acids
Evidence that DNA is the hereditary material
DNA structure
DNA replication
DNA in chromosomes

Research conducted in the first half of this century suggested erroneously that genes were made of proteins. It is now known that they are made of deoxyribonucleic acid (DNA). Each DNA molecule consists of two polynucleotide chains arranged in a coiled double helix. Nucleotides consist of a pentose sugar, a phosphate, and a nitrogenous base. The base is a purine or a pyrimidine. The two chains are joined together by hydrogen bonds between the pyrimidines on one chain and the purines on the other chain. The sequence of bases in DNA provides for the storage of genetic information. When DNA replicates, the hydrogen bonds between the two polynucleotide chains break, and the chains unwind and separate. Each half-helix then pairs with complementary nucleotides, replacing its missing partner. The result is two DNA double helices, each identical to the original and consisting of one original strand from the parent molecule and one newly synthesized complementary strand. DNA replication is complex requiring a number of different enzymes. One polynucleotide chain is synthesized continuously while the other is synthesized discontinuously in short pieces. DNA replication proceeds in both directions from the point at which replication was initiated. DNA is packaged in chromosomes in a highly organized way.

CHAPTER OUTLINE AND CONCEPT REVIEW (Fill in the blanks)

I. INTRODUCTION

II. MOST GENES CARRY INFORMATION FOR MAKING PROTEINS

■ 1 Experiments by _____ led to the "one gene one protein" concept.

III. GENES ARE MADE OF DEOXYRIBONUCLEIC ACID (DNA)

IV. EVIDENCE THAT DNA IS THE HEREDITARY MATERIAL WAS FIRST FOUND IN MICROORGANISMS

■ 2 Evidence that genes are made of nucleic acids was provided by _____ experiments, a process whereby genetic characteristics may pass from one strain of bacteria to another.

■ 3 _____Virus_____ can reproduce by injecting only their DNA into cells, indicating that DNA is the genetic material.

V. THE STRUCTURE OF DNA ALLOWS IT TO CARRY INFORMATION AND TO SELF-REPLICATE

A. Nucleotides can be covalently linked in any order to form long polymers

■ 4 Nucleotides consist of a __phosphate group, pentose sugar, nitrogen base__.

B. DNA is made of two polynucleotide chains intertwined to form a double helix

■ 5 _____ studies by Franklin and Wilkins showed that DNA has a helical structure with nucleotide bases stacked like rungs of a ladder.

■ 6 Watson and Crick devised a DNA model based for the most part on existing data. Their model suggested that DNA was formed from two __polynucleotide strands__ arranged in a coiled double helix.

C. In double-stranded DNA, hydrogen bonds form between adenine and thymine and between guanine and cytosine

■ 7 In DNA, the following are found in ratios of about 1: purines to (a)__pyrimidine__, guanine to (b)__cytosine__, and adenine to (c)__thymine__.

VI. DNA REPLICATION IS SEMICONSERVATIVE: EACH DOUBLE HELIX CONTAINS AN "OLD" STRAND AND A NEWLY SYNTHESIZED STRAND

■ 8 Replication of DNA is considered semiconservative because each "old" strand serves as a __template__ for the formation of a new strand.

■ 9 DNA synthesis proceeds in a (a)_5_' --> (b)_3_' direction. One strand is copied continuously, the other adds (c)__Okazaki__ fragments discontinuously.

■10 ____DNA polymerase____ serve as catalysts in the construction of a new DNA strand from deoxyribonucleoside triphosphates.

■11 DNA replication usually starts at one or more points known as the "__origin of replication__," and proceeds in both directions from that/those point(s).

VII. DNA IN CHROMOSOMES IS PACKAGED IN A HIGHLY ORGANIZED WAY

■12 Most chromosomes in __prokaryotic__ cells are single strands of circular DNA.

■13 The structural unit of eukaryotic chromosomes is the (a)__nucleosome__. It consists of a segment of DNA coiled around (b)__histones__ and an adjacent linker DNA/protein complex.

■14 The loops that comprise nucleosomes are held together by nonhistone __scaffolding__ proteins.

KEY TERMS

Frequently Used Prefixes and Suffixes: Use combinations of these prefixes and suffixes to generate terms for the definitions below.

Prefixes	The Meaning
a-	without, not, lacking
anti-	against, opposite of

Prefix	Suffix	Definitions
_a_	-virulent	1. Not lethal; lacking in virulence.
anti	-parallel	2. The arrangement of the two polynucleotide chains in a DNA molecule, viz., "running" in opposite directions to each other.

Other Terms You Should Know

alkaptonuria
bacteriophage
Chargaff's rules
chromatin
complementary
complete medium (-ia)
conidium (-ia)
density gradient
   centrifugation
deoxyribonucleic acid
DNA
DNA helicase
DNA ligase
DNA polymerase

DNA replication
double helix (-ices)
helix-destabilizing protein
histone
lagging strand
leading strand
mutation
nucleoside triphosphate
nucleosome
nucleotide
Okazaki fragment
one gene, one enzyme
  hypothesis
origins of replication

phosphodiester bond
primosome
purine
pyrimidine
replication fork
ribonucleic acid
RNA
RNA primer
scaffolding protein
semiconservative
topoisomerase
transformation
triphosphate
virulent

SELF TEST

Multiple Choice: Some questions may have more than one correct answer.

1. Which of the following base pairs is/are correct?
   - a. A – T
   - b. C – A
   - c. G – C
   - d. T – A
   - e. T – C

2. A large quantity of homogentistic acid in urine
   - a. is associated with alkaptonuria.
   - b. indicates a kidney disease.
   - c. is caused by a block in the metabolic reactions that break down phenylalanine and tyrosine.
   - d. is due to absence of an oxidizing enzyme.
   - e. indicates that a diuretic has been consumed.

3. Chargaff's rules state or infer that ([ ] = concentration of)
   - a. [A] = [T].
   - b. [G] = [C].
   - c. ratio of purines to pyrimidines = 1.
   - d. ratio of T to A = 1.
   - e. ratio of G to C = 1.

4. The investigators credited with elucidating the structure of DNA are
    a. Hershey and Chase.   d. Watson and Crick.
    b. Meselson and Stahl.   e. Franklin and Wilkens.
    c. Avery, MacLeod, and McCarty.

5. Which of the following is/are true of linkages in DNA?
    a. 5' phosphate to 3' sugar carbon.   d. One strand ends with a 5' phosphate.
    b. The backbone has 5', 3' phosphodiester bonds.   e. Both strands end with a 3' hydroxyl group.
    c. The two strands are joined by covalent bonds.

6. *Neurospora* was a good choice for the Beadle and Tatum studies because
    a. it is diploid.   d. it had no known mutant forms.
    b. its sexual phase enables genetic analysis.   e. the control strain could not use arginine.
    c. it can be cultivated on minimal medium.

7. Deoxyribose and phosphate are joined in the DNA backbone by
    a. one of four bases.   d. phosphodiester bonds.
    b. purines.   e. the 1' carbon of the sugar.
    c. pyrimidines.

8. A mutation is
    a. a change in an enzyme.   d. an alteration of mRNA.
    b. a change in a gene.   e. sometimes inheritable.
    c. a change in DNA.

9. Beadle and Tatum concluded that
    a. one gene affects one polypeptide.   d. mutations have no effect on enzymes.
    b. one gene affects one protein.   e. proteins mutate when exposed to radiations.
    c. a mutation directly affects an enzyme.

10. The 3' end of one DNA fragment is linked to the 5' end of another fragment by means of
    a. DNA ligase.   d. hydrogen bonds.
    b. helicase enzymes.   e. complementary base pairing.
    c. a DNA polymerase.

11. The molecule(s) thought to be responsible for untying knots in replicating DNA is/are
    a. histones.   d. topoisomerases.
    b. chromatin.   e. protosomes.
    c. scaffolding proteins.

12. The following can be said about the leading strand and lagging strand, respectively:
    a. Both form continuously.   d. Forms in short pieces, forms continuously.
    b. Both form in short pieces.   e. Forms continuously, assembles Okazaki fragments.
    c. Forms continuously, forms in short pieces.

13. Nucleosomes are held together in large coiled loops by
    a. histones.   d. topoisomerases.
    b. chromatin.   e. protosomes.
    c. scaffolding proteins.

14. The coiled double helix structure of DNA was suggested as a result of X-ray diffraction data collected by
    a. Hershey and Chase.   d. Watson and Crick.
    b. Griffith.   e. Franklin and Wilkens.
    c. Avery, MacLeod, and McCarty.

15. The direction for synthesis of DNA is
    a. 5' —> 3'.
    b. 3' —> 5'.
    c. 5' —> 5'.
    d. 3' —> 3'.
    e. variable.

16. The proteins intimately associated with DNA in eukaryotic chromosomes are
    a. nucleosomes.
    b. chromatin.
    c. topoisomerases.
    d. absent in prokaryotes.
    e. histones.

17. The genetic code is carried in the
    a. DNA backbone.
    b. sequence of bases.
    c. arrangement of 5', 3' phosphodiester bonds.
    d. Okazaki fragments.
    e. histones.

18. In one molecule of DNA one would expect the composition of the two strands to be
    a. both either old or new.
    b. both all new.
    c. both partly new fragments and partly old parental fragments.
    d. one old, one new.
    e. unpredictable.

VISUAL FOUNDATIONS

Color the parts of the illustration below as indicated.

RED	❑	thymine
GREEN	❑	adenine
YELLOW	❑	cytosine
BLUE	❑	guanine
ORANGE	❑	sugar
BROWN	❑	phosphate
TAN	❑	5', 3' phosphodiester bond

RNA and Protein Synthesis: The Expression of Genetic Information

O U T L I N E

The flow of genetic information
Transcription
Translation
Transcription and translation in eukaryotes
The genetic code
The gene
Mutations

The manufacture of proteins by cells involves two major steps: the transcription of base sequences in DNA into the base sequences in RNA, and the translation of the base sequences in RNA into amino acid sequences in proteins. Messenger RNA (mRNA) is synthesized by DNA-dependent RNA polymerase enzymes using nucleoside triphosphate precursors. Transcription begins when an RNA polymerase recognizes a specific base sequence at the beginning of a gene. mRNA contains coding sequences that directly code for specific proteins and noncoding sequences that do not directly code for protein. Protein synthesis involves the enzymatic linkage of transfer RNA (tRNA) molecules to their respective amino acids. Ribosomes then couple these tRNAs to their codons on mRNA, catalyze peptide bonding of the amino acids, and move the mRNAs so the next codon can be read. The first stage of protein synthesis involves the formation of the protein-synthesizing complex. The second stage involves the addition of amino acids to a growing polypeptide chain. The third stage involves the release of the polypeptide and the dissociation of the ribosomal subunits which, sooner or later, are used with another mRNA molecule. Whereas prokaryotic mRNAs are translated as they are transcribed, eukaryotic mRNAs undergo posttranscriptional modification and processing prior to translation. Many eukaryotic genes differ from prokaryotic genes by having their protein-coding regions interrupted by noncoding regions. Of 64 possible codons, 61 code for amino acids and three serve as signals that specify the end of the coding sequence for a polypeptide chain. The genetic code is nearly universal, suggesting that it was established early in evolutionary history. A gene is a transcribed nucleotide sequence that yields a product with a specific cellular function. A mutation is a change in a DNA nucleotide sequence. Genes can be altered by mutation in a number of ways.

CHAPTER OUTLINE AND CONCEPT REVIEW (Fill in the blanks)

I. INTRODUCTION

II. BASE SEQUENCES IN DNA ARE TRANSCRIBED AS BASE SEQUENCES IN RNA, WHICH ARE THEN TRANSLATED INTO AMINO ACID SEQUENCES IN PROTEINS

■ 1 Sequencing of amino acids in a protein involves two major steps: first is (a)_transcription_____, wherein DNA codes are read into a special messenger RNA, and second is (b)_____translation_____, involving the actual construction of polypeptide chains.

■ 2 Each _codon___ in mRNA consists of three-bases that specify one amino acid in a polypeptide chain.

■ 3 Translation of messages in mRNA is accomplished by (a)_anticodons____ in tRNA. Each tRNA is specific for only one (b)_amino acid_.

III. TRANSCRIPTION IS THE SYNTHESIS OF RNA FROM A DNA TEMPLATE

■ 4 _rRNA mRNA tRNA_____ are the three principal types of RNA transcribed from DNA that are involved directly in protein synthesis.

■ 5 mRNA synthesis is catalyzed by DNA-dependent (a)_RNA polymerase____ enzymes, and it is formed from (b)_nucleoside triphosphate_____ precursors.

A. Messenger RNA contains base sequences that code for protein

■ 6 Transcription is initiated at sites containing specific base sequences called the _promoter___ regions of DNA.

■ 7 Any point closer to the 3' end of transcribed DNA (and therefore toward the 5' end of mRNA) is said to be (a)_upstream____ relative to a given reference point. Conversely, areas toward the 5' end of DNA and the 3' end of mRNA are (b)_downstream_.

B. Messenger RNA contains additional base sequences that do not directly code for protein

■ 8 Well before the protein-coding sequences at the 5' end of mRNA, a segment called the (a)_leader sequence___ contains recognition signals for ribosome binding. In addition, coding sequences are followed by (b)_termination signals___ that specify the end of the protein.

IV. THE NUCLEIC ACID MESSAGE IS DECODED DURING TRANSLATION

A. An amino acid must be attached to its specific transfer RNA before it can become incorporated into a polypeptide

■ 9 The specific enzymes that catalyze the formation of covalent bonds between amino acids and their respective tRNAs are the _aminoacyl-tRNA synthetases____.

B. Transfer RNA molecules have specialized regions with specific functions
C. Ribosomes bring together all the components of the translational machinery

■ 10 Ribosomes couple the (a)_anticodons____ in tRNAs to their appropriate codons on mRNA. They also catalyze the formation of the (b)_peptide___ bonds that link amino acids, and they move mRNA so the next codon can be read.

D. Translation includes initiation, elongation, and termination

■ **11** The codon _AUG_ initiates the formation of the protein-synthesizing complex.

■ **12** The addition of amino acids to a growing polypeptide chain is called _elongation_.

■ **13** Protein synthesis proceeds from the (a) _amino_ end to the (b) _carboxyl_ terminal end of the growing peptide chain.

■ **14** Newly formed polypeptides are released and ribosome subunits dissociate as the result of sequences in mRNA called _termination codons_.

E. A polyribosome is a complex of one mRNA and many ribosomes

V. TRANSCRIPTION AND TRANSLATION ARE MORE COMPLEX IN EUKARYOTES THAN IN PROKARYOTES

■ **15** _methyl – guanylate_ serves as a protective cap on the 5' end of some eukaryotic mRNA chains.

A. Both noncoding nucleotide sequences (introns) and coding sequences (exons) are transcribed from eukaryotic genes

■ **16** Intron is an abbreviation for (a) _intervening sequences_, and exon stands for (b) _expressed sequences_. Introns must be removed in order to form a continuous string of "readable" exons.

B. Evidence exists that new combinations of exons have arisen during evolution

VI. THE GENETIC CODE IS READ AS A SERIES OF CODONS

A. The genetic code is virtually universal
B. The genetic code is redundant

■ **17** There are (a) _61_ (#?) codons for about (b) _40_ (#?) different tRNAs.

VII. THE GENE IS DEFINED AS A FUNCTIONAL UNIT

✗ nucleotide sequences

■ **18** A gene consists of regulatory sequences and the sequence of _polypeptides_ that can be transcribed to yield a product with a specific cellular function.

VIII. MUTATIONS ARE CHANGES IN DNA

■ **19** A mutation is a change in the _base sequences_ in DNA.

■ **20** (a) _Point / base_ mutations involve a change in only one pair of nucleotides. These can lead to the substitution of one amino acid for another, a so called (b) _missense_ mutation or a (c) _nonsense_ mutation involving the conversion of an amino acid specifying codon to a termination codon.

■ **21** _Hot spots_ are regions of DNA that are particularly susceptible to mutations.

KEY TERMS

Frequently Used Prefixes and Suffixes: Use combinations of these prefixes and suffixes to generate terms for the definitions below.

Prefixes	The Meaning	Suffixes	The Meaning
anti-	against, opposite of	-gen	production of
poly-	much, many	-some	body

Prefix	Suffix	Definitions
anti	-codon	1. A sequence of three nucleotides in tRNA that is complementary to ("opposite of"), and combines with, the three-nucleotide codon on mRNA.
ribo-	some	2. An organelle (microbody) composed of RNA and proteins that functions in protein synthesis.
poly	some	3. A submicroscopic complex consisting of many ribosomes attached to an mRNA molecule during translation.
muta-	gen	4. A substance capable of producing mutations.
carcino-	gen	5. A substance capable of producing cancer.

Other Terms You Should Know

aminoacyl tRNA
base-substitution mutation
cap
coding sequence
codon
colinear
complex
DNA-dependent RNA
domain
downstream
elongation
exon
exon shuffling
frameshift mutation
genetic code
heterogeneous nuclear RNA
initiation
initiation factor
initiation tRNA
interrupted coding sequence

intron
leader sequence
messenger RNA
missense mutation
mRNA
mutation
N-formyl methionine
nonsense mutation
nuclear ribonucleoprotein
point mutation
poly-A tail
polyadenylated tail
polyadenylation signal
polymerase
polyribosome
posttranscriptional
   modification and processing
pre-mRNA
promoter
reading frame

ribonucleic acid
ribosomal RNA
ribosome recognition
sequence
ribozyme
rRNA
snRNPs
stop codon
termination
termination codon
transcription
transfer RNA
translation
translocation
transposon
tRNA
upstream
wobble hypothesis

SELF TEST

Multiple Choice: Some questions may have more than one correct answer.

1. DNA-dependent RNA polymerases are enzymes that
   a. add groups of polynucleotides to mRNA.
   b. require a primer.
   c. use nucleoside triphosphates as substrates.
   d. recognize promoters on a gene.
   e. are present in all cells.

2. A recognition sequence for ribosome binding to mRNA is/are
   a. aminoacyl-tRNA.
   b. coding sequence.
   c. initiator.
   d. upstream from the coding sequence.
   e. leader sequence.

3. Which of the following is/are termination codons?
   a. UAG
   b. UUA
   c. UAA
   d. GUA
   e. UGA

4. An "adapter" molecule containing an anticodon and a region to which an amino acid is bonded is a/an
   a. aminoacyl-tRNA.
   b. coding sequence.
   c. initiator.
   d. peptidyl transferase.
   e. leader sequence.

5. The A site accepts
   a. aminoacyl-tRNA.
   b. coding sequence.
   c. initiator.
   d. peptidyl transferase.
   e. leader sequence.

6. The elongation stage of protein synthesis involves
   a. the code AUG.
   b. initiation tRNA.
   c. loading initiation tRNA on the large ribosomal subunit.
   d. adding amino acids to a growing polypeptide chain.
   e. GTP.

7. Ribosomes attach to
   a. 5' end of mRNA.
   b. 3' end of mRNA.
   c. an mRNA recognition sequence.
   d. tRNA.
   e. RNA polymerase.

8. Translation involves
   a. hnRNA.
   b. decoding of codons.
   c. adapter molecules.
   d. copying codes into codons.
   e. copying codons into codes.

9. Energy for transfer of a peptide chain from the A site to the P site is provided by
   a. ATP.
   b. enzymes.
   c. GTP.
   d. ADP.
   e. guanosine triphosphate.

10. Transcription involves
    a. mRNA synthesis.
    b. decoding of codons.
    c. copying DNA information.
    d. copying codes into codons.
    e. peptide bonding.

11. Introns probably function to
    a. separate structural domains of proteins.
    b. integrate special coding sequences.
    c. provide for recombination.
    d. organize assembly of amino acids in proteins.
    e. indirectly provide an evolutionary advantage.

12. Initiation of protein synthesis involves
    a. the code AUG.
    b. initiation tRNA.
    c. loading initiation tRNA on the large ribosomal subunit.
    d. adding amino acids to a growing polypeptide chain.
    e. peptidyl transferase.

13. Post-transcriptional modification and processing of mRNA takes place in
    a. cytoplasm.
    b. the nucleus.
    c. prokaryotes only.
    d. eukaryotes only.
    e. both prokaryotes and eukaryotes.

14. A string of dozens to hundreds of adenine nucleotides thought to prevent mRNA degradation is a description of a
    a. cap on the 5' end of mRNA.
    b. poly-A tail.
    c. sequence added to mRNA about one minute after transcription.
    d. portion of the activated ribosomal complex.
    e. cap on the 3' end of mRNA.

VISUAL FOUNDATIONS

Color the parts of the illustration below as indicated.

RED   ❑   formylated methionine tRNA

GREEN   ❑   small ribosome subunit

YELLOW   ❑   large ribosome subunit

BLUE   ❑   P site

ORANGE   ❑   A site

BROWN   ❑   mRNA binding site

TAN   ❑   initiation codon

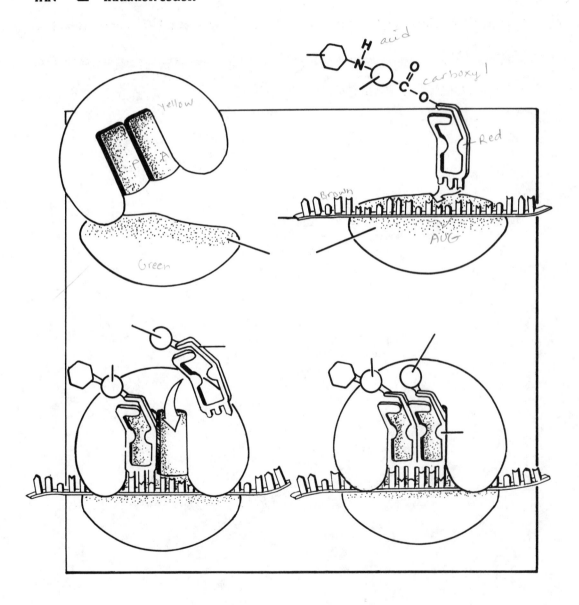

CHAPTER 13
□

Gene Regulation: The Control of Gene Expression

Economical controls in prokaryotes
Multifaceted gene regulation in eukaryotes

Genes are regulated and only part of the genetic information in any given cell is expressed. In prokaryotes, most regulated genes are organized into units called operons which may encode several proteins. The transcription of an operon is initiated at its promoter site, a sequence of DNA bases located upstream from the protein-coding region. Protein synthesis by the operon is controlled by the operator, a sequence of DNA bases also located upstream from the operon and overlapping part of the promoter site. Transcription is blocked when a repressor protein binds to the operator, thereby preventing RNA polymerase from binding to the promoter site. Inducible operons are normally turned off and are only turned on when a metabolite binds to the repressor protein so it cannot bind to the operator. Repressible operons are normally turned on and are only turned off when a metabolite binds to the operator and turns the operon off. Groups of operons may be organized into multigene systems called regulons that are controlled by a single regulatory gene. Constitutive genes are neither inducible nor repressible, but are active at all times. DNA-binding regulatory proteins recognize and bind to specific sequences of DNA bases, thereby inactivating functional groups of base pairs. The control of gene expression in prokaryotes also occurs during and after translation. Regulation of gene expression in eukaryotes occurs at the level of transcription, mRNA processing, translation, and the protein product. Eukaryotic transcription of both constitutive and inducible genes requires a promoter consisting of an RNA polymerase-binding site and short DNA sequences that affect the strength of the promoter. The rate of RNA synthesis of an inducible gene is increased by an enhancer, a regulatory element that can be located large distances away from the actual coding region of a gene. Some genes whose products are required in large amounts exist as multiple copies in the chromosome. Some genes are inactivated by changes in chromosome structure.

CHAPTER OUTLINE AND CONCEPT REVIEW (Fill in the blanks)

I. INTRODUCTION
II. GENE REGULATION IN PROKARYOTES EMPHASIZES ECONOMY

■ 1 _____ are genes that encode essential proteins that are in constant use.

A. Operons in prokaryotes allow for coordinated control of functionally related genes

■ 2 A group of functionally related genes may be controlled by one _____ _____ that is located upstream from the protein-coding region.

■ 3 The _____ is a sequence of bases that switches mRNA synthesis "on" or "off."

■ 4 Transcription of an operon is blocked when a _____ binds to the operator sequence and covers part of the promoter.

■ 5 An (a)_____ inactivates a repressor in order to turn "on" a gene or operon. Such a system is called an (b)_____ system.

■ 6 Repressible operons are turned "off" when a repressor is activated by binding to a product of the system called the _____.

■ 8 A _____ is a set of operons that is controlled by one regulatory gene.

B. Some posttranscriptional regulation occurs in prokaryotes

■ 9 _____ is when a product blocks its own production by binding to an enzyme that is required to generate that product.

III. GENE REGULATION IN EUKARYOTES IS MULTIFACETED

A. Eukaryotic transcription is controlled at many sites and by many different regulatory molecules

■ 10 RNA polymerase in multicellular eukaryotes binds to a portion of the promoter known as the _____.

■ 11 The efficiency of a eukaryotic promoter depends largely on the number and type of

_____.

■ 12 The structure of eukaryotic chromosomes varies with the degree of gene activity. Chromosomes containing inactive genes are called (a)_____, while chromosomes containing genes in an active state are referred to as (b)_____.

B. The long-lived, highly processed mRNAs of eukaryotes provide many opportunities for posttranscriptional control

■ 13 As a result of _____ _____, the same gene that produces calcitonin in the thyroid gland can produce a neurotransmitter in the brain.

C. The activity of eukaryotic proteins may be altered by posttranslational chemical modifications

■ 14 The addition or removal of phosphate groups is an example of _____ _____, a mechanism used by eukaryotic cells to regulate protein activity.

KEY TERMS

Frequently Used Prefixes and Suffixes: Use combinations of these prefixes and suffixes to generate terms for the definitions below.

Prefixes	The Meaning
co-	with, together, in association
eu-	good, well, "true"
hetero-	different, other
homo-	same

Prefix	Suffix	Definitions
_____	-chromatin	1. Chromatin that appears loosely coiled. Because it is the only chromatin capable of transcription, it is considered to be the "good" or "true" chromatin.
_____	-chromatin	2. The inactive chromatin that appears highly coiled and compacted, and is generally not capable of transcription. (Chromatin that is "different from" or "other than" the euchromatin.)
_____	-repressor	3. A substance that, together with a repressor, represses protein synthesis in a specific gene.
_____	-dimer	4. A dimer in which the two component polypeptides are different.
_____	-dimer	5. A dimer in which the two component polypeptides are the same.

Other Terms You Should Know

allosteric binding site
AMP
beta galactosidase
cAMP
CAP
catabolite activator protein
constitutive gene
differential nuclear RNA processing
dimer
enhancer

feedback inhibition
gene amplification
heterodimer
homodimer
inducer molecule
inducible system
lactose operon
lactose repressor
negative control
operator
positive control

posttranscriptional control
proteolytic processing
regulon
repressible enzyme
selective degradation
TATA box
temporal regulation
tissue-specific regulation
translational control
UPE
upstream promoter element

SELF TEST

Multiple Choice: Some questions may have more than one correct answer.

1. The basic way(s) that cells control their metabolic activities is/are by
   a. regulating enzyme activity.
   b. mutations.
   c. developing different genes for different purposes.
   d. controlling the number of enzyme molecules.
   e. transduction.

2. Codes for repressor and activator proteins are
   a. "off" usually.
   b. always "on".
   c. facultative.
   d. a constitutive gene.
   e. sometimes "off," sometimes "on".

3. Feedback inhibition is an example of
   a. transcriptional control.
   b. pretranscriptional control.
   c. a control mechanism affecting events after translation.
   d. an inducible system.
   e. a repressible system.

4. The lactose repressor
   a. can convert to an operator.
   b. is several bases upstream from the operator.
   c. is downstream from RNA polymerase coding sequences.
   d. becomes an activator when lactose is present.
   e. is always "on".

5. The genes and genetic information in different cells of multicellular organisms are
   a. all slightly different.
   b. distinctly different.
   c. identical in the same tissues, different in different tissues.
   d. identical.
   e. almost all identical.

6. Duplication of specific genes in cells that need more of them is an example of
   a. magnification.
   b. amplification.
   c. induction.
   d. a positive control system.
   e. a regulon.

7. The lactose operon promoter
   a. contains a translation termination codon.
   b. is several bases downstream from the operator.
   c. is downstream from RNA polymerase coding sequences.
   d. is the recognition site for three genes.
   e. is sometimes covered by a repressor.

8. One would expect a gene involved in the manufacture of ATP to be
   a. "off" usually.
   b. always "on".
   c. facultative.
   d. a constitutive gene.
   e. sometimes "off," sometimes "on".

9. The tryptophan operon in *E. coli* is
   a. usually on.
   b. on in the absence of corepressor.
   c. repressed only when tryptophan binds to repressor.
   d. an inducible system.
   e. a repressible system.

10. UPEs
    a. are in DNA near promoters.
    b. are amino acids.
    c. seem to determine the strength of a repressor.
    d. are sequences of DNA bases.
    e. affect promoter activity.

11. The lactose operon in *E. coli*
    a. contains three structural genes.
    b. is turned on by a repressor.
    c. contains an operator that overlaps a promoter.
    d. is transcribed as part of a single RNA molecule.
    e. is an inducible system.

12. "Zinc fingers" seems to be involved in
    a. unwinding DNA.
    b. activating transcription.
    c. insertion of a regulator domain into grooves of DNA.
    d. binding a regulatory protein to DNA.
    e. blocking posttranscriptional events.

13. Transcription of a gene is turned "on" instead of "off" in/when
    a. positive control systems.
    d. regulators stimulate transcription.
    b. negative control systems.
    e. regulons.
    c. systems that sense a metabolic substrate.

14. A group of operons controlled by one regulator may be
    a. embedded in the lactose operon.
    d. responding to changing environments.
    b. a regulon.
    e. a lactose repressor.
    c. involved in nitrogen metabolism.

15. A tightly coiled portion of DNA that contains inactive genes is
    a. called heterochromatin.
    d. found only in prokaryotes.
    b. called euchromatin.
    e. found only in eukaryotes.
    c. found in most eukaryotes and few prokaryotes.

Visual Foundations on next page ☞

VISUAL FOUNDATIONS

Color the parts of the illustration below as indicated.

RED ☐ operator
GREEN ☐ promoter
YELLOW ☐ repressor
BLUE ☐ mRNA
ORANGE ☐ metabolic derivative
BROWN ☐ inactivated repressor
TAN ☐ structural gene

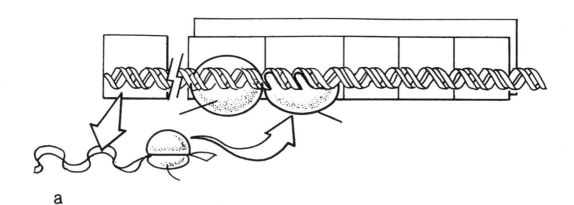

a

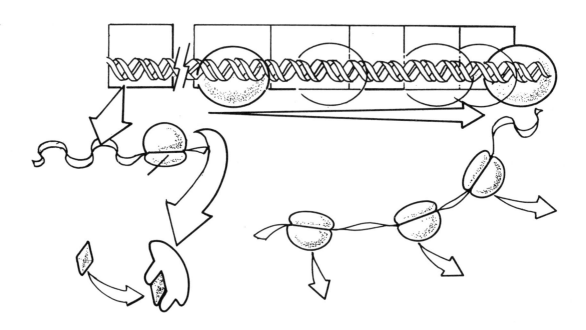

b

CHAPTER 14

Genetic Engineering

OUTLINE

Recombinant DNA methods
Practical applications of genetic engineering

Recombinant DNA technology involves introducing foreign DNA into cells of microorganisms where it replicates and is transmitted to daughter cells. In this way, a particular DNA sequence can be amplified to provide millions of identical copies that can be isolated in pure form. Bacteria manufacture specific enzymes that cut DNA molecules at specific base sequences. The resulting fragments are incorporated into vector molecules, forming recombinant DNA molecules. In this manner, an entire library of recombinant molecules can be formed which contains all of the fragments of an organism's genome. In eukaryotes, libraries of recombinant molecules can also be formed by making copies of mRNA which can then be incorporated into vector molecules. The recombinant molecules can be cloned by introduction into, and multiplication of, host cells. Once a piece of DNA is cloned, its functions can be studied and possibly engineered for a particular application. DNA can also be amplified in vitro by means of the polymerase chain reaction. Once a DNA fragment is cloned, a restriction map of it is constructed which provides the information needed to isolate (subclone) smaller DNA fragments. The subcloned regions of the fragment can then be DNA sequenced or used as DNA probes for analytical purposes. Sequencing DNA enables identification of the parts that contain protein-coding and regulatory regions. Recombinant DNA technology has led to genetic engineering -- the modification of the DNA of an organism to produce new genes with new characteristics. Engineered genes now produce improved pharmaceutical and agricultural products. Expression of eukaryotic genes in bacteria is often difficult because the gene must be linked to regulatory elements that the bacterium may not have. The expression of such genes in eukaryotic organisms is promising since eukaryotes already have the machinery to process and modify eukaryotic proteins. Transgenic organisms, i.e., plants and animals that have incorporated foreign genes, are useful in research and commerce.

CHAPTER OUTLINE AND CONCEPT REVIEW (Fill in the blanks)

I. INTRODUCTION
II. RECOMBINANT DNA METHODS GREW OUT OF RESEARCH IN MICROBIAL GENETICS

A. Restriction enzymes are "molecular scissors" that cleave DNA reproducibly

■ 1 A _____ sequence reads the same as its complement, but in the opposite direction.

B. Recombinant DNA is formed when DNA is spliced into a vector (DNA carrier)

■ 2 _____ cut isolated DNA at specific loci, producing precise fragments that can be incorporated into vector molecules.

■ 3 Recombinant DNA vectors are usually constructed from _____ _____ .

C. Cloning techniques provide the means for replicating and isolating many copies of a specific recombinant DNA molecule

■ 4 _____ are radioactive strands of RNA or DNA that are complementary to specific targeted cloned fragments. They are used to identify a particular fragment in a large population of clones.

■ 5 A _____ is a population of recombinant plasmids that collectively contain a fragmented genome.

■ 6 In order to avoid cloning introns, the enzyme _____ is used to create complementary DNA (cDNA) on a mRNA template.

D. The polymerase chain reaction is a technique for amplifying DNA in vitro

■ 7 The polymerase chain reaction (PCR) technique for amplifying DNA is advantageous because it side-steps the cumbersome, time-consuming process of cloning DNA. PCR is an *in vitro* process that alternately uses (a)_____ _____ to replicate DNA with the use of (b)_____ to dissociate the replicated DNA strands for further replications.

E. A cloned gene sequence is usually first analyzed by restriction mapping, followed by DNA sequencing

■ 8 _____ are detailed "charts" that help identify the specific sites cut by selected restriction enzymes.

■ 9 "Riflips," an acronym for _____ _____, are used to elucidate the degree of variability among genes in a population. "Riflips" are especially useful to study genetic relatedness of individuals.

III. GENETIC ENGINEERING HAS MANY PRACTICAL APPLICATIONS

A. Additional engineering is required for a recombinant eukaryotic gene to be expressed in bacteria

■10 Once a eukaryotic gene is incorporated into a bacterial genome, in order to produce the object product (an encoded protein), the gene must be linked to _____ _____ sequences that bacterial RNA polymerase can recognize.

B. Transgenic organisms incorporate foreign DNA into their cells

■11 Transgenic organisms (plants or animals) are generally produced by injecting the DNA of a gene into a recipient's (a)_____ , or by using (b)_____ as recombinant DNA vectors.

KEY TERMS

Frequently Used Prefixes and Suffixes: Use combinations of these prefixes and suffixes to generate terms for the definitions below.

Prefixes	The Meaning
poly-	much, many
retro-	backward
trans-	across

Prefix	Suffix	Definitions
_____	-morphism	1. The existence within a species of more than one (often many) different phenotypes; the presence of more than one allele for a given locus.
_____	-genic	2. Pertains to organisms that have had their genome altered by recombinant DNA technology. (Genes are taken from one organism and transferred "across" into another organism.)
_____	-virus	3. An RNA virus that makes DNA copies of itself by reverse transcription.

Other Terms You Should Know

bacteriophage
cDNA library
DNA ligase
gel electrophoresis
gene therapy
genetic engineering
genetic probe
genome

genomic library
hybridize
palindromic
plasmid
recombinant DNA
technology
restriction enzyme
restriction map

restriction polymorphism
reverse transcriptase
sticky end
transformation
transgenic animal
vector molecule

SELF TEST

Multiple Choice: Some questions may have more than one correct answer.

1. Creating a eukaryotic gene bank involves the use of
   a. eukaryotic DNA fragments.
   b. bacteriophages.
   c. plasmids treated with restriction enzymes.
   d. clones.
   e. identical plasmid and eukaryotic fragments.

2. Restriction enzymes are normally used by bacteria to
   a. defend against viruses.
   b. remove extraneous introns.
   c. denature antibiotics.
   d. cut plasmids from the large chromosome.
   e. insert operators in active DNA.

3. A restriction map of a DNA fragment involves
   a. plasmid markers.
   b. isolating "subfragments."
   c. reconstructing a gene from its fragments.
   d. identifying sites attacked by restriction enzymes.
   e. sizing "subfragments."

4. The Ti used to insert genes into plant cells is
   a. an RNA fragment.   d. a plasmid.
   b. only effective in dicots. e. a vector.
   c. useful to increase yields of grain foods.

5. A base sequence that is palindromic to 3'-AGCTTAA-5' would read
   a. 3'-AGCTTAA-5'.   d. 5'-UCGAAUU-3'.
   b. 5'-TCGAATT-3'.   e. 3'-UCGAAUU-5'.
   c. 3'-TCGAATT-5'.

6. Restriction fragment length polymorphisms are
   a. complementary.   d. possibly due to mutations.
   b. RNA variants.   e. variants paired in order to obtain different clones.
   c. the result of changes in DNA.

7. The vector(s) commonly used to incorporate a DNA fragment into a carrier is/are
   a. prophages.   d. translocation microphages.
   b. E. coli.   e. bacteriophages.
   c. plasmids.

8. Palindromic sequences are base sequences from one DNA strand that
   a. produce RNA polymerase.   d. cut specific fragments from DNA.
   b. form RNA complementary to DNA.   e. read in the opposite direction as its complement.
   c. can be cut by restriction enzymes.

9. cDNA libraries are compiled
   a. from introns.   d. from introns and exons.
   b. DNA copies of mRNA.   e. with the use of reverse transcriptase.
   c. from multiple copies of DNA fragments.

10. When a virus is used as a vector in mammalian cells, it is first
    a. transformed.   d. disabled.
    b. stripped of foreign DNA. e. coated with activating protein.
    c. incorporated into calcium phosphate crystals.

11. After "sticky ends" pair with other DNA molecules cut by the same enzyme, the two fragments can be covalently linked to form recombinant DNA by treating them with
    a. plamozymes.   d. DNA ligase.
    b. restriction enzymes. e. vector enzymes.
    c. DNA polymerase.

12. DNA fragments for recombination are produced by cutting DNA
    a. and introducing a cleaved plasmid.   d. with bacteriophages.
    b. with restriction enzymes.   e. with DNA ligase.
    c. at specific base sequences.

13. Obstacles that stand in the way of developing gene products of higher organisms in bacteria include
    a. obtaining eukaryotic fragments.   d. prokaryotic processing of eukaryotic introns.
    b. identifying a vector.   e. association of engineered gene with a promoter.
    c. association of engineered gene with a regulator.

14. RNA viruses that use reverse transcriptase to make DNA copies of themselves
    a. include the AIDS virus.   d. are the choice for plant vectors.
    b. are bacteriophages.   e. are retroviruses.
    c. can be used to eliminate troublesome introns.

15. Transgenic organisms are
    a. dead animals.
    b. vectors for retroviruses.
    c. used to produce recombinant proteins.
    d. animals with foreign genes.
    e. plants with foreign genes.

16. A radioactive strand of RNA or DNA that is used to locate a cloned DNA fragment is
    a. a genetic probe.
    b. the genome.
    c. complementary to the targeted gene.
    d. a restriction enzyme.
    e. complementary to the ligase used.

VISUAL FOUNDATIONS

Color the parts of the illustration below as indicated.

RED   ☐   exon

GREEN   ☐   intron

YELLOW   ☐   reverse transcriptase

BLUE   ☐   DNA polymerase

ORANGE   ☐   complementary DNA

Also label DNA, pre-mRNA, and mRNA.

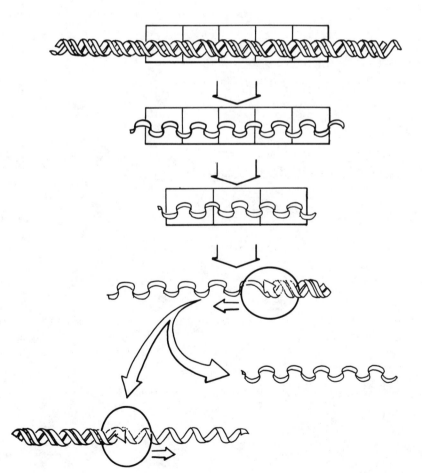

CHAPTER 15

Human Genetics

OUTLINE

Alternative methods used by human geneticists
Genetic and environmental interactions
Birth defects
Chromosomal abnormalities
Genetic diseases with simple inheritance patterns
Prenatal detection of genetic abnormalities
Genetic counseling
Gene replacement therapy
Natural variation in the human population
The Human Genome Initiative
Genetics and society

Human geneticists, unable to make specific crosses of pure strains, must rely on studies of populations, analyses of family pedigrees, and molecular studies. The development of human characteristics is regulated by a large number of interacting genes. Some birth defects are inherited; others are produced by environmental factors. Studies of the number and kinds of chromosomes in the nucleus permit detection of various chromosomal abnormalities, most of which are lethal or cause serious defects. Autosomal abnormalities are usually more severe than sex chromosome abnormalities. Abnormal alleles at a number of loci cause many inherited diseases. Most human genetic diseases with simple inheritance patterns are transmitted as autosomal recessive traits. Autosomal dominant disorders and x-linked recessive disorders also occur. Some genetic diseases can be diagnosed before birth by amniocentesis or chorionic villus sampling. Genetic counselors advise prospective parents with a history of genetic disease about the probability of their having an affected child. Gene replacement therapy is currently being actively researched. A great deal of the natural variation that exists in human beings has a genetic basis. Much of our knowledge of human variation is based on studies of ABO blood group and Rh alleles. Many human characteristics, such as height and intelligence, show continuous variation because a number of genes are involved in the expression of the characteristic. A major research project is underway to determine the total informational content of the human genome. Many people are misinformed about genetic diseases and their effect on society. A high frequency of one abnormal allele in a given group does not mean that the group has a higher frequency of abnormal alleles in general. Virtually everyone is heterozygous for several abnormal alleles. Alleles that cause a genetic disease when homozygous may confer an advantage when heterozygous. Matings of close relatives produce a greater frequency of abnormal offspring.

CHAPTER OUTLINE AND CONCEPT REVIEW (Fill in the blanks)

I. INTRODUCTION

■ **1** Human geneticists can rarely make specific crosses of _____; that is, pure strains of individuals that are homozygous at nearly all loci.

II. ANALYSIS OF INHERITANCE PATTERNS IN HUMANS REQUIRES ALTERNATIVE METHODS

III. MOST HUMAN TRAITS ARE DUE TO COMPLEX GENETIC AND ENVIRONMENTAL INTERACTIONS

IV. SOME BIRTH DEFECTS ARE INHERITED

■ **2** Defects that are present at birth are called birth defects, or _____ defects.

V. CHROMOSOME ABNORMALITIES ARE RESPONSIBLE FOR SOME BIRTH DEFECTS

■ **3** _____ is the study of chromosomes and their role in inheritance.

A. Karyotyping is the analysis of chromosomes

■ **4** The karyotype of an individual is a composite _____ that shows his/her chromosome composition.

B. Most chromosome abnormalities are lethal or cause serious defects

■ **5** _____, the presence of multiple sets of chromosomes, is usually lethal in humans.

■ **6** The abnormal presence or absence of one chromosome in a set is called (a)_____. For the affected chromosome, the normal condition, (b)_____, becomes a (c)_____ condition with the addition of an extra chromosome, and a (d)_____ condition when one member of a pair is missing.

■ **7** (a)_____ occurs when chromosomes do not separate at anaphase. (b)_____ is an example of a condition that results from such an occurrence in chromosome 21.

C. Sex chromosome abnormalities are usually less severe than autosomal abnormalities

■ **8** (a)_____ attempts to determine the gender of an individual based on the presence (female) or absence (male) of a (b)_____.

■ **9** Persons with _____ are nearly normal males, except for an extra X chromosome and Barr bodies.

■ **10** Persons with _____ are sterile females without Barr bodies.

D. Chromosome abnormalities are relatively common at conception but usually result in prenatal death

VI. MANY GENETIC DISEASES SHOW SIMPLE INHERITANCE PATTERNS

A. Most genetic diseases are inherited as autosomal recessive traits

■ **11** PKU and akaptonuria are examples of disorders involving enzyme defects, collectively known as _____.

■ 12 An infliction called _____ expresses itself in the form abnormal hemoglobin in RBCs.

■ 13 _____ results from a defective protein, the one that evidently controls the transport of chloride ions across cell membranes.

B. Huntington's disease is an autosomal dominant disorder that affects the nervous system
C. Hemophilia A is an X-linked recessive disorder that affects blood clotting

VII. SOME GENETIC ABNORMALITIES AND OTHER BIRTH DEFECTS CAN BE DETECTED BEFORE BIRTH

■ 14 _____ may be used to diagnose prenatal genetic diseases. It analyzes cells in amniotic fluid withdrawn from the uterus of a pregnant woman.

■ 15 One technique designed to detect prenatal genetic defects is _____ _____, which involves inspecting fetal cells involved in forming the placenta.

VIII. GENETIC COUNSELORS EDUCATE PEOPLE ABOUT GENETIC DISEASES AND GIVE THEM INFORMATION TO MAKE REPRODUCTIVE DECISIONS

IX. THE FEASIBILITY OF GENE REPLACEMENT THERAPY IS BEING EXPLORED

■ 16 Some victims of _____, a disorder affecting the immune system, have been successfully treated by introducing normal bone marrow cells or injecting the enzyme ADA.

X. A GREAT DEAL OF NATURAL VARIATION EXISTS IN THE HUMAN POPULATION

A. Studies on blood contribute to our understanding of genetic diversity in humans

■ 17 Genotypes (a)_____ produce blood type A, (b)_____ produce type B, (c)_____ produces AB, and (d)_____ results in type O.

■ 18 _____ may result when a pregnant Rh-negative woman's sensitized WBCs are producing large quantities of anti-D antibodies.

B. Quantitative traits are controlled by polygenes
C. Many common physical characteristics are inherited

XI. THE HUMAN GENOME INITIATIVE IS A SYSTEMATIC STUDY OF ALL HUMAN GENES

XII. BOTH HUMAN GENETICS AND OUR BELIEFS ABOUT GENETICS HAVE AN IMPACT ON SOCIETY

■ 19 Autosomal recessive genetic diseases appear with greater frequency in the offspring resulting from _____ than from random matings among individuals that are not related.

KEY TERMS

Frequently Used Prefixes and Suffixes: Use combinations of these prefixes and suffixes to generate terms for the definitions below.

Prefixes	The Meaning
cyto-	cell
iso-	equal, "same"

poly-   much, many
trans-   across, beyond, through
tri-   three

Prefix	Suffix	Definitions
_____	-genic	1. Pertaining to homozygosity (having the same alleles) at all loci.
_____	-genetics	2. The branch of biology that uses the methods of cytology to study genetics.
_____	-ploidy	3. The presence of multiples of complete chromosome sets.
_____	-somy	4. Condition in which a chromosome is present in triplicate instead of duplicate (i.e., the normal pair).
_____	-location	5. An abnormality in which a part of a chromosome breaks off and attaches to another chromosome.

Other Terms You Should Know

amniocentesis
aneuploidy (-dies)
chorionic villus sampling
codominant
congenital defect
consanquineous mating
CVS
cystic fibrosis
deletion
disomic
Down syndrome
duplication
erythroblastosis fetalis
hemophilia A

Huntington's chorea
Huntington's disease
inborn errors of metabolism
karyotype
Klinefelter syndrome
lectin
maternal PKU
mongolism
monosomic
monosomy
nondisjunction
nuclear sexing
pedigree
phenylketonuria

PKU
quantitative trait
Rh factor
Rh incompatibility
sickle cell anemia
sickle cell trait
syndrome
Tay-Sachs disease
tissue typing
trisomy 21
Turner syndrome
ultrasound imaging
XYY karyotype

SELF TEST

Multiple Choice: Some questions may have more than one correct answer.

1. An ideal organism for genetic studies would be one that
   a. is polygenic.
   b. is heterozygous.
   c. has a short generation time.
   d. produces one offspring per mating.
   e. can be controlled.

2. Consanguineous matings produce more abnormal offspring because of the increased possibility of
   a. heterozygosity.
   b. dominant homozygosity.
   c. recessive homozygosity.
   d. nondisjunctions.
   e. monosomy.

3. If a couple has one child with cystic fibrosis, their chance of having another affected child is
   a. 10%.
   b. 25%.
   c. 50%.
   d. 75%.
   e. almost 100%.

4. A person with Down syndrome will likely
   a. have 47 chromosomes.
   b. have trisomy 21.
   c. be mentally retarded.
   d. be born of a mother in her teens.
   e. have a father with Down syndrome.

5. The study of human genetics relies mainly on
   a. pedigrees.
   b. Punnett squares.
   c. fetal karyotypes.
   d. distribution of a trait in a population.
   e. nondisjunction.

6. Amniocentesis involves
   a. examination of maternal blood.
   b. insertion of a needle into the uterus.
   c. examination of fetal cells.
   d. examination of karyotypes.
   e. appraisal of paternal abnormalities.

7. A person with Klinefelter syndrome will
   a. have small testes.
   b. be female.
   c. have 45 chromosomes.
   d. have two X chromosomes.
   e. have a Barr body.

8. Congenital defects
   a. are acquired in infancy.
   b. may be inherited.
   c. may be produced by the environment.
   d. are present at birth.
   e. are only expressed when alleles are homozygous.

9. Translocations may involve
   a. loss of genes.
   b. acquisition of extra genes.
   c. duplication of genes.
   d. loss of a chromosome.
   e. acquisition of an extra chromosome.

10. A trisomic individual
   a. is missing one of a pair of chromosomes.
   b. may result from nondisjunction.
   c. has at least three X chromosomes.
   d. is a male.
   e. has an extra chromosome.

11. Nondisjunction of one chromosome in the second meiotic division would result in which of the following combination(s) of numbers of chromosomes and ratios in human gametes?
   a. 1-24:1-22
   b. 1-22:2-23:1-24
   c. all 23
   d. all 22
   e. 1-23:1-22

12. The most accurate statement about the frequency of abnormal genes is that they are found in
   a. certain ethnic groups.
   b. certain cultural groups.
   c. college professors.
   d. college students.
   e. everyone.

VISUAL FOUNDATIONS

Questions 13-15 show pedigrees of the type that are used to determine patterns of inheritance. Use the
following list to answer questions 13-15:
        a. sex-linked recessive
        b. sex-linked dominant
        c. autosomal dominant
        d. autosomal recessive
        e. cannot determine the pattern of inheritance

13. What is the pattern of inheritance for the following pedigree?

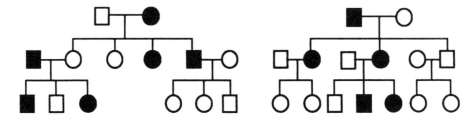

14. What is the pattern of inheritance for the following pedigree?

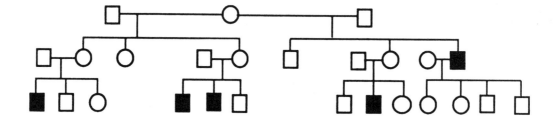

15. What is the pattern of inheritance for the following pedigree?

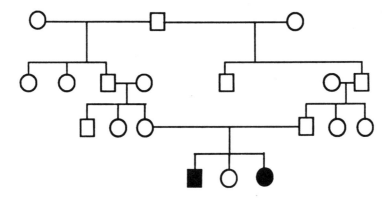

CHAPTER 16

❑

Genes and Development

OUTLINE

Cellular differentiation
Molecular genetic studies of development
Future challenges in developmental biology

Development is the process by which cells specialize and organize into a complex organism. Groups of cells become gradually committed to specific patterns of gene activity and progressively organized into recognizable structures. Cellular specialization is not due to the loss of genes during development, but rather to differential gene activity. A variety of organisms and methodologies are used today to identify genes that control development and to determine how those genes work. The earliest stages of fruit fly development are controlled by maternal genes. Then, following 13 cell divisions, the genes in the developing embryo begin to be expressed, extending the developmental program beyond the pattern established by the maternal genome. Still later-acting genes are responsible for specifying the identity of each body segment. Many of these genes contain a sequence of 180 DNA base pairs; homologous sequences have been found in a wide range of other organisms, including humans, and are thought to control development. Early development in the nematode worm is very rigid; if a particular cell is destroyed or removed, the structures that would normally develop from that cell are missing. Genes have been identified in the nematode that control developmental processes, such as interactions with neighboring cells, programmed cell death, and developmental timing. Mouse development, in contrast to that of the nematode, is not rigid. An embryo with extra or missing cells will still develop normally. Foreign DNA injected into fertilized mouse eggs can be incorporated into the chromosomes and expressed. The resulting mice provide information about gene activation during development. Genes affecting development have also been identified in certain plants. Although uncommon, one type of developmental regulation involves structural changes in the DNA that result in new coding sequences, an important mechanism for the development of the immune system. Another uncommon type of regulation involves making multiple copies of DNA sequences that code for products in high demand.

CHAPTER OUTLINE AND CONCEPT REVIEW (Fill in the blanks)

I. INTRODUCTION

■ 1 Development involves cell specialization, a developmental process involving a gradual commitment by each cell to a specific pattern of gene activity. This process is called (a)_____, and the final step in the process is referred to as (b)_____.

■ **2** The distinctive organizational pattern, or form, that characterizes an organism is acquired through a process known as (a)_____. It encompasses a series of steps known as (b)_____, by which groups of cells organize into identifiable structures.

II. CELLULAR DIFFERENTIATION USUALLY DOES NOT INVOLVE CHANGES IN THE DNA

■ **3** A body of data supports the assertion that, in at least some organisms, almost all nuclei of differentiated cells are identical to each other and to the nucleus of the single cell from which they are descended, which has given rise to the concept of

_____.

■ **4** _____ cells are fully differentiated, yet, since they can still develop, they apparently still contain all of the original genetic equipment of their common single ancestral cell.

A. A totipotent nucleus contains all the information required to direct normal development

B. Most differences among cells are due to differential gene expression

III. MOLECULAR GENETICS IS REVOLUTIONIZING THE STUDY OF DEVELOPMENT

A. Certain organisms are particularly well suited for studies on the genetic control of development

B. *Drosophila melanogaster* provides researchers with a wealth of developmental mutants

■ **5** Large, multistranded interphase chromosomes called _____ chromosomes result when DNA replicates repeatedly without mitosis occurring.

■ **6** _____ are "decondensed" segments of multistranded chromosomes that are actively synthesizing RNA.

■ **7** The initial stages of development in *Drosophila* are under the control of _____ genes.

■ **8** Once maternal genes have initiated pattern formation in the embryonic fruit fly, a set of (a)_____ begin to act on development. Among these are the (b)_____ that begin to organize segments and the (c)_____ _____ that specify the ultimate identity of each imaginal disc and segment.

■ **9** The _____ is a short DNA sequence that is consistently found in association with known developmental genes. It is useful as a probe for locating and cloning developmental mutants.

C. *Caenorhabditis elegans* is a roundworm with a very rigid early developmental pattern

■ **10** The fates of cells in adult roundworms are predetermined by a few _____ _____ that form in the early embryo.

■ **11** Development in *Caenorhabditis elegans* is rigidly fixed in a predetermined pattern; that is, specific cells throughout the early embryo are already irrevocably "programmed" to develop in a particular way. An embryo with this pattern of development is said to be _____.

D. **The mouse is a model for mammalian development**

■ 12 Early mice embryos can be fused to produce a _____, a term for offspring that exhibit a diversity of characteristics derived from two or more kinds of genetically dissimilar cells from different zygotes.

■ 13 A self-regulating embryo exhibits a form of development referred to as _____. Such an embryo can still develop normally when it has extra cells or missing cells .

E. **Homeotic-like mutants are found in plants**
F. **Some exceptions to the principle of nuclear equivalence have been found**

■ 14 _____ are physical changes in gene structure that affect gene activity as, for example, in cases of recombination of coding sequences or replacement of an active gene with a "silent" gene.

IV. THE STUDY OF DEVELOPMENTAL BIOLOGY PRESENTS MANY FUTURE CHALLENGES

KEY TERMS

Frequently Used Prefixes and Suffixes: Use combinations of these prefixes and suffixes to generate terms for the definitions below.

Prefixes	The Meaning		Suffixes	The Meaning
chrono-	time		-gen(esis)	production of
morpho-	form			
poly-	much, many			

Prefix	Suffix	Definitions
_____	_____	1. The development and differentiation (production) of the form and structures of the body during embryonic development.
_____	-tene	2. Pertains to a large, "many-stranded" chromosome.
_____	-gene	3. A gene that is involved in the timing of development.

Other Terms You Should Know

cellular differentiation
chimera
chromosome walking
determination
development
differential gene activity
differentiation
expression site
founder cell
gap gene
genomic rearrangement
hermaphrodite

homeobox
homeodomain
homeotic gene
imaginal disc
imago
induction
larva (-ae)
maternal effect gene
morphogen
mosaic
nuclear equivalence
pair rule gene

pattern formation
programmed cell death
puff
regulative development
segment polarity
segmentation gene
somatic cell
stem cell
tissue culture
transformation
zygotic gene

SELF TEST

Multiple Choice: Some questions may have more than one correct answer.

1. Morphogens are
   a. hypothetical chemicals.
   b. derived from a homeobox.
   c. determined by embryonic segmentation genes.
   d. thought to exist in *Drosophila* eggs.
   e. unravelled portions of active DNA.

2. If destruction of a single cell early in development results in the absence of a structure in the adult, the embryo of that organism
   a. is mosaic.
   b. is a fruit fly.
   c. contains predetermined cells.
   d. contained founder cells.
   e. is controlled to some extent by homeotic genes.

3. A group of cells grown in liquid medium from a single root cell is
   a. formed by a predetermined pattern.
   b. a mosaic.
   c. totipotent.
   d. predetermined.
   e. an embryoid.

4. A parasitic organism that can defeat its host's immune system by switching glycoproteins in the host cell surfaces probably has
   a. genetically engineered genes.
   b. mosaic development.
   c. a regulative development pattern.
   d. founder cells that give rise to adult parts.
   e. many genomic rearrangements.

5. An embryo with an invariant developmental pattern in which the fate of cells is predetermined is said to be
   a. induced.
   b. mosaic.
   c. homeotic.
   d. transgenic.
   e. differentiated.

6. Embryos that act as a self-regulating whole and can accommodate missing parts
   a. are transgenic.
   b. are mosaics.
   c. have regulative development patterns.
   d. have founder cells that give rise to adult parts.
   e. contain many genomic rearrangements.

7. Gap genes in *Drosophila*
   a. are a type of segmentation gene.
   b. act on all embryonic segments.
   c. are the first zygotic genes to act.
   d. are complementary to maternal genes.
   e. determine segment polarity.

8. Genes in *Drosophila* that designate the adult structure formed by an imaginal disc are
   a. gap genes.
   b. segmentation genes.
   c. maternal genes.
   d. homeotic genes.
   e. morphogens.

9. The concept of nuclear equivalence holds that
   a. adult cells are identical to one another.
   b. embryonic cells are identical to one another.
   c. adult somatic nuclei are identical to the nucleus of the fertilized egg.
   d. adult cells are identical to egg cells.
   e. adult somatic nuclei are identical.

10. The process by which developing cells become committed to a particular pattern of gene activity is
    a. determination.
    d. morphogenesis.
    b. differentiation.
    e. the first step in cell specialization.
    c. transgenesis.

11. If differentiated cells can be induced to act like embryonic cells, the cells are
    a. formed by a predetermined pattern.
    d. predetermined.
    b. mosaics.
    e. embryoids.
    c. totipotent.

12. The process by which the differentiation of an embryonic cell is controlled or influenced by other embryonic cells is
    a. transformation.
    d. an example of programmed cell death.
    b. induction.
    e. determinate morphogenesis.
    c. mosaic development.

13. Pattern formation is a series of steps leading to
    a. determination.
    d. morphogenesis.
    b. differentiation.
    e. the final step in cell specialization.
    c. predetermination.

14. Differences in the molecular composition of different cells in a multicellular organism are due to
    a. loss of genes during development.
    d. nuclear equivalency.
    b. transformation.
    e. differential gene activity.
    c. regulation of the activities of different genes.

15. Polytene chromosomes
    a. form imaginal discs.
    d. form by DNA replication without mitosis.
    b. are formed during interphase.
    e. result from fusion of two or more chromosomes.
    c. are found in the nematode *Caenorhabditis elegans*.

PART IV

Evolution

□

CHAPTER 17

Evolution: Mechanism and Evidence

O U T L I N E

Ideas about evolution
Charles Darwin's concept of evolution
Natural selection
The synthetic theory of evolution
Evidence supporting evolution

All organisms that exist today evolved from earlier organisms. Evolution is a genetic change in a population over many generations. The concept of evolution is the cornerstone of biology; biologists do not question its occurrence. Although ideas about evolution existed before Charles Darwin, it remained for Darwin to discover natural selection – the actual mechanism of evolution. According to this mechanism, each species produces more offspring than will survive to maturity, genetic variation exists among the offspring, organisms compete for resources, and individuals with the most favorable characteristics are most likely to survive and reproduce. Favorable characteristics are thus passed to succeeding generations. Over time, changes occur in the gene pools of geographically separated populations that may lead to the evolution of new species. A subsequent modification of Darwin's theory, the synthetic theory of evolution, explains Darwin's observation of variation among offspring in terms of mutation and recombination. The concept of evolution is supported by an enormous body of scientific observations and experiments. The most direct evidence supporting evolution comes from the fossil record. Additional evidence comes from studies of comparative anatomy, mimicry, embryology, and biogeography. Especially compelling evidence supporting evolution comes from studies of biochemistry and molecular biology.

CHAPTER OUTLINE AND CONCEPT REVIEW (Fill in the blanks)

I. INTRODUCTION

■ 1 Evolution means change, or as Darwin described it with respect to organisms, in broad, general terms it is " _genetic change in a popu over time_ ."

■ 2 The genetic changes that bring about evolution do not occur in individual organisms, but rather in _populations_ .

■ 3 All of the genes in a population are referred to as the _gene pool_ for that population.

117

II. IDEAS ABOUT EVOLUTION ORIGINATED BEFORE DARWIN

■ 4 _____*Lamarck*_____ proposed a theory of evolution in 1809, coincidentally in the year of Darwin's birth. His theory held that traits *acquired* during an organism's lifetime could be passed along to their offspring.

III. CHARLES DARWIN WAS INFLUENCED BY HIS CONTEMPORARIES

■ 5 Charles Darwin served as the naturalist on the ship (a)_*HMS Beagle*_ during its five year cruise around the world. Central to the theory Darwin would propose were data about the similarities and differences among organisms in the (b)_*Galapagos Islands*_, a chain of islands west of mainland Ecuador.

■ 6 In the mid-1800s, most believed that organisms did not change, but some understood the significance of variations among domestic animals that were intentionally induced by breeders, a process called _*artificial select'n*_.

■ 7 Charles Lyell popularized the geological theory first proposed by Hutton called _____. This theory proposes that the physical features of the Earth were formed over long periods of time, and that geological processes and scientific laws operating today operated in the past.

■ 8 _____ made the important mathematical observation that populations increase in size geometrically until checked by factors in the environment.

■ 9 In 1858, both Darwin and _____*Wallace*_____ arrived at the conclusion that evolution occurred by natural selection.

■ 10 Darwin published his monumental book, _*Origin of Species*_ _____ in 1859.

IV. NATURAL SELECTION HAS FOUR PREMISES

■ 11 Darwin's mechanism of natural selection consists of four premises based on observations about the natural world. In summary, these are: _*Overproduction, variation, competition & survival to reproduce.*_

V. THE SYNTHETIC THEORY OF EVOLUTION COMBINES DARWIN'S THEORY WITH MENDELIAN GENETICS

■ 12 The synthetic theory of evolution explains variation in terms of _*mutation & recombination*_, both of which are inheritable and therefore potential explanations for *how* traits are passed from one generation to another.

VI. MANY TYPES OF EVIDENCE SUPPORT EVOLUTION

A. The fossil record indicates that organisms evolved in the past

■ 13 Theories of biological evolution draw heavily from the physical sciences, particularly from (a)_*geology*_, which is the study of the Earth and its history, and (b)_*paleontology*_, the study of fossils.

■ 14 Fossils include not only the remains of ancient plants and animals, but also any _*impressions*_ left by organisms.

■ **15** An impression of an organism in sediments may remain long after its organic components have "vaporized." ___compression___ are "imprints" of organisms that still contain organic material.

■ **16** _____ occur when minerals infiltrate an organism's tissues.

■ **17** ___index fossils___ are the remains or traces of organisms that were preserved in large numbers over a relatively short geological time. They are used to identify specific sedimentary layers and to arrange strata in chronological order.

B. Comparative anatomy of related species demonstrates similarities in their structures

■ **18** ___homologous___ organs have basic structural similarities, even though the organs may be used in different ways. They indicate evolutionary ties between the organisms possessing them.

■ **19** ___analogous___ organs have similar functions, whether or not structurally similar. They do not indicate close evolutionary ties.

■ **20** The occasional presence of "remnant organs" in organisms is to be expected as ancestral species evolve and adapt to different modes of life. Such organs or parts of organs, called ___vestigial___, are degenerate and have no apparent function.

C. Mimicry provides an evolutionary advantage

■ **21** In mimicry, one organism gains an selective advantage by resembling another organism. An organism with (a)___protective colouration___ resembles its immediate surroundings, making it difficult to perceive. In (b)___batesian___, a harmless species resembles a dangerous species, whereas in (c)___muellerian___ different harmful or dangerous species resemble one another.

D. Related species have similar patterns of development
E. The distribution of plants and animals supports evolution

■ **22** The study of the distribution of plants and animals is ___biogeography___.

■ **23** It is generally assumed that each species originated only once at a location called its ___centre of origin___. New species spread out from that point until halted by a barrier of some kind.

F. Molecules contain a record of evolutionary change

■ **24** Blood sera of closely related vertebrates are more similar than sera of distantly related vertebrates. Data used to determine the degree of relatedness based on similarities among blood proteins are obtained from ___Antibody–Antigen___ reactions and amino acid sequencing.

■ **25** A greater proportion of the sequence of nucleotides in DNA is identical in closely related organisms. Such relationships are derived from ___DNA hybridization___, a technique whereby single strands of DNA from different species are paired *in vitro*. The degree to which the strands pair infers relatedness.

KEY TERMS

Frequently Used Prefixes and Suffixes: Use combinations of these prefixes and suffixes to generate terms for the definitions below.

Prefixes	The Meaning
bio-	life
homo-	same

Prefix	Suffix	Definitions
_____	-logous	1. Pertains to having a similarity of form due to having the same evolutionary origin.
_____	-geography	2. The study of the geographical distribution of living things.

Other Terms You Should Know

adaptation	evolution	paleontology
analogous	fossil	petrifaction
artificial selection	gene pool	phylogenetic tree
Batesian mimicry	impression	protective coloration
cast	index fossil	species range
center of origin	mimicry	synthetic theory of evolution
compression	mold	uniformitarianism
convergent evolution	molecular clock	vestigial organ
DNA hybridization	Mullerian mimicry	
DNA sequencing	natural selection	

SELF TEST

Multiple Choice: Some questions may have more than one correct answer.

1. If phylogenetic trees were to be constructed independently from molecular characteristics and from morphological characteristics, the two trees would likely be
   a. basically dissimilar.   d. identical.
   b. totally dissimilar.   x   e. misleading.
   c. basically similar.

2. The first comprehensive theory of evolution, which proposed that the mechanism was an inner drive for improvement, was proposed by
   a. Aristotle.   d. Wallace.
   b. Lamarck.   e. da Vinci.
   c. Darwin.

3. The first to formally propose that organisms change over time as the result of natural phenomena rather than divine intervention was
   a. Aristotle.   d. Wallace.
   b. Lamarck.   e. Lyell.
   c. Darwin.

4. That tulip trees are found only in the eastern United States, Japan, and China is probably due to
   a. a past land bridge.
   b. similar climates.
   c. migrations.
   d. long range dispersal.
   e. two centers of origin.

5. The person(s) who suggested that populations increase geometrically and that their numbers must eventually be checked was
   a. Malthus.
   b. Lamarck.
   c. Darwin.
   d. Wallace.
   e. Lyell.

6. The individuals who simultaneously concluded that organisms evolve as nature selects the fittest were
   a. Aristotle.
   b. Lamarck.
   c. Darwin.
   d. Wallace.
   e. Lyell.

7. Organs of different organisms that have a similar form due to a common origin are
   a. analogous.
   b. artifacts.
   c. homologous.
   d. usually vestigial.
   e. found in fossils but never in contemporary organisms.

8. That Earth's physical features were formed over long periods of time by a series of gradual changes was proposed by
   a. Aristotle.
   b. Lamarck.
   c. Darwin.
   d. Wallace.
   e. Lyell.

9. The humerus in a bird and human are
   a. analogous but not homologous.
   b. homologous but not analogous.
   c. neither analogous nor homologous.
   d. both analogous and homologous.
   e. probably derived from a common ancestor.

10. When an organism's body part is trapped in sediments and is not completely decomposed, the resulting fossil is called a/an
    a. cast.
    b. petrifaction.
    c. impression.
    d. compression.
    e. index fossil.

11. The first to propose that the mechanism for the evolution of organisms is by natural selection was
    a. Malthus.
    b. Lamarck.
    c. Darwin.
    d. Wallace.
    e. Lyell.

12. The wings of a butterfly and a bat are
    a. analogous but not homologous.
    b. homologous but not analogous.
    c. neither analogous nor homologous.
    d. both analogous and homologous.
    e. probably derived from a common ancestor.

13. When a fossil consists of minerals that have replaced an organism's tissues, it is called a/an
    a. cast.
    b. petrifaction.
    c. impression.
    d. compression.
    e. index fossil.

14. Evolution of organisms occurs by means of
    a. changes in individual organisms.
    b. uniformitarianism.
    c. changes in gene frequencies in the gene pool.
    d. extinctions.
    e. changes in populations.

15. If Darwin had known about and employed current biochemical techniques, it most likely would have resulted in
    a. a different theory of evolution.
    b. no theory of evolution.
    c. his conviction of immutable species.
    d. confirmation of his theories.
    e. an argument with Wallace.

16. The similar color patterns in the harmless king snake and the poisonous coral snake could logically be considered an example of
    a. Mullerian mimicry.
    b. Batesian mimicry.
    c. interbreeding.
    d. overlapping ranges.
    e. divergent evolution.

VISUAL FOUNDATIONS

Color the parts of the illustration below as indicated.

RED ☐ humerus
GREEN ☐ phalanges
YELLOW ☐ radius
BLUE ☐ carpels
ORANGE ☐ ulna

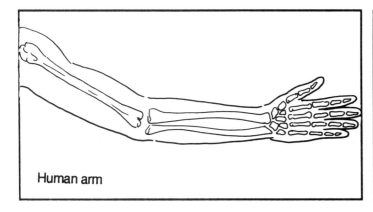

Human arm

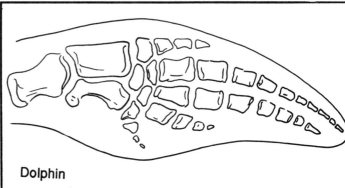

Dolphin

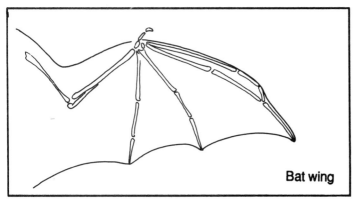

Bat wing

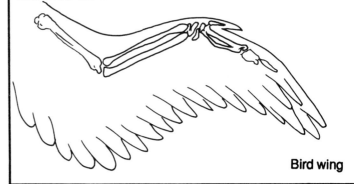

Bird wing

CHAPTER 18

Population Genetics

OUTLINE

The Hardy-Weinberg law
Changes in allele frequencies
Natural selection
Genetic variation

The gene pool of a population of organisms consists of all possible alleles at each locus of each chromosome in the population. Evolution has occurred when changes occur in allele frequencies in the gene pool over successive generations. Allele frequencies may change by genetic drift, migration, mutation, and natural selection. Genetic drift causes changes in allele frequencies due to chance events that result in a marked decrease in the number of individuals. Genetic drift also occurs when one or a few individuals extend beyond the population's normal range and establish a new colony. Migration of individuals between different populations of the same species causes a corresponding movement of alleles. Mutations contribute new genes to the gene pool. Natural selection checks the random effects of genetic drift, migration, and mutation and leads to a change in allele frequencies that bring the population into harmony with the environment. Natural selection is the most significant factor in changing allele frequencies in a population. It involves stabilizing, directional, and disruptive selection – processes that cause changes in the normal distribution of phenotypes in a population. Genetic variation in a gene pool is the raw material for evolutionary change. The gene pools of most populations contain a large reservoir of variability. When the heterozygote has a higher degree of fitness than either homozygote, both alleles tend to be maintained in the population. Often genetic variation is maintained in populations of prey by frequency-dependent selection. The predator catches and consumes the commoner phenotype, while ignoring the rarer phenotypes. Thus, frequency-dependent selection acts to decrease the frequency of the more common phenotypes and increase the frequency of the less common types. Some of the genetic variation seen in a population confers no selective advantage or disadvantage to the individuals possessing it. There is currently some disagreement among evolutionary biologists about the importance of genetic variation in natural selection.

CHAPTER OUTLINE AND CONCEPT REVIEW (Fill in the blanks)

I. INTRODUCTION

■ 1 The type of evolution involving changes *within* a population is referred to as
_____ micro evolution _____.

II. THE HARDY-WEINBERG LAW DEMONSTRATES THAT ALLELE FREQUENCIES DO NOT CHANGE IN A POPULATION THAT IS NOT EVOLVING

■ 2 Mendelian genetics predicts the frequency of genotypes expected among offspring from one mating, whereas the Hardy-Weinberg law describes the frequencies of genotypes in an entire ___population___. The Hardy-Weinberg law states that gene frequencies in a population tend to remain constant in successive generations unless certain factors are operating.

■ 3 When the distribution of genotypes in a population conforms to the binomial equation (a) $p^2 + 2pq + q^2$, the population is in a genetic equilibrium and is not evolving. This equilibrium is called the (b) ___Hardy-Weinberg equil'm___.

■ 4 The proportion of alleles in successive generations does not change in a population when certain conditions are met. Among them, briefly stated, are the following five: ___Large pop'n size, isolation, no selection, no mutation, random mating___.

III. EVOLUTION OCCURS WHEN THERE ARE CHANGES IN ALLELE FREQUENCIES IN A GENE POOL

A. Genetic drift causes changes in allele frequencies by random, or chance, events

■ 5 Genetic drift is the random change in ___allele frequencies___ of a small breeding population. The changes are usually not adaptive.

■ 6 Genetic drift may occur when the size of a population is suddenly reduced as a function of such temporary causes as a depleted food supply or disease. Such an event is called a ___bottleneck effect___.

■ 7 Genetic drift may occur among a few individuals that have established a colony outside the usual range of their population. This is known as the ___founder effect___.

B. Gene flow, which changes the amount of variation in the gene pool, is caused by the differential migration of organisms

■ 8 The localized, more or less isolated, populations within a species are (a) ___demes___. Occasionally, individuals from one local population migrate to another localized population, resulting in a movement of alleles between populations, a phenomenon called (b) ___gene flow___.

C. Mutation increases variation in the gene pool

■ 9 ___mutation___ is the ultimate source of all new alleles.

D. Natural selection changes allele frequencies in a way that leads to adaptation to the environment

■ 10 The most important reason for adaptive changes in gene frequencies is ___nat. selection___.

IV. SELECTION INCREASES THE ADAPTEDNESS OF A POPULATION FOR THE ENVIRONMENT IN WHICH IT LIVES

■ 11 A _____ is a group of genes that act synergistically to produce a better adapted phenotype.

A. Stabilizing selection selects against phenotypic extremes
B. Directional selection favors one phenotype over another
C. Disruptive selection selects for phenotypic extremes

V. GENETIC VARIATION IS NECESSARY IF EVOLUTION IS TO OCCUR

■ 12 _polymorphophism_ is the presence of more than one allele for a given locus.

A. Genetic variation can be maintained by heterozygote advantage
B. Genetic variation may be maintained by frequency-dependent selection

■ 13 Frequency-dependent selection decreases the frequency of (a)_common_____
phenotypes and their respective genotypes and increases the frequency of (b)_____
_____rare_____ types.

C. Neutral variation gives no selective advantage or disadvantage
D. The importance of selection in maintaining genetic variation is a major question in population genetics today

KEY TERMS

Frequently Used Prefixes and Suffixes: Use combinations of these prefixes and suffixes to generate terms for the definitions below.

Prefixes	The Meaning
allo-	other
micro-	small

Prefix	Suffix	Definitions
_____	-evolution	1. Changes in allele frequencies over successive generations; involves small changes within a population.
_____	-zymes	2. Different (other) versions of the same enzyme.

Other Terms You Should Know

adaptive value
Batesian mimicry
coadapted gene complex (-xes)
deme
directional selection
disruptive selection
fitness

founder effect
frequency-dependent selection
gene flow
genetic bottleneck
genetic drift
Hardy-Weinberg Law
heterozygote advantage

mimicry
mutation
neutral variation
neutralist-selectionist controversy
polymorphism
stabilizing selection

SELF TEST

Multiple Choice: Some questions may have more than one correct answer.

1. A cluster of members of a species that is separated spatially from other clusters is a/an
   a. deme.
   b. founder.
   c. subspecies.
   d. evolving unit.
   e. separate species.

2. A relatively quick, extreme environmental change that favors several phenotypes at the expense of the mean phenotype will likely result in
   a. stabilizing selection.
   b. no selection.
   c. a shift in gene frequencies.
   d. disruptive selection.
   e. directional selection.

3. An organism is likely to be exceptionally well adapted if it contains
   a. a stabilizing mutation.
   b. a coadapted gene complex.
   c. more than the average number of mutations.
   d. a large proportion of recessive alleles.
   e. neutral variations.

4. The source of all new alleles is
   a. gene flow.
   b. natural selection.
   c. mutations.
   d. genetic drift.
   e. the heterozygote advantage.

5. Natural selection associated with a well-adapted population is
   a. stabilizing selection.
   b. no selection.
   c. characterized by radical shifts in gene frequencies.
   d. disruptive selection.
   e. directional selection.

6. If undisturbed by other forces, random sexual reproduction among members of a large population in nature leads to
   a. new species.
   b. changes in gene frequencies.
   c. generations of unchanged gene frequencies.
   d. new traits.
   e. a change in the frequency of 2pq, but not p^2 or q^2.

7. When gradual environmental changes favor phenotypes at the extreme of the normal distribution curve, the result in time is likely to be
   a. stabilizing selection.
   b. no selection.
   c. a shift in gene frequencies.
   d. disruptive selection.
   e. directional selection.

8. Changes in allele frequencies within a population are referred to as
   a. evolution.
   b. macroevolution.
   c. microevolution.
   d. p^2 shifts.
   e. q^2 shifts.

9. The proportion of alleles in a population does not change if there is/are
   a. random mating.
   b. natural selection of advantageous alleles.
   c. no more than ten percent of the alleles mutating in each generation.
   d. mating with individuals in a similar population.
   e. a large number of individuals in the population.

10. Random evolutionary changes in a small breeding population resulting from random changes in gene frequencies are referred to as
    a. gene flow.
    b. natural selection.
    c. mutations.
    (d.) genetic drift.
    e. the heterozygote advantage.

11. The gene pool for a given species is
    a. fixed and unchanging.
    (b.) unique to that species.
    c. found in each individual in the species.
    (d.) isolated from other species.
    (e.) the sum of all genes in the species.

Use the following information, the Hardy-Weinberg equation, and the list below to answer questions 12-17. The phenotype coded for by the genotype "tt" is found among 400 individuals in a population of 10,000 randomly mating individuals.

a. 0.01	f. 0.16	k. 0.80	p. 4.00	u. 20.0
b. 0.02	g. 0.20	l. 1.00	q. 6.40	v. 32.0
c. 0.04	h. 0.32	m. 1.60	r. 9.60	w. 40.0
d. 0.08	i. 0.40	n. 2.00	s. 10.0	x. 64.0
e. 0.10	j. 0.64	o. 3.20	t. 16.0	y. 96.0

12. Frequency of q. g

13. Frequency of p. k

14. Frequency of heterozygous individuals. h

15. Frequency of homozygously recessive individuals. c

16. Frequency of homozygously dominant individuals. j

17. Percent of individuals containing one or more of the dominant alleles. y

$$p^2 + 2pq + q^2$$

freq of tt : $\frac{400}{10,000} = 0.04 = q^2$

$\therefore q = 0.2$

$p = 1 - 0.2 = 0.8$

$p^2 = (0.8)^2 = 0.64 \rightarrow$ freq of TT

$2(0.8)(0.2) = 0.32 \rightarrow$ freq of Tt

$\quad \hookrightarrow 0.64 \times 10,000 = 6400$ TT

$\quad 0.32 \times 10,000 = 3200$ Tt

$\begin{array}{r} 6400 \\ +3200 \\ \hline 9600 \end{array}$ $\frac{9600}{10,000} = 96\%$

Visual Foundations on next page ☞

VISUAL FOUNDATIONS

Color the parts of the illustration below as indicated.

RED	☐	axis indicating number of individuals
GREEN	☐	axis indicating phenotype variation
YELLOW	☐	normal distribution
BLUE	☐	unsuitable phenotypes
ORANGE	☐	directional selection
BROWN	☐	disruptive selection
TAN	☐	stabilizing selection

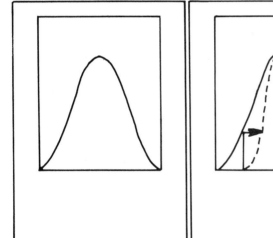 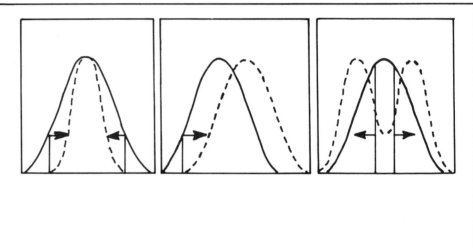

Speciation and Macroevolution

O U T L I N E

Mechanisms to achieve reproductive isolation
Speciation
The pace of evolution
Macroevolution
The fossil record
Relating microevolution to speciation and macroevolution

A species is a group of organisms with a common gene pool that is reproductively isolated from other organisms. Most species have two or more mechanisms that serve to preserve the integrity of the gene pool. Some mechanisms prevent fertilization from occurring in the first place. Others ensure reproductive failure should fertilization occur. Speciation, or the development of a new species, most commonly occurs when one population becomes geographically separated from the rest of the species and subsequently evolves. A new species can also evolve, however, within the same geographical region as its parent species. This latter type of speciation is especially common in plants and occurs as a result of hybridization and polyploidy. There are two theories about the pace of evolution. One says that there is a slow, steady change in species over time, and the other says there are long periods of little evolutionary change followed by short periods of rapid speciation. Macroevolution, or evolutionary change above the level of species, includes the origin of unusual features (e.g., wings with feathers), evolutionary trends (e.g., increase in body size), the evolution of many species from a single species, and the extinction of species. Although the synthetic theory of evolution (Chapter 17) appears adequate to explain macroevolution, it is essentially unproven.

CHAPTER OUTLINE AND CONCEPT REVIEW (Fill in the blanks)

I. INTRODUCTION

■ 1 The 18th century biologist, _____, is generally considered the founder of modern systematics. His system for separating plants into different species based on their morphological characteristics formed the basis for modern methods of systematics.

■ 2 Although still very important in describing species, morphological features alone are not enough to delineate species. Biologists now define a species as a group of (a)_____reproductively_____ isolated organisms with a common (b)__gene___ __pool____.

II. SPECIES HAVE VARIOUS MECHANISMS TO ACHIEVE REPRODUCTIVE ISOLATION FROM ONE ANOTHER

A. Prezygotic isolating mechanisms interfere with mating

■ 3 Prezygotic isolating mechanisms prevent __fertilization__.

■ 4 Prezygotic isolating mechanisms include (a)__temporal__, which occurs because two groups reproduce at different times; (b)__ecological__ separation due to breeding in different habitats; (c)__behavioural__ involving different, incompatible courtship patterns; (d)__mechanical__, isolation due to anatomical differences that thwart successful matings; and (e)__gametic__, in which chemical differences between gametes prevents interspecific fertilization.

B. Postzygotic isolating mechanisms prevent successful reproduction if mating occurs

■ 5 Embryos are aborted due to __hybrid inviability__.

■ 6 __hybrid sterility__ occurs when the gametes produced by interspecific hybrids are abnormal and nonfunctional.

III. THE KEY TO SPECIATION IS THE DEVELOPMENT OF REPRODUCTIVE ISOLATING MECHANISMS

■ 7 __Speciation__ is the evolution of a new species.

A. Long physical isolation and different selective pressures result in allopatric speciation

■ 8 Allopatric speciation occurs when one population becomes __geographically isolated__ from the rest of the species and subsequently evolves.

B. In sympatric speciation two populations diverge in the same physical location

■ 9 Sympatric speciation usually occurs *within* a geographical region as a result of __hybridization & polyploidy__.

■ 10 (a)__polyploidy__ is the possession of more than two sets of chromosomes, a common phenomenon in plants. When it results from the pairing of chromosomes from different species, it is known as (b)__allopolyploidy__.

IV. EVOLUTION IS GRADUAL, OCCURS IN SPURTS, OR IS A COMBINATION OF BOTH PROCESSES

■ 11 According to proponents of the theory of __punctuated equilibrium__, evolution proceeds in spurts; that is, short periods of active evolution are followed by long periods of inactivity or stasis.

■ 12 According to proponents of the theory of __gradualism__, populations slowly and steadily diverge from one another by the accumulation of adaptive characteristics within a population.

V. MACROEVOLUTION INVOLVES CHANGES IN THE KINDS OF SPECIES

■ 13 Evolutionary "novelties" originate from mutations that alter developmental pathways. For example, (a)_____ allometric _____ occurs when developing body parts grow at different rates, and (b)_____ paedomorphosis _____ results from differences in the *timing* of development.

VI. MACROEVOLUTION IS DEMONSTRATED BY THE FOSSIL RECORD

A. Evolutionary trends are indicated in the fossil record by morphological changes in hard body parts

B. Adaptive radiation is the evolutionary diversification of an ancestral species into many species

■ 14 (a)_____ Adaptive zones _____ are new ecological roles made possible by an adaptive advancement. When an organism with a newly acquired evolutionary advancement assumes a new ecological role made possible by its advancement(s), its diversification is known as (b)_____ adaptive radiation _____.

C. Extinction of species is an important aspect of evolution

■ 15 The geological record seems to indicate that two types of extinction have occurred: (a)_____ background _____ extinction, which is a continuous, ongoing, relatively low frequency form, and (b)_____ mass _____ extinction, a relatively rapid and widespread loss of numerous species.

VII. CAN MICROEVOLUTIONARY CHANGE LEAD TO SPECIATION AND MACROEVOLUTION?

KEY TERMS

Frequently Used Prefixes and Suffixes: Use combinations of these prefixes and suffixes to generate terms for the definitions below.

Prefixes	The Meaning
allo-	other
macro-	large, long, great, excessive
poly-	much, many

Prefix	Suffix	Definitions
poly	-ploidy	1. The presence of multiples of complete chromosome sets.
allo	-patric	2. Originating in or occupying different (other) geographical areas.
allo	-polyploidy	3. Polyploidy following the joining of chromosomes from two different species (half of the chromosomes come from one parent and half from the other parent).
macro	-evolution	4. Large-scale evolutionary change.
allo	-metric	5. Varied rates of growth for different parts of the body during development (some parts grow at rates different from other parts).

Other Terms You Should Know

adaptive radiation
adaptive zones
allometric growth
background extinction
behavioral isolation
biological species concept
ecological isolation
extinction
gametic isolation
gradualism

hybrid breakdown
hybrid inviability
hybrid sterility
hybridization
mass extinction
mechanical isolation
polyploidy
postzygotic isolating
   mechanism

prezygotic isolating
   mechanism
punctuated equilibria
reproductive isolation
speciation
species
stasis
subspecies
sympatric speciation
temporal isolation

SELF TEST

Multiple Choice: Some questions may have more than one correct answer.

1. A population that evolves within the same range as its parent species is an example of
   a. cladogenesis.
   b. character displacement.
   c. anagenesis.
   d. sympatric speciation.
   e. allopatric speciation.

2. If two distinct species produce a fertile hybrid, the hybrid resulted from
   a. anagenesis.
   b. phyletic evolution.
   c. cladogenesis.
   d. diversifying evolution.
   e. allopolyploidy.

3. The fact that dogs come in many sizes, shapes, and colors proves that
   a. dogs have one gene pool.
   b. dogs can produce hybrids.
   c. morphology alone is not enough to define a species.
   d. reproduction within a species produces sterile offspring.
   e. all dogs belong to one species.

4. If two distinct populations of organisms occasionally interbreed in the wild, they are considered to be
   a. separate species.
   b. one species.
   c. reproductively isolated.
   d. one gene pool.
   e. evolving.

   only members of SAME SPECIES can interbreed.

5. If a taxonomist discovered a species with two distinct populations, each population with its own characteristic gene complex, she/he would likely consider them to be
   a. two species.
   b. subspecies.
   c. two populations.
   d. varieties of one species.
   e. demes.

    *find demes 2 distinct local populations ?*

6. If flower-loving female botany majors were required to smell a flower before male botany majors would mate with them, and if male zoology majors would only mate with females that did not sniff flowers, zoology majors and botany majors would likely become
   a. separate species.
   b. one species.
   c. reproductively isolated.
   d. one gene pool.
   e. behaviorally isolated.

7. The largest unit within which gene flow can occur is the
   a. species.
   b. subspecies.
   c. population.
   d. individual organism.
   e. deme.

8. Hybrid breakdown affects the
   a. P₁ generation.
   b. F₁ generation.
   c. F₂ generation.
   d. F₃ generation.
   e. course of evolution.

9. A population that evolves as the result of its separation from the rest of the species is an example of
   a. cladogenesis.
   b. character displacement.
   c. anagenesis.
   d. sympatric speciation.
   e. allopatric speciation.

10. If college-educated persons mated only at night and noncollege-educated persons mated only at the lunch hour, and no amount of laboratory experimentation could change these habits, these two populations would be considered
   a. separate species.
   b. one species.
   c. reproductively isolated.
   d. members of one gene pool.
   e. temporally isolated.

11. A species typically has
   a. a common gene pool.
   b. an isolated gene pool.
   c. the capacity to reproduce with other species in the laboratory.
   d. members with different morphological characteristics.
   e. become reproductively isolated.

VISUAL FOUNDATIONS

Color the parts of the illustration below as indicated.

RED	☐	original bird
GREEN	☐	second birds to evolve
YELLOW	☐	time
BLUE	☐	first birds to evolve
ORANGE	☐	gradualism
BROWN	☐	punctuated equilibrium
TAN	☐	third birds to evolve

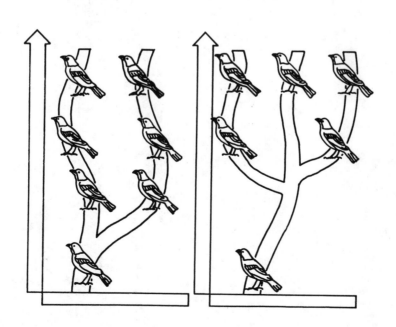

CHAPTER 20

□

The Origin and Evolutionary History of Life

OUTLINE

Chemical evolution on early Earth
The history of life

Life on earth developed from nonliving matter. Small organic molecules formed, accumulated, and became organized into complicated assemblages. Cells evolved from these macromolecular assemblages. This process was enabled by the absence of oxygen and the presence of energy, the chemical building blocks of organic molecules, and sufficient time for the molecules to accumulate and react. The first cells to evolve were anaerobic and prokaryotic. Two crucial events in the evolution of cells was the origin of molecular reproduction and the development of metabolism. The earliest cells are believed to have obtained their energy from organic compounds in the environment. Later, cells evolved that could obtain energy from sunlight. Photosynthesis produced enough oxygen to change significantly the atmosphere and thereby alter the evolution of early life. Organisms evolved that had the ability to use oxygen in cell respiration. Eukaryotes are believed to have evolved from prokaryotes. Mitochondria, chloroplasts, and certain other organelles may have originated from symbiotic relationships between two prokaryotic organisms. In Precambrian times, life began and evolved into bacteria, protists, fungi, and animals. During the Paleozoic era, all major groups of land plants evolved except for flowering plants, and fish and amphibians flourished. The Mesozoic era was characterized by the evolution of flowering plants and reptiles; insects flourished, and birds and mammals appeared. The Cenozoic era has been characterized by the diversification of flowering plants and mammals, including humans and their ancestors.

CHAPTER OUTLINE AND CONCEPT REVIEW (Fill in the blanks)

I. INTRODUCTION
II. EARLY EARTH PROVIDED THE CONDITIONS FOR CHEMICAL EVOLUTION

■ 1 The Big Bang theory holds that a densely compacted universe exploded some (a)_____ years ago. The Earth is about (b)_____ years old.

■ 2 The four basic requirements for the chemical evolution of life are: _____ _____.

135

A. Organic molecules formed on primitive Earth before cells existed

■ **3** The idea that organic molecules could form from nonliving components on primitive Earth originated in the early 20th century with the Russian biochemist (a)_____ and the Scottish biologist (b)_____.

■ **4** In the 1950s, (a)_____ constructed conditions in the laboratory that were thought to prevail on primitive Earth. They started with an atmosphere of water (H_2O) and the three gases (b)_____ _____. These and other experiments have produced a variety of organic molecules.

■ **5** _____ are assemblages of organic polymers that form spontaneously, are organized and to some extent resemble living cells.

B. The first cells probably assembled from organic molecules

■ **6** The first cells were (aerobic or anaerobic?) (a)_____ and (prokaryotic or eukaryotic?) (b)_____.

■ **7** (a)_____ may have been the first polynucleotide to carry "hereditary" information. Interestingly, some forms of this molecule, called (b)_____, can catalyze the formation of more forms.

■ **8** _____, the term for all of the biochemical reactions taking place in a living organism, developed in a gradual, stepwise fashion.

C. Heterotrophy may have evolved before autotrophy

■ **9** Organisms like ourselves that require preformed organic molecules (food), because they cannot synthesize them *de novo*, are called (a)_____. On the other hand, (b)_____ possess the chemical apparatus required to synthesize food.

■ **10** The evolution of photosynthesis ultimately changed early life and afforded vastly new opportunities for diversity because it generated _____.

D. As oxygen increased in the atmosphere, aerobes evolved that could use it
E. Eukaryotic cells evolved after prokaryotic cells

■ **11** The _____ theory proposes that mitochondria, chloroplasts, and possibly centrioles and flagella evolved from prokaryotes engulfed by other cells.

III. THE FOSSIL RECORD PROVIDES US WITH CLUES TO THE HISTORY OF LIFE

A. Evidence of living cells is found in precambrian times

■ **12** Signs of the first life on Earth appeared in the (a)_____ era, about (b)_____ years ago. In the second era, the (c)_____ era, which began about (d)_____ years ago, there are clear signs of bacteria, fungi, protists, and animals.

B. A considerable diversity of life forms evolved during the Paleozoic era

■ **13** The Paleozoic era began about (a)_____ years ago with a burst of evolution so profound that it has been nicknamed the (b)_____ _____. Later, in the (c)_____ period, the first vertebrate, the jawless, bony-armored fish called (d)_____ appeared.

C. Dinosaurs and other reptiles dominated the Mesozoic era

■ 14 The Mesozoic era began about (a)_____ years ago. It consisted of the three periods: (b)_____.
Mammals appeared in the (c)_____ period and birds appeared in the (d)_____ period.

D. The Cenozoic era is known as the "age of mammals"

■ 15 The Cenozoic era began about _____ years ago and extends into the present.

KEY TERMS

Frequently Used Prefixes and Suffixes: Use combinations of these prefixes and suffixes to generate terms for the definitions below.

Prefixes	The Meaning		Suffixes	The Meaning
lipo-	fat		-bio(nt)	life
Pre-	before, prior to		-oid	resembling, like
proto-	first, earliest form of		-some	body

Prefix	Suffix	Definitions
protein-	_____	1. Polypeptides obtained by heating dry amino acids; compounds "resembling" protein.
_____	_____	2. Spontaneous assemblages of organic polymers that may have been involved in the evolution of the earliest forms of life.
_____	_____	3. A protobiont made from lipids.
_____	-cambrian	4. The geologic time period prior to the Cambrian.

Other Terms You Should Know

Archean era	epoch	Permian period
autotroph	era	Pleistocene epoch
Cambrian explosion	heterotrophic	plesiosaur
Cambrian period	ichthyosaur	Pliocene epoch
Carboniferous period	Jurassic period	Proterozoic era
Cenozoic era	Mesozoic era	Quaternary period
chemical evolution	microsphere	Recent epoch
coacervate	Miocene epoch	ribozyme
coelacanth	Oligocene epoch	saurischian
cotylosaur	Ordovician period	Silurian period
creodont	ornithischian	sink
Cretaceous period	ostracoderm	stromatolite
Devonian period	ozone	therapsid
dinosaur	Paleocene epoch	Tertiary period
endosymbiont theory	Paleozoic era	Triassic period
Eocene epoch	period	ungulate

SELF TEST

Multiple Choice: Some questions may have more than one correct answer.

1. The atmosphere of early Earth contained large quantities of
   a. oxygen.   d. carbon monoxide.
   b. water vapor.   e. helium.
   c. hydrogen.

2. The basic requirements for chemical evolution include
   a. DNA.   d. water.
   b. oxygen.   e. time.
   c. energy.

3. Initially, Earth's atmosphere became oxygenated through the photosynthetic activity of
   a. cyanobacteria.   d. water-splitting autotrophs.
   b. purple sulfur bacteria.   e. hydrogen-sulfide bacteria.
   c. green sulfur bacteria.

4. Oparin's coacervates were
   a. proteinoids.   d. cells.
   b. advanced liposomes.   e. protobionts.
   c. capable of replicating and dividing.

5. In a sequence of simplest (primitive) to most complex (advanced), which one of the following items is out of order: monomer, coacervate, proteinoid, prokaryote, eukaryote?
   a. microsphere   d. coacervate
   b. proteinoid   e. monomer
   c. prokaryote

6. According to the endosymbiont theory of the origin of eukaryotic cells, mitochondria were likely once
   a. free-living prokaryotes.   d. cells or proteinoids.
   b. coacervates.   e. eukaryotes.
   c. viruses.

7. Aerobes are generally better competitors than anaerobes because aerobic respiration
   a. is more efficient.   d. has had a longer period of evolution.
   b. splits water.   e. extracts more energy from a molecule.
   c. is confined to symbiotic mitochondria.

8. Experimenters that simulated conditions thought to have existed on early Earth found that these conditions could produce
   a. DNA.   d. proteins.
   b. RNA.   e. amino acids.
   c. nucleotide bases.

9. The consensus among scientists is that the first cells were
   a. aerobic autotrophic prokaryotes.   d. anaerobic heterotrophic prokaryotes.
   b. anaerobic autotrophic prokaryotes.   e. anaerobic heterotrophic eukaryotes.
   c. aerobic heterotrophic prokaryotes.

10. Strong support for the endosymbiotic theory of eukaryotic cell origins is derived from the fact that mitochondria and chloroplasts have
    a. unique DNA.
    b. two membranes.
    c. ribosomes.
    d. tRNA.
    e. endosymbionts.

11. Which of the following era/period combinations is/are mismatched?
    a. Cenozoic/Tertiary
    b. Paleozoic/Miocene
    c. Mesozoic/Jurassic
    d. Cenozoic/Cambrian
    e. Paleozoic/Devonian

12. The first large organic polymers (e.g., proteins, nucleic acids) most likely formed
    a. in the atmosphere.
    b. in shallow seas.
    c. in coacervates.
    d. from monomers.
    e. on clay surfaces.

13. The Earth is thought to be about how many years old?
    a. 10-20 billion.
    b. 4,000.
    c. 3.5 billion.
    d. 4.6 billion.
    e. 10-20 million.

14. About how many years passed between the Earth's formation and the appearance of recognizable cells?
    a. 10 million
    b. 100 million
    c. one billion
    d. five billion
    e. 10 billion

Visual Foundations on next page ☞

VISUAL FOUNDATIONS

Color the parts of the illustration below as indicated.

RED	❑	DNA
GREEN	❑	chloroplast
YELLOW	❑	aerobic bacterium
BLUE	❑	nuclear envelope
ORANGE	❑	mitochondrion
BROWN	❑	endoplasmic reticulum
TAN	❑	photosynthetic bacterium

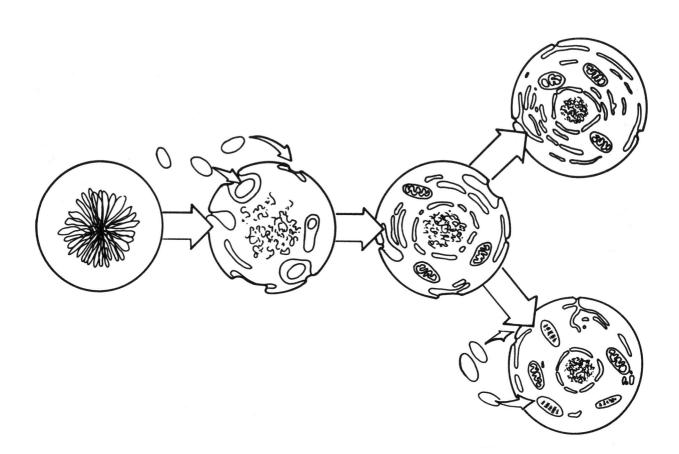

CHAPTER 21

The Evolution of Primates

OUTLINE

Evolution from shrewlike mammals
Trends in hominid evolution
Human cultural evolution

Primates evolved from small, tree-dwelling, shrewlike mammals during the "Age of Reptiles." The first primates to evolve were the prosimians. Then, during the Oligocene epoch, the prosimians gave rise to the anthropoids. The early anthropoids branched into two groups, the New World Monkeys and the Old World Monkeys. The latter gave rise to apes that, in turn, gave rise to the hominids. Early hominids evolved a two-footed posture that has resulted in distinct characteristics of the skeleton and skull. Hominid evolution began in Africa. The earliest hominids belong to the genus *Australopithecus*. The first hominid to have enough human features to be placed in the same genus as modern humans is *Homo habilis*. Although it may have been a contemporary of *Australopithecus*, it had a larger brain, and it made tools from stone. *Homo erectus* was taller than *H. habilis* and fully erect. Also, its brain size was larger, it made more sophisticated tools, and it discovered how to use fire. *Homo sapiens* appeared approximately 200,000 years ago. Since then its cranial capacity has increased by over 60 percent. This evolutionary increase in human brain size made cultural evolution possible. Two of the most significant stages in cultural evolution have been the development of agriculture and the industrial revolution.

CHAPTER OUTLINE AND CONCEPT REVIEW (Fill in the blanks)

I. INTRODUCTION
II. PRIMATES EVOLVED FROM SHREWLIKE MAMMALS

■ 1 There are two subgroups in the order Primates: the (a)_____ include lemurs, lorises, and tarsiers, and the (b)_____ include monkeys, apes, and humans.

A. Prosimians are primitive, arboreal primates
B. Anthropoids include monkeys, apes, and humans

■ 2 Anthropoids evolved from (a)_____ ancestors during the (b)_____ epoch about 38 million years ago.

■ 3 Apes and humans are members of the superfamily _____.

■ 4 The common names for the four genera of apes are _____
_____.

III. THE FOSSIL RECORD SUGGESTS GENERAL TRENDS IN HOMINID EVOLUTION

■ 5 The hominid line separated from the ape line approximately _____ years ago.

A. The earliest hominids belong to the genus *Australopithecus*

■ 6 Earliest hominids appeared about _____ years ago.

■ 7 "Lucy" was one of the most ancient hominids, a member of the species _____.

B. *Homo habilis* is the oldest member of the Genus *Homo*

■ 8 *Homo habilis* appeared about _____ years ago. *H. habilis* was the first primate to consciously design useful tools.

C. Numerous fossils of *Homo erectus* have been discovered

■ 9 *Homo erectus* is about _____ years old. It had a larger brain than *H. habilis*, made more sophisticated tools, and discovered how to use fire.

D. *Homo sapiens* appeared approximately 200,000 years ago

■10 Modern *Homo sapiens*' probably arose in _____ and migrated to Eurasia fairly recently.

IV. HUMANS UNDERGO CULTURAL EVOLUTION

■11 Cultural evolution is the progressive addition of _____ to the human experience. It is made possible by an evolutionary increase in brain size in humans.

■12 Three of the most significant advances in cultural evolution were the _____ _____ _____.

A. Development of agriculture resulted in a more dependable food supply
B. Cultural evolution has had profound impacts on the ecosphere

KEY TERMS

Frequently Used Prefixes and Suffixes: Use combinations of these prefixes and suffixes to generate terms for the definitions below.

Prefixes	The Meaning	Suffixes	The Meaning
anthrop(o)-	human	-oid	resembling, like
bi-	twice, two	-ped(al)	foot
pro-	"before"		
quadr(u)-	four		
supra-	above		

Prefix	Suffix	Definitions
_____	_____	1. Member of the Suborder Anthropoidea; animals that "resemble" humans.
_____	_____	2. Pertains to an animal that walks on four feet.

_____ _____		3. Pertains to an animal that walks on two feet.
_____	-simian	4. Belonging or pertaining to the Suborder Prosimii; the primates that evolved before the simians (apes and monkeys).
homin-	_____	5. Member of the Superfamily Hominoidea; a gibbon, orangutan, gorilla, chimpanzee, or human.
_____	-orbital	6. Situated above the eye socket.

Other Terms You Should Know

brachiate	endothermic	prehensile
cultural evolution	foramen magnum	primate
dentition	hominid	supraorbital ridge
diurnal	knuckle walking	viviparous

SELF TEST

Multiple Choice: Some questions may have more than one correct answer.

1. The Peking man and Java man are classified as
   a. *H. habilis.*   d. apes.
   b. *H. erectus.*   e. hominids.
   c. *H. sapiens.*

2. The "Neanderthal man" is classified as
   a. *H. habilis.*   d. an ape.
   b. *H. erectus.*   e. a hominid.
   c. *H. sapiens.*

3. *Homo sapiens* appeared about how many years ago?
   a. 2,000   d. 200,000
   b. 10,000   e. one million
   c. 20,000

4. The first primates
   a. were hominids.   d. were anthropoids.
   b. had opposable thumbs.   e. were prosimians.
   c. had an ancestor in common with insectivores.

5. Anthropoids evolved directly from
   a. hominids.   d. apes.
   b. primates.   e. prosimians.
   c. insectivores.

6. Based on molecular similarities and other characteristics, it is thought that the nearest living relative of humans is the
   a. gorilla.   d. chimpanzee.
   b. monkey.   e. ape.
   c. gibbon.

7. Both apes and humans
   a. are hominoids.
   b. are hominids.
   c. can brachiate.
   d. lack tails.
   e. have opposable thumbs.

8. The first hominid that migrated to Europe and Asia is
   a. *H. habilis*.
   b. *H. erectus*.
   c. *H. sapiens*.
   d. australopithecines.
   e. "Lucy" and her descendents.

9. The earliest hominids
   a. had short canines.
   b. are in the genus *Homo*.
   c. are in the genus *Australopithecus*.
   d. appeared about 3.8 million years ago.
   e. evolved in Africa.

10. The earliest hominid to be placed in the genus *Homo* is
   a. *H. habilis*.
   b. *H. erectus*.
   c. *H. sapiens*.
   d. australopithecines.
   e. "Lucy" and her descendents.

VISUAL FOUNDATIONS

Color the parts of the illustration below as indicated.

RED ❑ brain
GREEN ❑ jaw (mandible bone)
BLUE ❑ supraorbital ridge
YELLOW ❑ canine teeth
ORANGE ❑ other teeth

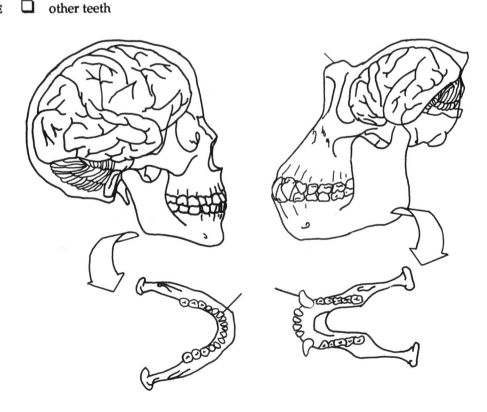

Color the parts of the illustration below as indicated.

RED   ☐   pelvis
GREEN   ☐   skull
YELLOW   ☐   big toe
BLUE   ☐   spine
ORANGE   ☐   arm
BROWN   ☐   leg

In each box, label the indicated anatomical difference.

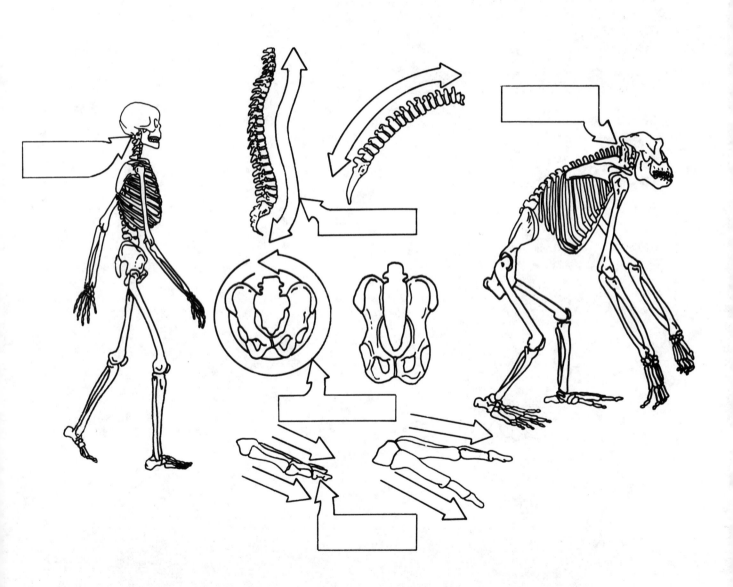

PART V

The Diversity of Life

CHAPTER 22

❑

The Classification of Organisms

OUTLINE

Modern taxonomy
Reconstructing phylogeny: Systematics
Three approaches to classification: Phenetics, cladistics, and
classical evolutionary taxonomy

Organisms are classified according to a system developed in the mid-18th century. In this system, each organism is given a two-part name; the first name is the genus and the second is the species. The basic unit of classification is the species. Classification is hierarchical, with one or more related genera constituting a family, and one or more related families constituting an order, and so on. Currently, five major categories of organisms are recognized: Prokaryotae, Protista, Fungi, Plantae, and Animalia. Classification is based on evolutionary relationships. In some groupings, all of the organisms have a common ancestor. In others, all of the organisms are derived from different ancestors. Two or more different organisms are grouped together when evidence suggests that they have evolved from a common ancestor. The evidence might be shared homologous structures, or similarities in DNA and protein structure. There are three principal approaches to the classification of organisms. In one, organisms are classified according to the number of characteristics they have in common, whether homologous or analogous. In the second, organisms in a grouping all have a common ancestor and are classified based on how long ago one group of organisms branched off from another. In the third and most widely accepted and used system of classification, organisms are classified based on their evolutionary relationships as well as the extent of their divergence.

CHAPTER OUTLINE AND CONCEPT REVIEW (Fill in the blanks)

I. INTRODUCTION

■ 1 _____ is the science of classifying and naming organisms.

II. MODERN TAXONOMY IS BASED ON THE WORK OF LINNAEUS

■ 2 The system developed by Carolus Linnaeus that assigns a two part name to each kind of organism is logically referred to as the (a)_____ of nomenclature. The name consists first of the (b)_____, which is capitalized, followed by the decapitalized (b)_____ name.

■ 3 The basic unit of classification is the _____.

A. Each taxonomic level is more general than the one below

■ **4** The classification system consists of several levels of organization. For example, a group of closely related species is placed in the same genus. In similar fashion, closely related genera are grouped together in the same (a)_____, which in turn are grouped and placed in orders. The next level in the hierarchy above orders is the (b)_____, followed in succession by the (c)_____ in plants or fungi, and the (d)_____ for animals and protists.

■ **5** In the system for classifying organisms, each grouping, or level, such as a particular species, or genus, or order, and so forth is called a _____.

B. Subspecies may become species
C. Taxonomists may "split" or "lump"
D. Organisms are classified in five kingdoms

■ **6** The five kingdoms proposed by R. H. Whitaker in 1969 are: _____
_____.

III. SYSTEMATICS IS CONCERNED WITH RECONSTRUCTING PHYLOGENY

■ **7** Modern classification is based on the reconstruction of _____
_____, or phylogeny, as it is called.

A. Taxa may be monophyletic or polyphyletic

■ **8** The organisms in a (a)_____ taxon have a common ancestor, and the taxon containing the common ancestor and all of its descendant species is a (b)_____.

■ **9** When the organisms in a given taxon have evolved from different ancestors, the taxon is said to be _____.

B. Biologists consider homologous structures
C. Derived characters have evolved more recently than primitive characters

■ **10** Shared <u>primitive</u> characters suggest (ancient or recent?) (a)_____ ancestry, whereas, shared <u>derived</u> characters indicate more (ancient or recent?) (b)_____ common ancestry.

D. Biologists carefully choose taxonomic criteria
E. Molecular biology provides new taxonomic tools

■ **11** Comparisons of the structure of the two molecules _____ are used to confirm evolutionary relationships.

IV. TAXONOMISTS USE THREE MAIN APPROACHES

■ **12** The three main approaches to taxonomy are _____
_____.

A. Phenetics is based on phenotypic similarities

■ **13** The phenetic system is a numerical taxonomy in which organisms are grouped according to the number of _____ they share.

B. Cladistics emphasizes phylogeny

■ 14 The cladistic approach emphasizes phylogeny, focusing on how long ago one group branched off from another. Each taxon contains a common ancestor and its descendants, and is therefore said to be _____.

C. Classical evolutionary taxonomy uses a phylogenetic tree

KEY TERMS

Frequently Used Prefixes and Suffixes: Use combinations of these prefixes and suffixes to generate terms for the definitions below.

Prefixes	The Meaning		Suffixes	The Meaning
eu-	good, well, "true"		-karyo(te)	nucleus
mono-	alone, single, one			
poly-	much, many			
sub-	under, below			

Prefix	Suffix	Definitions
_____	-species	1. A unit of classification below species.
_____	-phyletic	2. Refers to a taxon in which all of the subgroups have a common (single) ancestry.
_____	-phyletic	3. Refers to a taxon consisting of several evolutionary lines and not including a common ancestor.
pro-	_____	4. An organism that lacks a nuclear membrane.
_____	_____	5. An organism with a distinct nucleus surrounded by nuclear membranes.

Other Terms You Should Know

Animalia
clade
cladistics
class
classical evolutionary
taxonomy
derived character
division
Eukaryota
family (-lies)

Fungi
genus (-nera)
lumper
molecular clock
Monera
order
phenetics
phylogeny
phylum (-la)
Plantae

primitive character
Prokaryota
Protista
species
splitter
strain
systematic
taxon (-xa)
taxonomy

SELF TEST

Multiple Choice: Some questions may have more than one correct answer.

1. The systematist would likely be most interested in
   a. lumping taxa.
   b. splitting taxa.
   c. evolutionary relationships.
   d. ontogeny.
   e. ancestries.

2. Humans are placed in the phylum Chordata because they have
   a. hair.
   b. an embryonic notochord.
   c. mammary glands.
   d. an opposable digit.
   e. a backbone.

3. Which of the following would the pheneticist emphasize when classifying organisms?
   a. phylogenetic trees.
   b. amino acid sequences.
   c. the point in time that branching occurred.
   d. number of shared characteristics.
   e. antigen-antibody relationships.

4. Which one of the following differs most?
   a. Animals and plants.
   b. Fungi and animals.
   c. Algae and animals.
   d. Prokaryotes and eukaryotes.
   e. Plants and fungi.

5. The fundamental unit of classification is the
   a. epithet.
   b. binomial.
   c. variety or subspecies.
   d. species.
   e. phylum or division.

6. Which of the following would the cladist emphasize when classifying organisms?
   a. phylogenetic trees.
   b. amino acid sequences.
   c. the point in time that branching occurred.
   d. number of shared characteristics.
   e. antigen-antibody relationships.

7. The superficial similarities of two distantly related species probably result from
   a. convergent evolution.
   b. divergent evolution.
   c. interbreeding.
   d. fortuitous adaptation to similar environments.
   e. a common ancestor.

8. Which of the following is/are used extensively by the molecular biologist to determine relationships?
   a. carbohydrates.
   b. nucleic acids.
   c. lipids.
   d. proteins.
   e. hormones.

9. Fungi are generally not classified as plants because they do not
   a. photosynthesize.
   b. have a true nucleus.
   c. have cell walls.
   d. have mitochondria.
   e. reproduce sexually.

10. Which of the following is/are listed in order from taxa of greatest diversity to taxa of least diversity?
    a. Phylum, order, genus
    b. Family, class, order
    c. Class, family, genus
    d. Family, genus, species
    e. Class, order, division

VISUAL FOUNDATIONS

Color the parts of the illustration below as indicated. Inside each box, write the specific name of each category used in classifying the depicted organisms.

RED   ❑   specific epithet
GREEN   ❑   phylum
YELLOW   ❑   class
BLUE   ❑   kingdom
ORANGE   ❑   order
VIOLET   ❑   family
PINK   ❑   genus

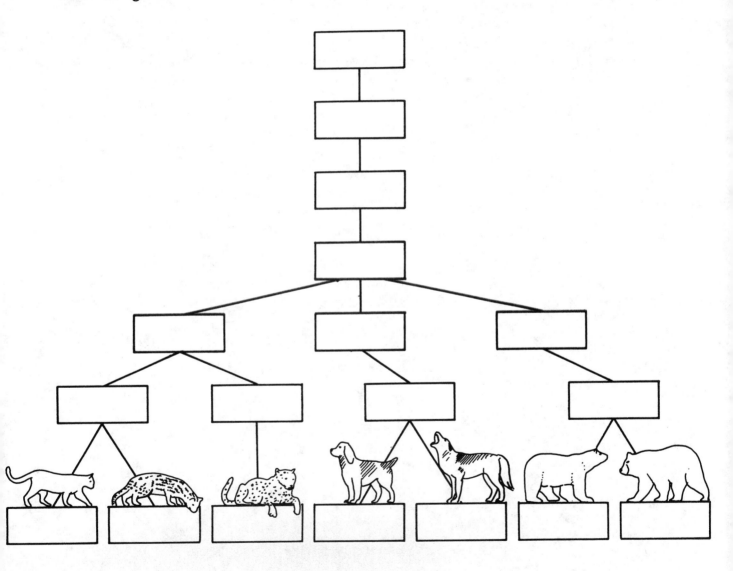

Viruses and the Kingdom Prokaryotae

O U T L I N E

Viruses — tiny, infectious agents
Bacteriophages
Viruses that infect animals
Viruses that infect plants
Origin of viruses
The kingdom Prokaryotae
Two different groups of bacteria:
archaebacteria and eubacteria
Archaebacteria
Eubacteria

Viruses are not true organisms. They consist of a core of DNA or RNA surrounded by a protein coat. Some are surrounded by an outer envelope too. Most can only be seen with an electron microscope. Whereas some viruses may or may not kill their hosts, others invariably do. Some integrate their DNA into the host DNA, conferring new properties on the host. Bacterial viruses released from host cells may contain some host DNA, which can then become incorporated into the genome of a new host. Some viruses infect humans and other animals, causing many diseases and cancers. Others infect plants, causing serious agricultural losses. Receptor molecules on the surface of a virus determine what type of cell it can infect. Viruses enter and exit cells in a variety of ways. Prions are virus-like particles associated with certain diseases in sheep, cattle, and humans. Viroids are infectious molecules consisting only of RNA. Viruses are believed to have their origin in bits of nucleic acid that "escaped" from cellular organisms. Bacteria are in the kingdom Prokaryotae. Most are unicellular, but some form colonies or filaments. They have ribosomes, but lack membrane-bounded organelles. Their genetic material is a circular DNA molecule, not surrounded by a nuclear membrane. Some have flagella. Whereas most bacteria get their nourishment either from dead organic matter or symbiotically from other organisms, some manufacture their own organic molecules. Bacteria respire in a variety of ways. Although most reproduce asexually by transverse binary fission, some exchange genetic material. There are two major groups of bacteria; the Archaebacteria and Eubacteria. The former are thought to be the original bacteria from which all cellular life descended. Although most Eubacteria are harmless, a few are notorious for the diseases they cause. The mycoplasmas are the only bacteria without cell walls. Bacteria are classified on the basis of their ability to absorb and retain crystal violet stain. Those that do are said to be gram-positive; those that don't are gram-negative.

CHAPTER OUTLINE AND CONCEPT REVIEW (Fill in the blanks)

I. INTRODUCTION

II. VIRUSES ARE TINY, INFECTIOUS AGENTS THAT ARE NOT ASSIGNED TO ANY OF THE FIVE KINGDOMS

■ 1 Viruses are tiny particles consisting of a (a)_____ core surrounded by a protein coat called a (b)_____.

■ 2 Some viruses are surrounded by a membranous layer called the (a)_____, which contains four major components: (b)_____ _____.

■ 3 The largest virus is the _____. It can be seen with the light microscope; all others require the electron microscope.

III. BACTERIOPHAGES ARE VIRUSES THAT ATTACK BACTERIA

■ 4 Bacteriophages that break apart host cells are said to be (a)_____ _____, while those that do not destroy the host cell are referred to as (b)_____.

A. In lytic infections, many new phages are produced and the host bacterial cell undergoes lysis

■ 5 The five steps that almost all lytic viral infections follow are _____ _____.

B. In lysogenic infections, phage DNA is integrated into the host bacterial chromosome

■ 6 Temperate (lysogenic) viruses may or may not destroy their host cells; some integrate their nucleic acid into the host's DNA and multiply whenever the host does. Viral DNA that is integrated into host DNA is called a (a)_____, and host cells carrying integrated viral DNA are said to be (b)_____.

■ 7 _____ occurs when bacteria exhibit new properties as a result of integration of temperate viral DNA.

■ 8 Phages released from lysogenic cells may carry a portion of host DNA to a new host. This form of recombination in called _____.

IV. SOME VIRUSES INFECT ANIMALS

■ 9 Unenveloped animal viruses enter the host cell by a process similar to phagocytosis called _____.

■ 10 RNA viruses called _____ transcribe their RNA genome into a DNA strand that is then used as a template to produce more viral RNA. HIV is one such virus.

A. Viruses cause diseases in animals
B. Viruses are linked to certain types of cancer

■ 11 _____ are viral genes that can transform host cells into cancer cells.

■ 12 The _____ retrovirus enters human T cells and initiates events that lead to leukemia.

■13 _____ is a herpes virus found in almost everyone, and in most infected people it causes no detectable symptoms, while in others it can cause infectious mononucleosis or lymphoma.

C. Prions, which appear to consist only of protein molecules, are linked to certain animal diseases

V. SOME VIRUSES INFECT PLANTS

■14 _____ are disease-causing RNA particles that are smaller and simpler than viruses and infect plants.

VI. THE ORIGIN OF VIRUSES IS UNCERTAIN

■15 Viruses are thought to have had a multiple evolutionary origin; that is, some originated from plant cells, others from animal cells, and still others from bacteria. In all cases, regardless of their ancestral origins, they are thought to be _____ _____.

VII. BACTERIA ARE PLACED IN THE KINGDOM PROKARYOTAE

■16 The bacterial cell wall owes its rigidity to _____, a molecule found only in prokaryotes.

■17 Although membrane-bound organelles, such as a true nucleus, mitochondria, and chloroplasts, are entirely lacking in prokaryotes, some bacteria have _____, elaborate infoldings of the plasma membrane that are associated with cell division.

■18 The bacterial chromosome is a single, long, circular strand of DNA. Additional genetic information may also be found on small DNA loops called _____.

A. Bacterial diversity is evident in their varied metabolism

■19 Most bacteria are heterotrophic (a)_____, organisms that obtain nutrients from dead organic matter. Others are (b)_____ _____ that produce organic molecules by oxidizing simple inorganic molecules.

■20 Symbiotic bacteria may be (a)_____ which neither harm nor help their hosts, or (b)_____ that live at the expense of the host, or (c)_____ that both derive benefits from and provide benefits for the host.

■21 The five groups of photosynthetic bacteria are _____ _____ _____ _____ _____.

B. Bacteria reproduce by fission

■22 Bacteria usually reproduce asexually by simple transverse binary fission, but sometimes genetic material is exchanged in one of several ways. In (a)_____, fragments of DNA from a cell are taken in by another cell. In (b)_____, bacterial genes are carried from one cell to another by phages, and, finally, bacterial "mating types" may exchange genetic material during (c)_____.

VIII. THERE ARE TWO FUNDAMENTALLY DIFFERENT GROUPS OF BACTERIA, THE ARCHAEBACTERIA AND THE EUBACTERIA

IX. THE ARCHAEBACTERIA INCLUDE HALOPHILES, METHANOGENS, AND THERMOACIDOPHILES

■ 23 Halophiles ("salt loving" bacteria) live only in salty environments. Some of them photosynthesize using the purple photosynthetic pigment _____ _____.

■ 24 Methanogens are anaerobes that produce methane from _____ _____.

X. THE EUBACTERIA ARE DIVIDED INTO THREE GROUPS BASED ON THEIR CELL WALL COMPOSITION

■ 25 Eubacteria cells are spherical, rod-shaped, or spirals. Spherical bacteria are called (a)_____. They may occur singly, in pairs called (b)_____, or in chains called (c)_____, or in clumps or bunches called (d)_____. Rod-shaped bacteria are known as (e)_____, and spiral-shaped bacteria are called (f)_____.

A. Eubacteria may be grouped on the basis of their staining properties

■ 26 Gram-positive bacteria have a thick-layered cell wall of (a)_____ that retains (b)_____ stain. Gram-negative bacteria have a thick outer layer of lipid compounds surrounding the peptidoglycan layer in the cell wall.

B. Mycoplasms lack a rigid cell wall
C. Gram-negative bacteria have thin cell walls

■ 27 Spirochetes are spiral-shaped bacteria with flexible cell walls; they move by means of _____.

■ 28 Most of the gram-negative (a)_____ are photosynthetic autotrophs that contain chlorophyll *a* and accessory pigments on internal membranes called (b)_____.

■ 29 Rickettsias are obligate intracellular parasites. Most of them parasitize members of the taxon _____.

D. Gram-positive bacteria have thick cell walls of peptidoglycan

■ 30 Representatives of the group of gram-positive bacteria known as _____ can cause scarlet fever, "strep throat," and other infections.

■ 31 _____ are opportunistic pathogens that may cause boils, skin infections, food poisoning, and possibly toxic shock syndrome.

■ 32 Certain species of the anaerobic _____ cause tetanus, gas gangrene, and botulism.

■ 33 (a)_____ form branching filaments and spores called (b)_____. Some of them cause tuberculosis and leprosy.

KEY TERMS

Frequently Used Prefixes and Suffixes: Use combinations of these prefixes and suffixes to generate terms for the definitions below.

Prefixes	The Meaning
an-	without, lacking
arch(ae)-	primitive
bacterio-	bacteria
cyano-	dark blue
endo-	within
eu-	good, well, "true"
halo-	salt
onco-	"tumor"
patho-	suffering, disease, feeling
phyco-	seaweed, algae
vir-	virus

Suffixes	The Meaning
-cyan(in)	dark blue
-erythr(in)	red
-gen	production of
-oid	resembling, like
-phage	eat, devour
-phile	loving, friendly, lover

Prefix	Suffix	Definitions
_____	_____	1. A virus that infects and destroys bacteria.
_____	_____	2. Infective agent resembling a virus, but smaller and simpler.
_____	-gene	3. A gene that can cause cancer.
_____	-bacteria	4. Blue-green bacteria; formerly known as the "blue-green algae."
_____	-bacteria	5. Bacteria that inhabit extreme environments and do not have peptidoglycan in their cell walls; primitive bacteria.
_____	_____	6. Bacteria that live only in extremely salty environments; "salt lovers."
methano-	_____	7. A bacterium that produces methane from carbon dioxide and water.
_____	_____	8. A blue pigment found only in cyanobacteria (formerly called blue-green algae) and red algae.
_____	_____	9. A red pigment found in cyanobacteria and red algae.
_____	-spore	10. A thick-walled spore that forms within a bacterium in response to adverse conditions.
_____	-bacteria	11. The so-called "true bacteria."
_____	_____	12. Any disease-producing organism.
_____	-aerobe	13. An organism that metabolizes only in the absence of (without) molecular oxygen.

Other Terms You Should Know

absorptive endocytosis	assembly	bacillus (-li)
actinomycete	attachment	bacteriorhodopsin
aerobic	axial filament	Burkitt's lymphoma

capsid
capsule
chemoautotroph
chemosynthetic autotroph
chlamydia
clostridium (-ia)
coccus (-ci)
commensal
conjugation
enterobacterium (-ia)
envelope
Epstein-Barr virus
F pilus (-li)
facultative anaerobe
fermentation
fruiting body (-dies)
genome
gram-negative bacterium (-ia)
gram-positive bacterium (-ia)
heterotroph
lactic acid bacterium (-ia)

lysogenic
lysogenic conversion
lytic
mesosome
methanogen
mutualistic
mycoplasma
myxobacterium (-ia)
Neisseria gonorrhoeae
nitrogen-fixing aerobic
  bacterium (-ia)
obligate anaerobe
parasite
penetration
peptidoglycan
phage
photosynthetic autotroph
photosynthetic lamella (-ae)
pilus (-li)
plasmid
prion

prophage
release
replication
retrovirus (-ses)
reverse transcriptase
rickettsia
saprobe
spirillum (-la)
spirochete
staphylococcus (-ci)
streptococcus (-ci)
Streptomyces
temperate
thermoacidophile
trachoma
transduction
transformation
transverse binary fission
vector
virulent

SELF TEST

Multiple Choice: Some questions may have more than one correct answer.

1. Bacteria that retain crystal violet stain differ from those that do not retain the stain in having a
   a. thick lipoprotein layer.
   b. thick lipopolysaccharide layer.
   c. thicker peptidoglycan layer.
   d. cell wall.
   e. mesosome.

2. The coat surrounding the nucleic acid core of a virus is
   a. a capsid.
   b. a protein.
   c. a virion.
   d. a viral envelope.
   e. composed of capsomeres.

3. Viral genes that transform host cells into cancer cells are
   a. called oncogenes.
   b. the cause of AIDS.
   c. capable of coding for cellular growth factors.
   d. called prions.
   e. the cause of Hodgkin's disease.

4. The simplest form of life capable of independent growth and metabolism is probably the
   a. virus.
   b. spirochete.
   c. azotobacterium.
   d. myxobacterium.
   e. mycoplasma.

5. Bacteria that are adapted to extreme conditions and are perhaps most like the first organisms on the primitive Earth are the
   a. cyanobacteria.
   b. monera.
   c. eubacteria.
   d. archaebacteria.
   e. saprobes.

6. An important difference between virulent and temperate viruses is that only the temperate virus
   a. destroys the host cell.
   b. lyses the host cell.
   c. does not lyse host cells in the lysogenic cycle.
   d. contains DNA.
   e. infects eukaryotic cells.

7. Prokaryotic cells are distinguished from eukaryotic cells by
   a. absence of a nuclear membrane.
   b. absence of mitochondria.
   c. absence of DNA in the genetic material.
   d. absence of a plasma membrane.
   e. bacteriorhodopsin in cell walls.

8. Infoldings of the bacterial plasma membrane associated with cell division and metabolic processes are
   a. called plasmids.
   b. mesosomes.
   c. responsible for photosynthesis.
   d. found in archaebacteria and cyanobacteria, but not eubacteria.
   e. found in gram-negative, but not gram-positive bacteria.

9. Asexual reproduction in bacteria typically involves
   a. DNA replication.
   b. cytokinesis.
   c. exchange of genetic material (DNA) between donors and recipients.
   d. inward growth of the cell wall.
   e. expansion of a central plate.

10. When a host bacterium exhibits new properties because of a prophage the phenomenon is called
    a. lysogenic conversion.
    b. assembly.
    c. transduction.
    d. viroid induction.
    e. reverse transcription.

11. Viruses are usually grouped (classified) according to
    a. size.
    b. shape.
    c. evolutionary relationships.
    d. phylogeny.
    e. type of nucleic acid.

12. The majority of bacteria are
    a. autotrophs.
    b. heterotrophs.
    c. pathogens.
    d. saprobes.
    e. photosynthesizers.

13. The phenomenon by which a virus introduces DNA from a bacterium it previously infected into a new host bacterium is called
    a. lysogenic conversion.
    b. assembly.
    c. transduction.
    d. viroid induction.
    e. reverse transcription.

14. A prion
    a. is a subviral pathogen.
    b. contains DNA.
    c. is composed of glycoprotein.
    d. contains RNA.
    e. is larger than a typical virus.

15. The group of bacteria that have molecular traits so different from other prokaryotes as to be considered a separate kingdom by some biologists is the
    a. cyanobacteria.
    b. monera.
    c. eubacteria.
    d. archaebacteria.
    e. saprobes.

16. A layer in some viruses that contains lipid, carbohydrate, and protein is a
    a. capsid.
    b. viral core.
    c. virion.
    d. viral envelope.
    e. capsomere.

17. The characteristic(s) of life that viruses do not exhibit is/are
   a. presence of nucleic acids.
   b. independent movement.
   c. reproduction.
   d. a cellular structure.
   e. independent metabolism.

18. Exchange of genetic material between bacteria sometimes occurs by
   a. fusion of gametes.
   b. conjugation.
   c. incorporation by a bacterium of DNA fragments from another bacterium.
   d. transfer of genes in bacteriophages.
   e. transduction.

19. A small piece of DNA that is separate from the main chromosome is
   a. called a plasmid.
   b. a mesosome.
   c. responsible for photosynthesis.
   d. found in archaebacteria and cyanobacteria, but not eubacteria.
   e. found in gram-negative, but not gram-positive bacteria.

VISUAL FOUNDATIONS

Color the parts of the illustration below as indicated.

RED   ❑ genetic material
GREEN   ❑ outer membrane
YELLOW ❑ inner (plasma) membrane
BLUE   ❑ cell wall
ORANGE ❑ capsule

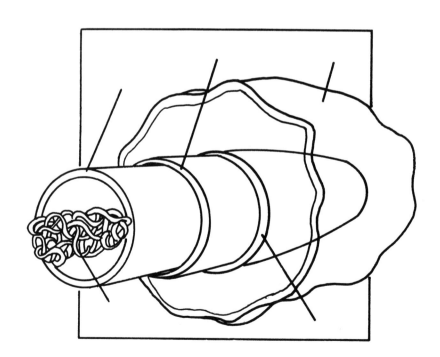

Color the parts of the illustration below as indicated.

RED	❑	DNA
GREEN	❑	RNA
YELLOW	❑	capsid
BLUE	❑	spikes
ORANGE	❑	fibers
BROWN	❑	tail

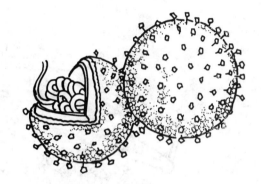

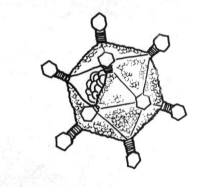

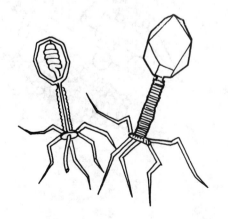

Color the parts of the illustration below as indicated.

RED	❑	protein
GREEN	❑	lipoprotein
YELLOW	❑	outer membrane
BLUE	❑	lipopolysaccharide
ORANGE	❑	cell wall
BROWN	❑	plasma membrane

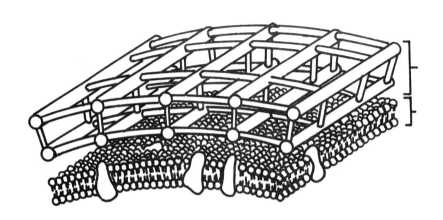

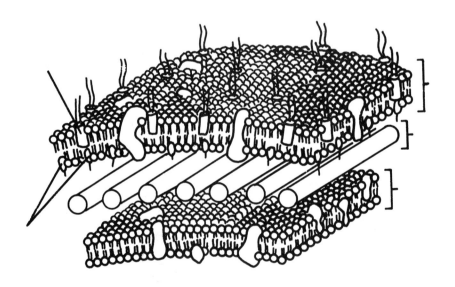

The Protist Kingdom

O U T L I N E

Characteristics of protists
Protozoa: animal-like protists
Algae: plant-like protists
Slime molds and water molds: fungus-like protists
The earliest eukaryotes

The protist kingdom, an unnatural assemblage of mostly aquatic organisms, includes three major groupings -- protozoa, algae, and slime and water molds. Although eukaryotic, the cell structure of protists is more complex than that of either animal or plant cells. Protists range in size from microscopic single cells to meters-long multicellular organisms. They do not have specialized tissues. Whereas some protists are free-living, others form symbiotic associations ranging from mutualism to parasitism. Reproductive strategies vary considerably among the different groups of protists; most reproduce both sexually and asexually, others only asexually. Whereas some are nonmotile, most move by either flagella, pseudopods, or cilia. The protozoa are the animal-like protists, so considered because they ingest their food like animals. Included in this group are the amoebas, foraminiferans, actinopods, flagellates, ciliates, and sporozoans. The algae are the plant-like protists, so considered because most are photosynthetic. Algae include the dinoflagellates, diatoms, euglenoids, and the green, red and brown algae. The slime and water molds are the fungus-like protists. There are two groups of slime molds -- the plasmodial slime molds and the cellular slime molds. Protists are believed to be the first eukaryotes. Flagellates appear to be the most primitive protists, and are thought to have given rise to animals. Green algae are believed to have given rise to plants, and red algae to fungi. Multicellular organisms may have evolved from colonial forms.

CHAPTER OUTLINE AND CONCEPT REVIEW (Fill in the blanks)

I. INTRODUCTION
II. PROTISTS ARE THE "SIMPLE" EUKARYOTES

■ 1 Organisms are placed in the kingdom Protista for convenience. They do not comprise a naturally occurring phylogenetic group, but they are all eukaryotes and most of them are unicellular. Others are colonial, or multicellular, or single multinucleated cells, the latter a condition known as _____.

■ 2 Although most protists are unicellular, their cell structure is more complex than those of animal or plant cells, because in most protists, a single cell contains all of the "equipment" required for life's activities. For example, water regulation is often accomplished by specialized intracellular organelles called _____ _____.

■ 3 Most protists are aquatic, and many of them are floating microscopic forms called _____.

III. PROTOZOA ARE ANIMAL-LIKE PROTISTS

A. Single-celled amoebas are in the phylum Rhizopoda

■ 4 Amoebas glide along surfaces by flowing into cytoplasmic extensions called _____, meaning "false feet."

■ 5 The parasitic species _____ causes amoebic dysentery.

B. Species in the phylum Foraminifera have tests made of organic materials strengthened with minerals

C. The phylum Actinopoda contains organisms with slender cytoplasmic projections

■ 6 Actinopods utilize long, filamentous cytoplasmic projections called _____ to entrap prey.

D. Organisms in the phylum Zoomastigina move by means of flagella

■ 7 The parasitic flagellate _____ causes African sleeping sickness.

■ 8 The _____, or "collared flagellates," as their name implies, resemble sponge cells.

E. Members of the phylum Ciliophora use cilia for locomotion

■ 9 Many ciliates possess _____, which are small organelles that eject filaments to trap their prey.

■ 10 Ciliates are distinct in having one or more small (a)_____ that function during the sexual process and a larger (b)_____ that regulates other cell functions.

■ 11 "Cross fertilization" occurs among many ciliates during the sexual process known as _____.

F. Phylum Apicomplexa contains spore-forming parasites of animals

■ 12 A sporozoon in the genus _____ can enter human RBCs and cause malaria.

IV. ALGAE ARE PLANT-LIKE PROTISTS

A. Most species in the phylum Dinoflagellata are marine planktonic forms

■ 13 The dinoflagellates are mostly unicellular, biflagellate, photosynthetic organisms with a cellulose shell-like covering. The _____, however, are photosynthetic endosymbionts that lack cellulose "shells" and flagella.

B. Members of the phylum Bacillariophyta have shells composed of two parts

■ 14 Diatoms are mostly single-celled, with silica impregnated in their cell walls. When they die, their shells accumulate, forming a sediment called _____ _____.

C. Most of the phylum Euglenophyta are freshwater unicellular flagellates

■ 15 Euglenoids are single-celled, flagellated protists containing (a)_____ _____, the same three pigments as those found in green algae and higher plants. They store food in the form of the polysaccharide (b)_____.

D. Phylum Chlorophyta contains the green algae

■ 16 Green algae exhibit a wide diversity in size, complexity, and reproduction. They resemble plants, but some are endosymbionts, and others are symbiotic with fungi in a mutually obligatory "biorganismic" entity known as a _____.

■ 17 Green algae exhibit three kinds of sexual reproduction. In (a)_____ reproduction, the two gametes that fuse are apparently identical, whereas (b)_____ reproduction involves gametes of different sizes. Fusion of a nonmotile "egg" and a motile sperm-like gamete is called (c)_____ reproduction.

E. Red algae are classified in the phylum Rhodophyta

■ 18 Most red algae are attached to a surface by means of a root-like structure called the (a)_____. Their chloroplasts contain chlorophyll *a* and carotenoids, and in addition, the red pigment (b)_____ and the blue pigment (c)_____.

F. The brown seaweeds are classified in the phylum Phaeophyta

■ 19 All brown algae are multicellular and produce flagellated cells during reproduction. Kelp are the largest brown algae; in addition to the holdfast, they possess differentiated leaf-like (a)_____ and stem-like (b)_____.

V. SLIME MOLDS AND WATER MOLDS ARE FUNGUS-LIKE PROTISTS

A. The plasmodial slime molds are classified in the phylum Myxomycota

■ 20 Plasmodial slime molds form intricate stalked reproductive structures called _____, within which meiosis occurs. The resulting haploid nuclei are then surrounded by a layer of chitin or cellulose, forming an extremely resistant spore.

■ 21 When environmental conditions are suitable, spores open and one-celled haploid gametes of two types, one a flagellated form called a (a)_____, the other an amoeboid (b)_____, emerge and fuse.

B. Phylum Acrasiomycota contains the cellular slime molds

■ 22 The cellular slime molds live vegetatively as individual, single-celled organisms, except during reproduction they form a multicellular aggregate called a _____, which then differentiates and forms spores.

C. Members of the phylum Oomycota produce flagellated zoospores

■ 23 The water molds, like the fungi, have a coenocytic vegetative (a)_____, and reproduce asexually by biflagellate (b)_____ and sexually by means of (c)_____.

VI. THE EARLIEST EUKARYOTES WERE PROTISTS

■ 24 _____ are probably the most primitive protists. They are thought to have given rise to animals.

■ 25 _____ are thought to have given rise to plants.

KEY TERMS

Frequently Used Prefixes and Suffixes: Use combinations of these prefixes and suffixes to generate terms for the definitions below.

Prefixes	The Meaning	Suffixes	The Meaning
cyto-	cell	-gamy	united
iso-	equal, "same"	-pod	foot, footed
macro-	large, long, great, excessive	-zoa	animal
micro-	small		
proto-	first, earliest form of		
pseudo-	false		
syn-	with, together		
tricho-	hair		

Prefix	Suffix	Definitions
_____	_____	1. Sexual reproduction; the coming together and the uniting of the gametes.
_____	_____	2. Single-celled, animal-like protists including amoebae, ciliates, flagellates, and sporozoans; members of this group are believed to have given rise to the earliest form of animal life.
_____	-pharynx	3. The gullet of a protozoan, a single-celled organism.
_____	_____	4. A temporary protrusion of the cytoplasm of an amoeboid cell that the cell uses for feeding and locomotion; a "false foot."
_____	-cyst	5. A cellular organelle of some ciliates that can discharge a hair-like filament that may aid in trapping and holding prey.
_____	-nucleus	6. A small nucleus found in ciliates.
_____	-nucleus	7. A large nucleus found in ciliates.
_____	-plasmodium	8. The aggregation of cells for reproduction in cellular slime molds; not a true plasmodium (multinucleate amoeboid mass) as occurs in the plasmodial slime molds.
_____	_____	9. Reproduction resulting from the union of two gametes that are identical ("the same") in size and structure.

Other Terms You Should Know

agar
algin
anisogamous
axopod
blade
blooms
carotenoid
carrageenan
coenocytic
conjugation
contractile vacuole
diatom
diatomaceous earth
flagella

fucoxanthin
gametangium (-ia)
holdfast
hypha (-ae)
isogamous
mycelium (-ia)
myxamoeba (-ae)
oogamous
oospore
oral groove
pellicle
phycocyanin
phycoerythrin
plankton

plasmodium (-ia)
protist
red tide
siphonous
sporangium (-ia)
spore
sporozoan (-oa)
stipe
swarm cell
test
zoospore
zooxanthella (-ae)

SELF TEST

Multiple Choice: Some questions may have more than one correct answer.

1. The parasite that causes malaria is in the phylum
   a. Ciliophora.
   b. Apicomplexa.
   c. Sarcomastigophora.
   d. Dinoflagellata.
   e. Euglenophyta.

2. Which of the following does the text include among protists?
   a. fungi.
   b. slime molds.
   c. azobacteria.
   d. protozoa.
   e. algae.

3. A protozoan whose cells bear a striking resemblance to cells in sponges is the
   a. foraminiferan.
   b. diatom.
   c. choanoflagellate.
   d. amoeba.
   e. radiolarian.

4. The amoeboid protozoans that form the grey mud on the bottom of oceans and that are used as indicators of geophysical changes in the environment are the
   a. foraminiferans.
   b. dinoflagellates.
   c. choanoflagellates.
   d. amoebae.
   e. radiolarians.

5. A protist rich in mineral nutrients and containing an agent used to thicken ice cream is a
   a. foraminiferan.
   b. Phaeophyta.
   c. Rhodophyta.
   d. red alga.
   e. brown alga.

6. A protist covering a large area of a piece of decaying tree bark would most likely be a
   a. Myxomycota.
   b. fungus.
   c. plasmodium.
   d. alga.
   e. slime mold.

7. One of the reasons that protists are placed in a separate kingdom is that many of them possess both plantlike and animal-like characteristics, which is particularly well-illustrated in the genus
   a. *Paramecium.*
   b. *Euglena.*
   c. *Plasmodium.*
   d. *Amoeba.*
   e. *Didinium.*

8. Ciliates are
   a. all motile.
   b. all sessile.
   c. sometimes motile, sometimes sessile.
   d. protozoa.
   e. binucleated or multinucleated.

9. Which of the following is/are characteristic of protists and clearly distinguish(es) them from monerans?
   a. Membrane-bound nucleus.
   b. Plasma membrane.
   c. Cell walls.
   d. Mitosis and meiosis.
   e. Eukaryotic cell structure.

10. Red algae are placed in the phylum
    a. Phaeophyta.
    b. Rhodophyta.
    c. Acrasiomycota.
    d. Oomycota.
    e. Apicomplexa.

11. A protozoan with two nuclei and projectile filaments is classified as
    a. Ciliophora.
    b. Apicomplexa.
    c. Sarcomastigophora.
    d. Dinoflagellata.
    e. Euglenophyta.

VISUAL FOUNDATIONS

Color the parts of the illustration below as indicated.

RED	❑	micronucleus
GREEN	❑	trichocyst
YELLOW	❑	oral groove
BLUE	❑	contractile vacuole
ORANGE	❑	cilia
BROWN	❑	food vacuole
TAN	❑	anal pore
VIOLET	❑	macronucleus

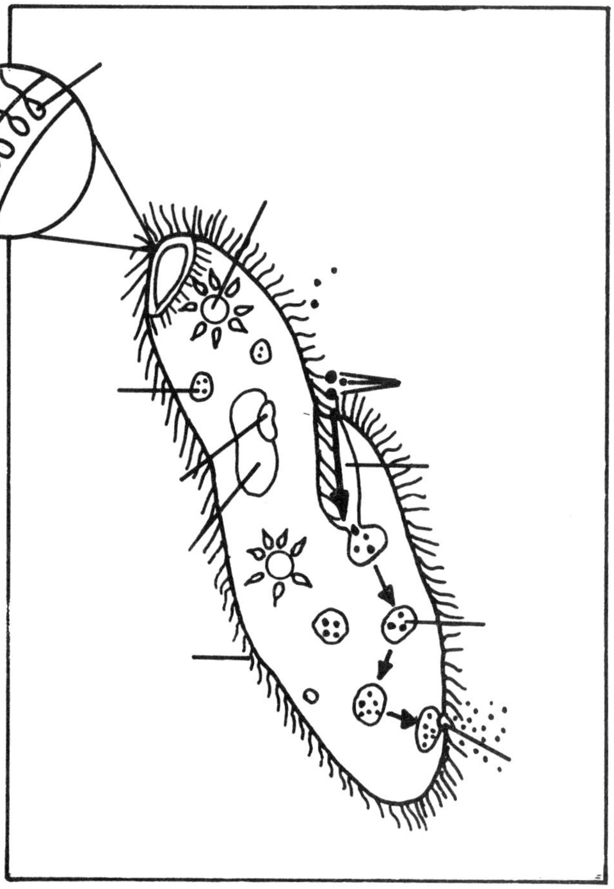

Color the parts of the illustration below as indicated.

RED ❑ nucleus
GREEN ❑ chloroplast
YELLOW ❑ eyespot
BLUE ❑ contractile vacuole
ORANGE ❑ external flagellum
BROWN ❑ short flagellum
VIOLET ❑ paramylon body

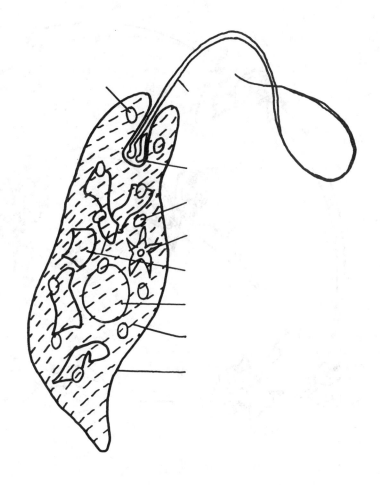

Color the parts of the illustration below as indicated. Label the portions of the illustration depicting sexual reproduction, and the portions depicting asexual reproduction.

RED ☐ positive strain
GREEN ☐ negative strain
YELLOW ☐ zygospore

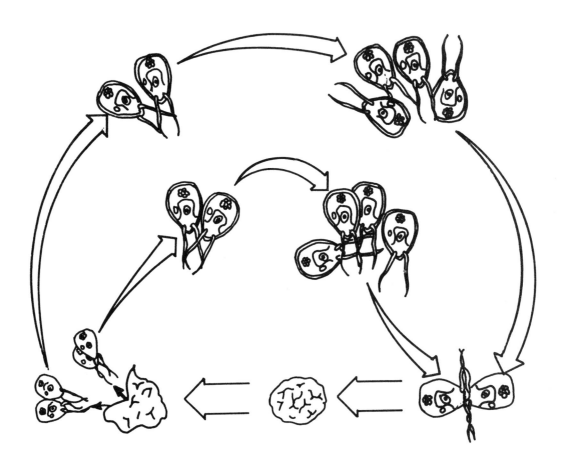

CHAPTER 25

□

Kingdom Fungi

OUTLINE

Ecological importance of fungi
Filamentous body plan
Fugal reproduction
Fungal classification
Lichens
Economic importance of fungi

Fungi are eukaryotes with cell walls. They digest their food outside their body and absorb the nutrients through the cell wall and cell membrane. Most fungi are decomposers, and as such play a critical role in the cycle of life. Others however form symbiotic relationships with other organisms. Some, for example, are parasites, causing diseases in animals and plants. And some are involved in mutualistic relationships, such as that between fungi and the roots of plants known as fungus-roots or mycorrhizae, and that between fungi and certain algae and cyanobacteria known as lichens. Fungi are nonmotile and reproduce by means of spores that are formed either sexually or asexually. Some fungi are unicellular (the yeast form) and some are multicellular, having a filamentous body plan. Fungi are classified into four divisions: Zygomycota, Ascomycota, Basidiomycota, and Deuteromycota. Fungi are of both positive and negative economic importance. Some damage stored goods and building materials. Some provide food for humans. Some function in brewing beer and in baking. Some produce antibiotics and other drugs and chemicals of economic importance, and some cause serious diseases in economically important animals, humans, and plants.

CHAPTER OUTLINE AND CONCEPT REVIEW (Fill in the blanks)

I. INTRODUCTION
II. FUNGI ARE ECOLOGICALLY IMPORTANT

- ■ 1 Most fungi are heterotrophic decomposers, more specifically they are (a)_____ that absorb nutrients from externally digested nonliving organic matter. In the process, they release (b)_____ to the atmosphere and return (c)_____ to the soil.

- ■ 2 Some fungi are parasites. Others are mutualistic symbionts, notable among which are the _____ that live on the roots of most plants.

III. MOST FUNGI HAVE A FILAMENTOUS BODY PLAN

■ 3 Body structures vary in complexity from single-celled (a)_____ to filamentous, sometimes elaborately complex multicellular (b)_____.

■ 4 Cell walls of most fungi contain _____, which is highly resistant to microbial breakdown.

IV. MOST FUNGI REPRODUCE BY SPORES

■ 5 Fungal spores are usually produced on (a)_____ that project into the air above their food source. In some fungi, these reproductive structures are elaborate fruiting bodies or (b)_____.

■ 6 Some form of sexual conjugation often occurs, sometimes producing hyphal cells with two distinctly different unfused nuclei. Such cells are said to be (a)_____, whereas cells containing only one haploid nucleus are (b)_____.

V. FUNGI ARE CLASSIFIED INTO FOUR DIVISIONS

■ 7 Based mainly on the characteristics of the (a)_____ _____, fungi are classified into four divisions: (b)_____ _____.

A. Zygomycetes reproduce sexually by forming zygospores

■ 8 The zygomycete species (a)_____ is the well known black bread mold. It is (b)_____, meaning a fungal mycelium will only mate with a *different* mating type, not with its own type.

B. Ascomycetes (sac fungi) reproduce sexually by forming ascospores

■ 9 Ascomycetes produce sexual spores in sacs called (a)_____, and asexual spores called (b)_____ at the tips of (c)_____.

■ 10 Sexual spore sacs (asci) develop on specialized dikaryotic (n + n) fruiting bodies known as _____.

C. Basidiomycetes (club fungi) reproduce sexually by forming basidiospores

■ 11 Basidiomycetes develop an enlarged, club-shaped hyphal cell called a (a)_____, on the surface of which four spores called (b)_____ develop.

■ 12 The mushroom we eat is actually a compact mass of hyphae that form a stalk and cap called the (a)_____. The mushroom is technically a sporocarp, or (b)_____.

D. Deuteromycetes (imperfect fungi) are fungi with no known sexual stage

■ 13 Deuteromycetes are called "imperfect fungi" simply because no one has observed them to have a _____ in their life cycles. If one is discovered in a species, it is moved to another division.

VI. LICHENS ARE DUAL "ORGANISMS" COMPOSED OF A FUNGUS AND PHOTOTROPH

■ 14 A lichen is a symbiotic combination of a fungus and a phototroph in which the fungus benefits from the photosynthetic activity of the phototroph. Lichens typically assume one of three forms: (a)_____, a low, flat growth, or (b)_____, a moderately flat growth with leaf-like lobes, or (c)_____, an erect and branching shrub-like growth.

■ 15 The phototroph portion of the lichen is usually a (a)_____ _____. The fungus is sometimes a basidiomycete, but usually an (b)_____.

■ 16 Lichens reproduce asexually by fragmentation, wherein a piece of dried lichen called a _____ breaks off and begins growing when it lands on a suitable substrate.

VII. FUNGI ARE ECONOMICALLY IMPORTANT

A. Fungi provide food for humans

■ 17 (a)_____ are the fungi used to make wine and beer, and to produce baked goods. Wine is produced when the fungus ferments (b)_____, beer results from fermentation of (c)_____, and (d)_____ bubbles cause bread to rise.

■ 18 Roquefort and Camembert cheeses are produced using the genus (a)_____, and the imperfect fungus (b)_____ is used to make soy sauce from soybeans.

■ 19 There are about 200 edible mushrooms and about 70 poisonous ones (toadstools). The most toxic substances in toadstools are the (a)_____, one of which inhibits mRNA synthesis. Some of the most toxic mushrooms, such as the "destroying angel" and "death angel," belong to the genus (b)_____.

■ 20 The chemical _____ found in some mushrooms causes hallucinations and intoxication.

B. Fungi produce useful drugs and chemicals

■ 21 Alexander Fleming noticed that bacterial growth is inhibited by the mold (a)_____, and this discovery eventually led to the development of the most widely used of all antibiotics, (b)_____.

■ 22 An ascomycete that infects cereal plant flowers produces a structure called an (a)_____ where seeds would normally form. When livestock or humans eat contaminated grain or grain products, they may develop the disease (b)_____, a condition characterized by convulsions, delusions, and sometimes gangrene and death.

C. Fungi cause many important diseases of plants

■ 23 All plants are susceptible to fungal diseases. Fungi enter plants through stomata, or wounds, or by dissolving a portion of cuticle with the enzyme (a)_____. Parasitic fungi often extend specialized hyphae called (b)_____ into host cells in order to obtain nutrients from host cell cytoplasm.

D. Fungi cause certain diseases of animals

■ 24 Ringworm, athlete's foot, and candidiasis ("yeast infection") are infections caused by members of the division _____.

KEY TERMS

Frequently Used Prefixes and Suffixes: Use combinations of these prefixes and suffixes to generate terms for the definitions below.

Prefixes	The Meaning		Suffixes	The Meaning
coeno-	common		-cyt(ic)	cell
hetero-	different		-karyo(tic)	nucleus
homo-	same		-phore	bearer
mono-	alone, single, one			
sapro-	rotten			
sporo-	spore			

Prefix	Suffix	Definitions
_____	-phyte	1. An organism that absorbs its nutrients through its cell membrane following extracellular digestion of dead and decaying life forms.
_____	_____	2. Pertains to an organism made up of a multinucleate, continuous mass of cytoplasm enclosed by one cell wall (all nuclei share a common cell).
_____	-carp	3. Large complex reproductive structure in which spores are produced. It is formed by the aerial hyphae of some fungi (e.g., mushrooms).
_____	_____	4. Pertains to hyphae that contain only one nucleus per cell.
_____	-thallic	5. Pertains to an organism that has two different mating types.
conidio-	_____	6. Hyphae in the ascomycetes that bear conidia.
_____	-thallic	7. Pertains to an organism that can mate with itself.

Other Terms You Should Know

ascocarp	deuteromycetes	mycorrhizae
Ascomycota	dikaryotic	parasite
ascospore	ergot	phototroph
ascus (-ci)	foliose	primary mycelium (-ia)
basidiocarp	fruticose	sac fungi
Basidiomycota	gametangium (ia)	secondary mycelium (-ia)
basidiospore	haustorium (-ia)	septum (-ta)
basidium (-ia)	hypha (-ae)	soredium (-ia)
budding	imperfect fungi	sporangium (-ia)
chitin	lichen	yeast
conidium (-ia)	mutualism	Zygomycota
crustose	mycelium (-ia)	zygospore

SELF TEST

Multiple Choice: Some questions may have more than one correct answer.

1. Yeast participates in the brewing of beer by
   - a. adding vital amino acids.
   - b. fermenting grain.
   - c. fermenting fruit sugars.
   - d. producing ethyl alcohol.
   - e. converting barley to hops.

2. The course of history was changed when Peter the Great's cavalry was killed by the fungus-caused disease known as
   - a. autoimmune response.
   - b. St. Anthony's fire.
   - c. histoplasmosis.
   - d. ergotism.
   - e. candidiasis.

3. A fungus infection throughout the body obtained by exposure to bird droppings is likely
   - a. an autoimmune response.
   - b. St. Anthony's fire.
   - c. histoplasmosis.
   - d. ergotism.
   - e. candidiasis.

4. The fungus that makes minerals available to a tree through mutualism with the tree's roots is a
   - a. water mold.
   - b. mycorrhizae.
   - c. lichen.
   - d. saprophyte.
   - e. yeast.

5. A collection of filamentous hyphae is called a
   - a. hypha.
   - b. mycelium.
   - c. conidium.
   - d. thallus.
   - e. ascocarp.

6. A hypha containing two genetically distinct nuclei can result from
   - a. conjugation.
   - b. meiosis.
   - c. fusion of hyphae from two genetically different mating types.
   - d. polynucleate mitosis.
   - e. dikaryosis.

7. If you eat just any mushroom that you find in the wild, there's a chance that you will
   - a. die.
   - b. become intoxicated.
   - c. see colors that aren't really there.
   - d. not become nauseated or die.
   - e. ingest the hallucinogenic drug psilocybin.

8. A common fungal infection of the mucous membranes of the mouth or vagina is
   - a. an autoimmune response.
   - b. St. Anthony's fire.
   - c. histoplasmosis.
   - d. ergotism.
   - e. candidiasis.

9. The black fungus growing on a piece of bread
   - a. is heterothallic.
   - b. has male and female strains.
   - c. has coenocytic hyphae.
   - d. has sporangia on the tips of stolons.
   - e. is in the division Zygomycota.

10. When you eat a mushroom, you are eating
    - a. a club fungus.
    - b. a sporocarp.
    - c. the bulk of the vegetative body.
    - d. a Basidiomycota.
    - e. an ascocarp.

11. A lichen can be composed of a/an
    a. alga and fungus.
    b. phototroph and fungus.
    c. cyanobacterium and ascomycete.
    d. alga and basiodiomycete.
    e. alga and ascomycete.

12. The genus *Penicillium*
    a. is an ascomycete.
    b. is a sac fungus.
    c. produces the flavor in Roquefort cheese.
    d. produces the antibiotic penicillin.
    e. is an imperfect fungus.

13. Lichens can be used as indicators of air pollution because they
    a. cannot excrete absorbed elements.
    b. tolerate sulfur dioxide.
    c. can endure large quantities of toxins.
    d. do not grow well in polluted areas.
    e. overgrow polluted areas.

14. Fungi can reproduce
    a. sexually.
    b. asexually.
    c. by spore formation.
    d. by fission.
    e. by budding.

VISUAL FOUNDATIONS

Color the parts of the illustration below as indicated.

RED ☐ haustorium
GREEN ☐ hypha
YELLOW ☐ epidermal cell
BLUE ☐ cuticle

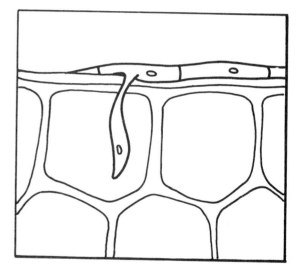

Color the parts of the illustration below as indicated.

RED	❑	released basidiospore
GREEN	❑	mycelium
YELLOW	❑	gills
BLUE	❑	basidium
ORANGE	❑	base
BROWN	❑	stalk
TAN	❑	cap

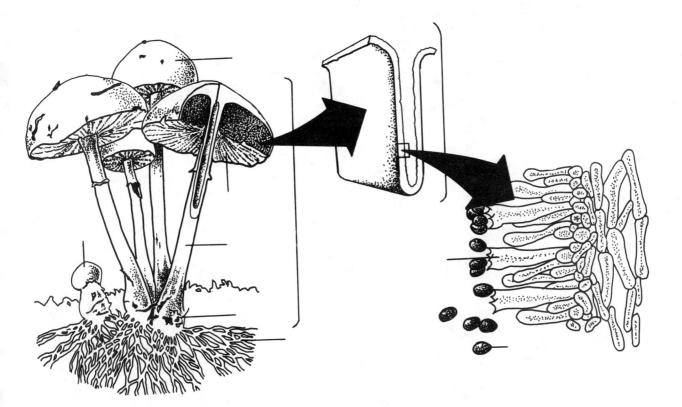

The Plant Kingdom: Seedless Plants

O U T L I N E

The plant kingdom
Alternation of generations
Mosses and other bryophytes
Ferns and their allies

Plants, believed to have evolved from green algae, are multicellular and exhibit great diversity in size, habitat, and form. They photosynthesize to obtain their energy. To survive on land, plants evolved 1) a waxy surface layer that protects against water loss, 2) tiny openings in the surface layer that allow for the gas exchange that is essential for photosynthesis, and 3) a strengthening polymer in the cell walls that enables them to grow tall and dominate the landscape. Plants spend part of their life cycle in a haploid stage that produces gametes (egg and sperm) by mitosis. The gametes fuse in a process known as fertilization, forming the diploid stage of the life cycle. To complete the life cycle, the diploid stage produces haploid spores by meiosis that divide mitotically producing the haploid stage. The seedless plants include the bryophytes and the ferns and their allies. The bryophytes, i.e., the mosses, liverworts, and hornworts, are the only land plants without vascular tissues, and are therefore restricted in size. Most of the bryophytes require a moist environment. Ferns and their allies, i.e., the whisk ferns, club mosses, and horsetails, have a vascular system and are therefore more advanced than the bryophytes. A vascular system allows plants to achieve larger sizes because water and nutrients can be transported over great distances to all parts of the plant. The ferns and their allies also have a larger and more dominant diploid stage, which is a trend in the evolution of land plants. Most of the seedless plants produce only one kind of spore as a result of meiosis. A few however produce two types of spores, an evolutionary development that led to the evolution of seeds.

CHAPTER OUTLINE AND CONCEPT REVIEW (Fill in the blanks)

I. INTRODUCTION

■ 1 Because of the numerous characteristics that they share, plants are thought to have evolved from ancient forms of the (a)_____. For example, both groups have the same photosynthetic pigments, which are (b)_____ _____; they both store excess carbohydrates as (c)_____; (d)_____ is the major component of their cell walls; and a (e)_____ forms during cytokinesis in both groups.

■ **2** There are four major groups of plants. Of these, only the (a)_____ _____ lack a vascular (conducting) system. Of the remaining three groups, only the (b)_____ reproduce entirely by spores. The (c)_____ have "naked seeds," often in cones, while the (d)_____ produce seeds within a fruit.

II. COMPLEX PHOTOSYNTHETIC ORGANISMS ARE PLACED IN THE PLANT KINGDOM

■ **3** The adaptation of plants to land environments involved a number of important adaptations. Among them was the (a)_____ that covers aerial parts, protecting the plant against water loss. Openings in this covering also developed. These (b)_____ facilitate the gas exchange that is necessary for photosynthesis.

■ **4** The sex organs of plants, or (a)_____, as they are called, are multicellular structures containing gametes. Both the male sex organ, the (b)_____, and the female sex organ, the (c)_____, contain gametes, sperm and egg respectively, that are protected from desiccation by an outer layer or jacket of "sterile" cells.

■ **5** _____ is a polymer in the cell walls of large, vascular plants that strengthens and supports the plant and its conducting tissues.

III. THE PLANT LIFE CYCLE ALTERNATES BETWEEN A GAMETOPHYTE GENERATION AND A SPOROPHYTE GENERATION

■ **6** Plants have alternation of generations, spending part of their life cycle in the haploid (gametophyte) stage and part in the diploid (sporophyte) stage. The first stage in the sporophyte generation is the (a)_____, and the haploid (b)_____ are the first stage of the gametophyte generation.

IV. MOSSES AND OTHER BRYOPHYTES ARE NONVASCULAR PLANTS

■ **7** Bryophytes can be separated into three classes, known in the vernacular as the _____. All of them are nonvascular and have similar life cycles.

A. Mosses have a dominant gametophyte generation

■ **8** Mosses are land plants without _____. They are therefore restricted in size and most of them require a moist environment for growth and reproduction.

■ **9** Among the most important mosses are those with large water storage vacuoles in their cells. These mosses are in the genus (a)_____, and are known by their common name, the (b)_____.

B. Liverworts are either thallose or leafy

■ **10** Liverworts reproduce sexually and asexually. In sexual reproduction, the thallus may simply branch and grow. Another form of asexual reproduction involves the formation of small masses of cells called _____ on the thallus, which can grow into a new liverwort thallus.

C. Hornworts are inconspicuous thalloid plants

V. SEEDLESS VASCULAR PLANTS INCLUDE FERNS AND THEIR ALLIES

■ 11 Ferns and fern allies have several advancements over the bryophytes, including the possession of vascular tissue and a dominant sporophyte generation. They also have two basic forms of leaves: the small (a)_____ with its single vascular strand and the larger (b)_____ with multiple vascular strands.

A. Ferns have a dominant sporophyte generation

■ 12 In ferns, the dominant (conspicuous), diploid sporophyte generation is composed mainly of an underground stem called the (a)_____, from which extend roots and leaves called (b)_____.

■ 13 (a)_____, the spore cases on the leaves of ferns, often occur in clusters called (b)_____. The haploid spore develops into a tiny, heart-shaped mature gametophyte plant called a (c)_____.

B. Whisk ferns are the simplest vascular plants

■ 14 The main organ of photosynthesis in sporophyte whisk ferns is the (a)_____, which exhibits (b)_____ branching (dividing into two *equal* halves). The small haploid gametophyte is a nonphotosynthetic subterranean plant that apparently obtains nourishment through a symbiotic relationship with a (c)_____.

C. Horsetails have hollow, jointed stems

■ 15 The few surviving species of horsetails are all in the genus _____.

D. Club mosses are small plants with rhizomes and short, erect branches

E. More advanced plants are less dependent on water as a transport medium for reproductive cells

F. Some ferns and club mosses are heterosporous

■ 16 The bryophytes, horsetails, whisk ferns, and most ferns and club mosses produce only one type of spore, a condition known as (a)_____. Some ferns and club mosses exhibit (b)_____, the production of two different types of spores.

■ 17 The strobilus of the club moss *Selaginella* has two kinds of sporangia. The small (a)_____ produce small "mother cells" or (b)_____ that undergo meiosis to form tiny, haploid (c)_____ that develop into sperm-producing male gametophytes. Egg-producing female gametophytes develop from (d)_____ that in turn result from the meiosis of (e)_____.

KEY TERMS

Frequently Used Prefixes and Suffixes: Use combinations of these prefixes and suffixes to generate terms for the definitions below.

Prefixes	The Meaning	Suffixes	The Meaning
arch(e)-	primitive	-angi(o)(um)	vessel, container
bryo-	moss	-gon(o)(ium)	sexual, reproductive

gamet(o)-	sex cells, eggs and sperm	-phyte plant
hetero-	different, other	-spor(e)(y) spore
homo-	same	
mega-	large great	
micro-	small	
spor(o)-	spore	
xantho-	yellow	

Prefix	Suffix	Definitions
_____	-phyll	1. Yellow plant pigment.
_____	_____	2. Special structure (container) of plants, protists, and fungi in which gametes are formed.
_____	_____	3. The female reproductive organ in primitive land plants.
_____	_____	4. The gamete-producing stage in the life cycle of a plant.
_____	_____	5. The spore-producing stage in the life cycle of a plant.
_____	-phyll	6. A small leaf that contains one vascular strand.
_____	-phyll	7. A large leaf that contains multiple vascular strands.
_____	_____	8. Special structure (container) of certain plants and protists in which spores and sporelike bodies are produced.
_____	_____	9. Production of one type of spore in plants (all spores are the same type).
_____	_____	10. Production of two different types of spores in plants, microspores and megaspores.
_____	_____	11. Large spore formed in a megasporangium.
_____	_____	12. Small spore formed in a microsporangium.
_____	_____	13. Mosses, liverworts, and their relatives.

Other Terms You Should Know

alternation of generations	gemmae cup	rhizome
antheridium (-ia)	lignin	seta (-ae)
capsule	megasporangium (-ia)	sorus (-ri)
carotene	megaspore mother cell	spore
dichotomous	microsporangium (-ia)	spore mother cell
embryo	microspore mother cell	sporophyte generation
fertilization	node	stoma (-ata)
fiddlehead	phloem	strobilus (-li)
frond	prothallus (-li)	thallus (-li)
gametophyte generation	protonema (-ata)	xylem
gemma (-ae)	rhizoid	zygote

SELF TEST

Multiple Choice: Some questions may have more than one correct answer.

1. A strobilus is
   a. on a diploid plant.
   b. on a haploid plant.
   c. on a vascular plant.
   d. found on horsetails.
   e. found on plants that do not have true leaves.

2. Mosses
   a. are in the class Bryopsida.
   b. are in the division Bryophyta.
   c. may have antheridia and archegonia on the same plant.
   d. contributed significantly to coal deposits.
   e. are vascular plants.

3. The sporophyte generation of a plant
   a. is diploid.
   b. is haploid.
   c. produces haploid gametes.
   d. produces haploid spores by mitosis.
   e. produces haploid spores by meiosis.

4. The spore cases on a fern are
   a. usually on the fronds.
   b. formed by the haploid generation.
   c. called sporangia.
   d. often arranged in a sorus.
   e. precursors to the fiddlehead.

5. Land plants are thought to have evolved from
   a. fungi.
   b. green algae.
   c. bryophytes.
   d. Euglena-like autotrophs.
   e. mosses.

6. Plants all have
   a. chlorophyll a.
   b. chlorophyll b.
   c. xanthophyll.
   d. carotene.
   e. yellow pigments.

7. The gametophyte generation of a plant
   a. is diploid.
   b. is haploid.
   c. produces haploid spores.
   d. produces haploid gametes by mitosis.
   e. produces haploid gametes by meiosis.

8. Plant sperm form in
   a. diploid gametophyte plants.
   b. haploid sporophyte plants.
   c. haploid gametophyte plants.
   d. antheridia.
   e. archegonia.

9. The leaves of vascular plants that evolved from several branches
   a. are megaphylls.
   b. are microphylls.
   c. evolutionarily derived from stem tissue.
   d. contain one vascular strand.
   e. contain more than one vascular strand.

10. Plants in the class Bryopsida
   a. are mosses.
   b. have rhizoids.
   c. do not have vascular systems.
   d. have true leaves and stems.
   e. have true roots.

11. The leafy green part of a moss is the
    a. sporophyte generation.
    b. gametophyte generation.
    c. rhizoid.
    d. product of buds from a protonema.
    e. thallus.

12. Liverworts
    a. are vascular plants.
    b. are bryophytes.
    c. contain a medicine that cures liver disease.
    d. can produce archegonia and antheridia on a haploid thallus.
    e. are in the same class as hornworts.

13. When a gamete produced by an archegonium fuses with a gamete produced by a male gametophyte plant, the result is
    a. an anomaly.
    b. a diploid zygote.
    c. the first stage in the sporophyte generation.
    d. the first stage in the gametophyte generation.
    e. called fertilization.

14. The conspicuous fern plant we often see in yards and homes is a
    a. Psilophyta.
    b. Pterophyta.
    c. gamete producer.
    d. haploid sporophyte.
    e. diploid sporophyte.

15. Spores grow
    a. into gametophyte plants.
    b. into sporophyte plants.
    c. into a haploid plant.
    d. to form a plant body by meiosis.
    e. to form a plant body by mitosis.

VISUAL FOUNDATIONS

Color the parts of the illustration below as indicated.

RED ❑ zygote

GREEN ❑ archegonium

YELLOW ❑ egg

BLUE ❑ sperm

ORANGE ❑ antheridium

BROWN ❑ capsule

TAN ❑ gametophyte (haploid) generation

PINK ❑ sporophyte (diploid) generation

VIOLET ❑ spore

Color the parts of the illustration below as indicated.

RED ☐ zygote
GREEN ☐ archegonium
YELLOW ☐ egg
BLUE ☐ sperm
ORANGE ☐ antheridium
BROWN ☐ sporangium
TAN ☐ gametophyte (haploid) generation
PINK ☐ sporophyte (diploid) generation
VIOLET ☐ spores

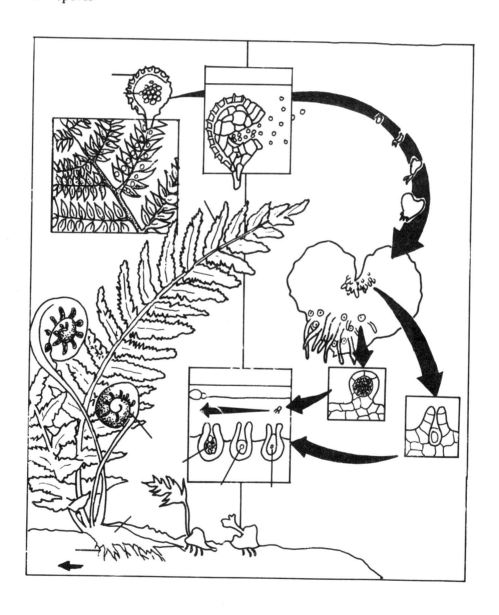

□

CHAPTER 27

The Plant Kingdom: Seed Plants

O U T L I N E

Gymnosperms: conifers, cycads, ginkgoes, and gnetophytes
Flowering plants
The evolution of seed plants

The most successful plants on earth are the seed plants, also known as the gymnosperms and angiosperms (flowering plants). They are the dominant plants in most habitats. Whereas gymnosperm seeds are totally exposed or borne on the scales of cones, those of angiosperms are encased within a fruit. All seeds are multicellular, consisting of an embryonic root, stem, and leaves. They are protected by a resistant seed coat that allows them to survive until conditions are favorable for germination, and they contain a food supply that nourishes the seed until it becomes self-sufficient. Seed plants have vascular tissue, and they exhibit alternation of generations in which the gametophyte generation is significantly reduced. They are heterosporous, producing both microspores and megaspores. There are four divisions of gymnosperms -- the Coniferophyta, the Ginkgophyta, the Cycadophyta, and the Gnetophyta. Although there is only one division of angiosperms -- the Magnoliophyta -- it is a huge one and is further divided into two major groups, the monocots and the dicots. Seeds and seed plants have been intimately connected with the development of human civilization. Human survival, in fact, is dependent on angiosperms, for they are a major source of human food. Angiosperm products play a major role in world economics. The organ of sexual reproduction in angiosperms is the flower. Flowering plants may be pollinated by wind or animals. Obligate relationships exist between some animal pollinators and the plants they pollinate. Flowering plants have several evolutionary advancements that account for their success. The fossil record indicates that seed plants evolved from seedless vascular plants.

CHAPTER OUTLINE AND CONCEPT REVIEW (Fill in the blanks)

I. INTRODUCTION

■ **1** The two groups of seed plants are the (a)_____, from the Greek "naked seed," and the (b)_____, meaning "seed in a case."

■ **2** Seed plants all have two types of vascular tissues: (a)_____ for conducting water and (b)_____ to conduct food.

II. GYMNOSPERMS ARE VASCULAR PLANTS WITH "NAKED SEEDS"

■ **3** The four divisions of plants with "naked seeds" are: _____
_____.

A. Conifers are woody plants that bear their seeds in cones

■ 4 The conifers are the largest group of gymnosperms. They are woody plants that bear needle leaves and produce their seeds in cones. Most of them are _____, which means that the male and female reproductive organs are in different places in the same plant.

■ 5 A few examples of conifers include _____
_____.

■ 6 The leaflike structures that bear sporangia on male cones of conifers are (a)_____, at the base of which are two microsporangia containing many "microspore mother cells" called (b)_____. These cells develop into the (c)_____.

■ 7 Female cones have (a)_____ at the base of bracts, within which are megasporocytes. Each of these cells produces four (b)_____, three of which disintegrate, while the fourth develops into the egg producing (c)_____.

B. Cycads are gymnosperms with compound leaves and simple seed cones

■ 8 The cycads are palmlike or fernlike gymnosperms that reproduce in a manner similar to pines, except cycads are _____, meaning that male cones and female cones are on *separate* plants. There are only a few extant members of this once large division.

C. *Ginkgo* is the only living species in its division

D. Gnetophytes include three unusual gymnosperms

■ 9 The gnetophytes share a number of advances over the rest of the gymnosperms, one of which is the presence of distinct _____ in their xylem tissues.

III. FLOWERING PLANTS ARE VASCULAR PLANTS THAT PRODUCE FLOWERS, FRUITS, AND SEEDS

■ 10 Angiosperms are all in the division _____.

A. There are two classes within the flowering plants, monocots and dicots

■ 11 The two groups of angiosperms are the monocots, in the class (a)_____ and the dicots, in the class (b)_____.

■ 12 The monocots have floral parts in multiples of (a)_____, and their seeds contain (b)_____(#?) cotyledons. The nutritive tissue in their mature seeds is the (c)_____.

■ 13 The dicots have floral parts in multiples of (a)_____, and their seeds contain (b)_____(#?) cotyledons. The nutritive tissue in their mature seeds is usually in the (c)_____.

■ 14 A few examples of monocots include (a)_____
_____. A few examples of dicots include (b)_____
_____.

B. Sexual reproduction in flowering plants involves the flower

■ 15 The four main organs in flowers are the (a)_____ _____, each of them arranged in whorls. The "male" organs are the (b)_____ and the "female" organs are the (c)_____. Flowers with both "male" and "female" parts are said to be (d)_____.

■ 16 Stamens are collectively known as the (a)_____, and carpels together comprise the (b)_____. Pollen forms in the (c)_____, a saclike structure on the tip of a stamen, and ovules form within the (d)_____ at the base of a carpel.

C. The life cycle of flowering plants includes a unique double fertilization process

■ 17 The megasporocyte in an ovule produces four haploid (a)_____, three of which disintegrate while the remaining one develops into the female gametophyte or (b)_____, as it is called. The female gametophyte contains eight nuclei, only three of which are directly involved in fertilization — these are the egg and the two (c)_____.

■ 18 Microsporocytes in the anther produce four haploid (a)_____, each of which develops into a (b)_____. A cell in this structure forms two male gametes called (c)_____.

■ 19 Double fertilization is a phenomenon that is unique to angiosperms. It results in the formation of three structures, namely, the _____ _____.

D. Flowering plants and their animal pollinators have affected one another's evolution

■ 20 _____ is the term for changes that occur in two species over time as a result of their very close, long term interactions. The evolution of flowers and their pollinators is an example.

E. The evolutionary advancements of the flowering plants account for their success

IV. THE FOSSIL RECORD PROVIDES VALUABLE CLUES ABOUT THE EVOLUTION OF SEED PLANTS

■ 21 (a)_____ is an extinct group of plants with features intermediate to ancient seedless plants and modern gymnosperms. This group gave rise to conifers and to another group of extinct plants called the (b)_____, which in turn are thought to be the ancestors of cycads and ginkgoes. Flowering plants evolved from (c)_____.

KEY TERMS

Frequently Used Prefixes and Suffixes: Use combinations of these prefixes and suffixes to generate terms for the definitions below.

Prefixes	The Meaning
andr(o)-	male
angio-	vessel, container
di-	two, twice, double

gymno-	naked, bare, exposed
gyn(o)-	female, woman
mon(o)-	alone, single, one

Prefix	Suffix	Definitions
_____	-sperm	1. A plant having its seeds enclosed in an ovary (a "vessel" or "container").
_____	-sperm	2. A plant having its seeds exposed or naked.
_____	-oecium	3. The female portion of a flower (all the carpels).
_____	-oecious	4. Pertains to plant species in which the male and female cones or flowers are on one plant.
_____	-oecious	5. Pertains to plant species in which the male and female flowers are on two different plants.
_____	-oecium	6. The male portion of a flower.

Other Terms You Should Know

anther
calyx (-xes, -yces)
carpel
complete flower
cone
conifer
Coniferophyta
corolla
cotyledon
cycad
Cycadophyta
deciduous
dicot
double fertilization
embryo sac
endosperm

fertilization
filament
fruit
Ginkgophyta
Gnetophyta
gnetophyte
imperfect flower
incomplete flower
Liliopsida
Magnoliophyta
Magnoliopsida
monocot
needle
ovary (-ries)
ovule
perfect flower

petal
pistil
polar nucleus (-ei)
pollen grain
pollination
pollination syndrome
progymnosperm
seed
seed fern
sepal
sporophyll
stamen
stigma (-ata)
style
triploid

SELF TEST

Multiple Choice: Some questions may have more than one correct answer.

1. Lilies are/have
   a. monocots.
   b. angiosperms.
   c. in the division Magnoliophyta.
   d. in the class Liliopsida.
   e. flower parts in multiples of three.

2. Dioecious plants with naked seeds, motile sperm, and the ability to affect fertilization without water are
   a. ginkgoes.
   b. cycads.
   c. extinct.
   d. gnetophytes.
   e. gymnosperms.

3. The sepals of a flower are collectively known as the
   a. corolla.
   b. calyx.
   c. gynoecium.
   d. androecium.
   e. florus perfecti.

4. In angiosperms, of the four haploid megaspores that result from meiosis of the megaspore mother cell, one of them becomes the
   a. male gametophyte generation.
   b. female gametophyte generation.
   c. embryo sac.
   d. archegonium.
   e. antheridium.

5. A traditional Christmas tree would likely be
   a. a gymnosperm.
   b. a conifer.
   c. the gametophyte generation.
   d. in the division Coniferophyta.
   e. heterosporous.

6. Which of the following is/are correct about a pine tree?
   a. Sporophyte generation is dominant.
   b. Male gametophyte produces an antheridium.
   c. Gametophyte is dependent on sporophyte for nourishment.
   d. Nutritive tissue in seed is gametophyte tissue.
   e. Pollen grain is an immature male gametophyte.

7. A pine tree has
   a. sporophylls.
   b. megasporangia on the female cones.
   c. separate male and female parts on the same tree.
   d. two sizes of spores in separate cones.
   e. female cones that are larger than male cones.

8. In gymnosperms, the pollen grain develops from
   a. microspore cells.
   b. spores.
   c. the male gametophyte.
   d. the gametophyte generation.
   e. meiosis of cells in the microsporangium.

9. The triploid endosperm of angiosperms develops from fusion of
   a. two sperm and one polar nucleus.
   b. two polar nuclei and one sperm.
   c. three polar nuclei.
   d. one diploid ovule and one haploid sperm.
   e. one haploid egg and one diploid sperm.

10. A perfect flower has
    a. stamens only.
    b. carpels only.
    c. both stamens and carpels.
    d. anthers.
    e. ovules.

Visual Foundations on next page ☞

VISUAL FOUNDATIONS

Color the parts of the illustration below as indicated.

RED   ❑ zygote

GREEN   ❑ female cone

YELLOW   ❑ megaspore mother cell, megaspore, egg

BLUE   ❑ pollen tube

ORANGE   ❑ microspore mother cell, microspore, pollen grain

BROWN   ❑ male cone

TAN   ❑ sporophyte (diploid) generation

PINK   ❑ gametophyte (haploid) generation

VIOLET   ❑ embryo

Color the parts of the illustration below as indicated.

RED   ❑   zygote

GREEN   ❑   female floral part

YELLOW   ❑   megaspore mother cell, megaspore, egg

BLUE   ❑   pollen tube

ORANGE   ❑   microspore mother cell, microspore, pollen grain

BROWN   ❑   male floral part

TAN   ❑   sporophyte (diploid) generation

PINK   ❑   gametophyte (haploid) generation

VIOLET   ❑   embryo sac

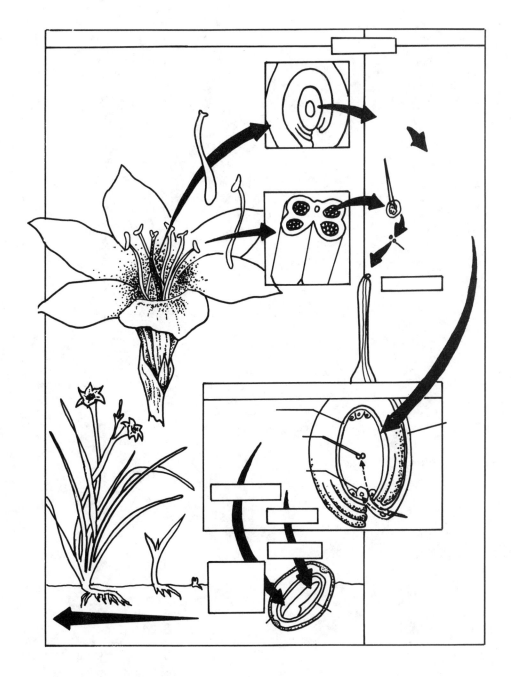

The Animal Kingdom: Animals Without a Coelom

O U T L I N E

Animal environments
Classification of animals according to body structure
or pattern of development
Phylum Porifera: Sponges
Phylum Cnidaria: Animals with stinging cells
Phylum Ctenophora: Comb jellies
Phylum Platyhelminthes: Flatworms
Phylum Nemertinea: Proboscis worms
Phylum Nematoda: Roundworms
Phylum Rotifera: Wheel animals

Animals are eukaryotic, multicellular, heterotrophic organisms. They are made up of localized groupings of cells with specialized functions. One grouping, for example, might function in locomotion, another might respond to external stimuli, and another might function in sexual reproduction. Animals inhabit virtually every environment on earth. They may be classified in a variety of ways -- as parazoa (sponges) or eumetazoa (all other animals); as radially symmetrical (the Radiata) or bilaterally symmetrical (the Bilateria); as those with no body cavity (the acoelomates), a body cavity between mesoderm and endoderm (the pseudocoelomates), or a body cavity within mesoderm (the coelomates); or those in which the first opening that forms in the embryonic gut becomes either the mouth (the protostomes) or the anus (the deuterostomes). This chapter briefly discusses the acoelomates and the pseudocoelomates. The acoelomates include the sponges, the cnidarians (hydras, jelly fish, and corals), the ctenophores (comb jellies), the platyhelminthes (flatworms), and the nemerteans (proboscis worms). The pseudocoelomates include the nematodes (roundworms), and the rotifers (wheel animals).

CHAPTER OUTLINE AND CONCEPT REVIEW (Fill in the blanks)

I. INTRODUCTION

II. ANIMALS INHABIT MOST ENVIRONMENTS OF THE ECOSPHERE

■ 1 Of the three major environments, the most hospitable to animals is
(a)_____. (b)_____ environments are hypotonic to
tissue fluids and (c)_____ environments tend to dehydrate
animals.

III. ANIMALS CAN BE CLASSIFIED ACCORDING TO BODY STRUCTURE OR PATTERN OF DEVELOPMENT

■ 2 There are two subkingdoms of animals: the (a)_____, containing only sponges, and (b)_____, containing all the rest.

A. Animals can be classified according to body symmetry

■ 3 Eumetazoans are grouped according to body types. Members of the branch (a)_____ possess repetitive body parts that radiate out from a central locus, and animals in the branch (b)_____ have right and left sides that are basically mirror images of one another.

■ 4 There are many anatomical terms that are used to describe the locations of body parts. Among them are: (a)_____ for front and (b)_____ for the rear; (c)_____ for a back surface and (d)_____ for the under or "belly" side; (e)_____ for parts closer to the midline and (f)_____ for parts farther from the midline, towards the sides; (g)_____ towards the head end and (h)_____ when away from the head, toward the tail.

B. Animals can be grouped according to type of body cavity

■ 5 Eumetazoans can also be grouped according to the presence or absence and, when present, the type of body cavity. For example, the "lowly" cnidarians and flatworms are called (a)_____ because they do not have a body cavity. The rest of the eumetazoans are either (b)_____, possessing a body cavity of sorts between the mesoderm and endoderm, or (c)_____ with a true body cavity within the mesoderm.

C. Animals can be classified as protostomes or deuterostomes

■ 6 "Higher" animals can also be grouped by shared developmental characteristics. In mollusks, annelids, and arthropods — all protostomes — the blastopore develops into the (a)_____, whereas in echinoderms and chordates — both deuterostomes — the blastopore becomes the (b)_____. Early cell divisions in deuterostomes are either parallel to or at right angles to the polar axis, which is called (c)_____, while protostomes early embryonic divisions are at oblique angles to the polar axis, a type of division referred to as (d)_____.

IV. PHYLUM PORIFERA CONSISTS OF THE SPONGES

■ 7 Because they have collar cells, sponges are thought to have evolved from the protozoan _____. They appear to be an evolutionary "dead end."

■ 8 Sponges are divided into three main classes on the basis of the type of skeleton they secrete. The (a)_____ secrete spicules of calcium carbonate, the spicules of (b)_____ (the glass sponges) contain silicon, and the (c)_____ which have protein fibers called (d)_____, or silicon spicules, or a combination of both.

■ 9 The sponge is a cellular sac perforated by numerous "pores." Water enters the internal cavity or (a)_____ through these openings, then exits through the (b)_____, which is the only open end in the body.

V. CNIDARIANS HAVE STINGING CELLS

■ **10** Cnidarians are characterized by radial symmetry; stinging cells called
(a)_____; two definite tissue layers, the outer (b)_____
and the inner (c)_____ separated by a gelatinous
(d)_____; and a nerve net.

■ **11** The phylum is divided into three classes: Hydras, hydroids, and the Portuguese
man-of-war are in the class (a)_____; jellyfish are in (b)_____;
and sea anemones, corals, and sea whips and sea fans are in (c)_____.

A. Class Hydrozoa includes solitary and colonial forms

■ **12** The epidermis of hydra contains specialized "stinging cells" called
(a)_____, within which are "thread capsules" called
(b)_____. These capsules release a long thread that entraps and
sometimes also paralyzes their prey.

B. The jellyfish belong to class Scyphozoa

■ **13** The familiar jellyfish is the (a)_____ stage of this scyphozoan. The
(b)_____ stage is a small, relatively inconspicuous larva.

C. The corals belong to class Anthozoa

VI. PHYLUM CTENOPHORA INCLUDES THE COMB JELLIES

■ **14** Phylum ctenophora consists of the comb jellies, which are fragile, luminescent,
(a)_____ symmetrical marine animals with (b)_____(#?) rows
of cilia resembling a comb. They have two tentacles with adhesive
(c)_____ that trap prey.

VII. FLATWORMS BELONG TO PHYLUM PLATYHELMINTHES

■ **15** Flatworms are characterized by (a)_____ symmetry, cephalization,
(b)_____(#?) definite tissue layers, well-developed organs, a simple brain and
nervous system, and specialized excretory structures called
(c)_____.

■ **16** The three classes of Platyhelminthes are (a)_____, the free-living
flatworms; (b)_____, the flukes; and (c)_____, the tapeworms.

A. Class Turbellaria includes planarians

■ **17** Planarians trap prey in a mucous secretion, then extend the _____ outward
to "vacuum" them into their digestive tube.

B. The flukes belong to class Trematoda

■ **18** Flukes are similar to free-living flatworms, except they are parasitic, they have
(a)_____ for clinging to their host and their mouths are (b)_____
instead of ventral.

C. The tapeworms belong to class Cestoda

■ **19** Most tapeworms have suckers and/or hooks on the (a)_____ for
attachment to their hosts. Each segment, or (b)_____, contains both
male and female reproductive organs. They absorb nutrients through the body
wall.

VIII. PHYLUM NEMERTINEA HAS EVOLUTIONARY IMPORTANCE

■ 20 Members of the phylum nemertinea have a tube-within-a-tube body plan, a complete digestive tract with mouth and anus, and a separate circulatory system. Their most distinctive feature is the _____, which is a long, hollow, muscular tube that can be everted from the anterior end of the body for use in seizing food or in defense.

IX. ROUNDWORMS BELONG TO PHYLUM NEMATODA

■ 21 Nematodes include species of great ecological importance and species that are parasitic in plants and animals. They are characterized by (a)_____(#?) definite tissue layers, a body cavity called a (b)_____, (c)_____ symmetry, and a complete digestive tract.

A. *Ascaris* is a parasitic roundworm
B. Several other parasitic roundworms infect humans

■ 22 The _____ larva enters the body through the skin, then migrates to the intestine where it matures.

■ 23 (a)_____ enter the intestines of humans who eat undercooked, infected pork. Their larvae migrate to skeletal muscles where they form (b)_____.

■ 24 Female _____ in the intestine migrate to the host's anal region to deposit eggs.

X. PHYLUM ROTIFERA ARE WHEEL ANIMALS

■ 25 "Wheel animals" are aquatic, pseudocoelomate, microscopic worms that exhibit _____, which means that each member of a given species is composed of exactly the same number of cells. Their nickname comes from the circular aggregate of (b)_____ that seems to be spinning like a wheel.

■ 26 Rotifers have a complete digestive tract, including a muscular organ for grinding food called the _____.

KEY TERMS

Frequently Used Prefixes and Suffixes: Use combinations of these prefixes and suffixes to generate terms for the definitions below.

Prefixes	The Meaning	Suffixes	The Meaning
deutero-	second	-coel(om)	cavity
ecto-	outer, outside, external	-cyte	cell
entero-	intestine	-stome	mouth
gastro-	stomach		
meso-	middle		
proto-	first, earliest form of		
pseudo-	false		
schizo-	split		

Prefix	Suffix	Definitions
_____	_____	1. A body cavity between the mesoderm and endoderm; not a true coelom.
_____	-derm	2. The outermost germ layer.
_____	-derm	3. The middle layer of the three basic germ layers.
_____	_____	4. Major division of the animal kingdom in which the mouth forms from the first opening in the embryonic gut (the blastopore); the anus forms secondarily.
_____	_____	5. Major division of the animal kingdom in which the mouth forms from the second opening in the embryonic gut; the first opening (the blastopore) forms the anus.
_____	_____	6. The process of coelom formation in which the mesoderm splits into two layers.
_____	_____	7. The process of coelom formation in which the mesoderm forms as "outpocketings" of the developing intestine, eventually separating and forming pouches which become the coelom.
spongo-	_____	8. The central cavity in the body of a sponge.
amoebo-	_____	9. Amoeba-like cell found in many invertebrates.
_____	-dermis	10. The tissue lining the gut cavity ("stomach") that is responsible for digestion and absorption in certain phyla.

Other Terms You Should Know

acoelomate
anterior
Anthozoa
Ascaris
auricle
bilateral symmetry
Bilateria
blastopore
Calcispongiae
caudal
cephalic
cephalization
Cestoda
choanocyte
cleavage
Cnidaria
cnidocyte
coelom
coelomate
collar cell
ctenophore
cuticle
Demospongia
determinate cleavage

dorsal
endoderm
enterocoelomate
epidermis
Eumetazoa
final host
flame cell
flatworm
fluke
frontal plane
ganglion (-ia)
gastrovascular cavity
germ layer
hermaphroditic
Hexactinellida
hookworm
Hydrozoa
indeterminate cleavage
inferior
intermediate host
invertebrate
lateral
mastax
medial

medusa (-as)(-ae)
mesoglea
nematocyst
Nematoda
Nemertinea
nerve net
osculum (-la)
Parazoa
pharynx (-nges)(-xes)
pinworm
planarian
plankton
planula (-ae)
Platyhelminthes
polyp
Porifera
posterior
proboscis (-ses)(-ides)
proglottid
protonephridion (-ia)
pseudocoelomate
radial cleavage
radial symmetry
Radiata

rostral	Scyphozoa	transverse section
Rotifera	sessile	Trematoda
roundworm	spiral cleavage	trichina worm
sagittal plane	sponge	Turbellaria
schizocoelomate	spongin	ventral
scolex (-eces)(-lices)	superior	vertebrate

SELF TEST

Multiple Choice: Some questions may have more than one correct answer.

1. The parasitic roundworms
   a. are pseudocoelomates.
   b. have a mastax.
   c. are in the phylum Platyhelminthes.
   d. include hookworms.
   e. are nematodes.

2. The taxon that has "thread capsules" in epidermal stinging cells is
   a. Scyphozoa.
   b. Hydrozoa.
   c. Porifera.
   d. Cnidaria.
   e. Ctenophora.

3. Deuterostomes characteristically have
   a. radial cleavage.
   b. spiral cleavage.
   c. indeterminate cleavage.
   d. schizocoely.
   e. enterocoely.

4. Sessile, marine animals with no medusa stage and a partitioned gastrovascular cavity include
   a. hydrozoans.
   b. scyphozoans.
   c. anthozoans.
   d. jellyfish.
   e. corals, sea anemones, and their close relatives.

5. Examples of acoelomate eumetazoans include
   a. sponges.
   b. insects.
   c. flatworms.
   d. those groups whose coelom developed from a blastocoel.
   e. cnidarians.

6. The simplest bilaterians are members of
   a. Ctenophora.
   b. Porifera.
   c. Platyhelminthes.
   d. the phylum that includes round worms.
   e. deuterostomes.

7. The phyla of animals that are placed in a separate branch of the Eumetazoa because they are radially symmetrical are the
   a. Parazoa.
   b. Radiata.
   c. sponges.
   d. cnidarians.
   e. ctenophores.

8. With the possible exception of sponges, all animals
   a. are heterotrophs.
   b. are eukaryotes.
   c. are multicellular.
   d. reproduce sexually.
   e. are capable of locomotion.

9. The group of animals in which every individual has the same number of cells is
   a. Ctenophora.
   b. Nematoda.
   c. Porifera.
   d. Rotifera.
   e. Trematoda.

10. When a group of cells moves inward to form a pore in early embryonic development, and that pore ultimately develops into a mouth, the group of animals is considered to be a
    a. pseudocoelomate.
    b. deuterostome.
    c. protostome.
    d. parazoan.
    e. schizocoelomate.

11. Animals that are parasitic and have a scolex and proglottids belong in the taxon/taxa
    a. Platyhelminthes.
    b. Nematoda.
    c. Turbellaria.
    d. Cestoda.
    e. Trichina.

12. The taxon that has biradial symmetry, a mesoglea, and eight rows of cilia that move the animal is
    a. Scyphozoa.
    b. Hydrozoa.
    c. Porifera.
    d. Cnidaria.
    e. Ctenophora.

13. The phylum/phyla that does/do not have specialized nerve cells is/are
    a. Porifera.
    b. Cnidaria.
    c. Platyhelminthes.
    d. Nemertinea.
    e. Echinodermata.

14. The one "lower" invertebrate phylum with radial symmetry and distinctly different tissues is
    a. Porifera.
    b. Cnidaria.
    c. Platyhelminthes.
    d. Nemertinea.
    e. Echinodermata.

Visual Foundations on next page ☞

VISUAL FOUNDATIONS

Color the parts of the illustration below as indicated.

RED ❑ mesoderm

GREEN ❑ endoderm

YELLOW ❑ body cavity

BLUE ❑ ectoderm

ORANGE ❑ mesoglea

Label the animals depicted below as either an acoelomate, a pseudocoelomate, or a coelomate.

CHAPTER 29

The Animal Kingdom: The Coelomate Protostomes

OUTLINE

Adaptations for life on land
Phylum Mollusca: Soft-bodied animals with a foot, visceral
mass, and mantle
Phylum Annelida: Segmented worms
Phylum Arthropoda: Animals with jointed
appendages and chitinous exoskeleton

The coelomate protostomes are animals with true coeloms in which the first opening that forms in the embryonic gut gives rise to the mouth. They have a complete digestive tract, and most have well-developed circulatory, excretory, and nervous systems. The coelom confers advantages to these animals over those without a coelom. For one thing, it separates muscles of the body wall from those of the digestive tract, permitting movement of food independent of body movements. Additionally, it serves as a space in which organs develop and function, and it can be used as a hydrostatic skeleton. Also, coelomic fluid helps transport food, oxygen, and wastes. Life on land requires many adaptations. To minimize water loss, terrestrial organisms have developed special body coverings as well as respiratory surfaces located deep inside the animal. To support the body against the pull of gravity, land animals have developed a supporting skeleton, located either within the body or externally covering the body. To meet the requirements of reproduction in a dry environment, some land animals return to water where their eggs and sperm are shed directly into the water. Most, however, copulate, depositing sperm directly into the female body. To prevent the developing embryo from drying out, it is either surrounded by a tough protective shell, or it develops within the moist body of the mother. The coelomate protostomes include the mollusks (chitons, snails, clams, oysters, squids, and octopuses), annelids (earthworms, leeches, and marine worms), onycophorans, and arthropods (spiders, scorpions, ticks, mites, lobsters, crabs, shrimp, insects, centipedes, and millipedes).

CHAPTER OUTLINE AND CONCEPT REVIEW (Fill in the blanks)

I. INTRODUCTION

■ 1 The coelomate protostomes include the mollusks, annelids, and arthropods, as well as some minor phyla. The blastopore develops into the (a)_____, and the coelom is formed within and is lined by the (b)_____ between the complete digestive tube and body wall.

■ **2** Among the many advantages derived from having a coelom are: _____

_____

_____

_____.

II. LIFE ON LAND REQUIRES MANY ADAPTATIONS

■ **3** Most invertebrate coelomates are still aquatic, however there are successful terrestrial groups as well. The first terrestrial, air-breathing animals were probably scorpion-like representatives of the phylum (a)_____.

■ **4** Terrestrial animals must overcome serious problems inherent in a land-based "life style." Three that are critical to their continued existence are: They must develop a body covering that prevents (a)_____, a support system (skeleton) that can withstand (b)_____, and reproductive adaptations that compensate for the lack of a supportive watery environment.

III. MOLLUSKS ARE SOFT-BODIED ANIMALS WITH A FOOT, VISCERAL MASS, AND MANTLE

■ **5** Mollusks are soft-bodied animals usually covered by a shell; they possess a ventral foot for locomotion and a (a)_____ that covers the visceral mass. With about 60,000 species, the phylum Mollusca is second only to (b)_____ in number of species.

■ **6** Most mollusks have an open circulatory system, meaning the blood flows through a network of open sinuses called the blood cavity or _____.

■ **7** Most marine mollusks have larval stages, the first a free-swimming, ciliated form called a (a)_____ larva, and the second, when present, a (b)_____ larva with a shell and foot.

A. Class Polyplacophora includes the chitons

■ **8** A distinctive feature of chitons is their shell, which is composed of _____(#?) overlapping plates.

B. Class Gastropoda includes snails and their relatives

■ **9** Class gastropoda, the largest and most successful group of mollusks, includes the snails, slugs, and whelks. It is a class second in size only to the _____.

■ **10** The gastropod shell (when present) is coiled, and the visceral mass is twisted, a phenomenon known as _____.

C. Class Bivalvia includes clams, oysters, and their relatives

■ **11** Foreign matter lodged between the shell and (a)_____ of bivalves may cause deposition of the compound (b)_____, forming a pearl.

■ **12** The _____ is the part of a scallop that we humans eat.

D. Class Cephalopoda includes squids and octopods

■ **13** Class cephalopoda includes the squids and octopods, which are active predatory animals. The foot is divided into (a)_____ that surround the mouth, (b)_____(#?) of them in squids, (c)_____(#?) of them in octopods.

IV. THE SEGMENTED WORMS BELONG TO PHYLUM ANNELIDA

■14 Phylum annelida, the segmented worms, includes many aquatic worms, earthworms, and leeches. Their body segments are separated from one another by partitions called _____.

■15 A pair of bristle-like structures called _____ project from the segments in annelids. These structures anchor portions of the worms' body during locomotion.

■16 _____ are pairs of excretory organs in each annelid segment.

A. Polychaetes have parapodia

■17 Class polychaeta consists of marine worms characterized by bristled _____, used for locomotion and gas exchange.

■18 The well-developed head of polychaetes is called the _____.

B. The earthworms belong to class Oligochaeta

■19 The earthworm is in the phylum (a)_____, the class (b)_____, and the genus and species (c)_____.

■20 The two parts of the earthworm's stomach are the (a)_____ where food is stored and the (b)_____ where ingested materials are ground into a mulch.

■21 The respiratory pigment in earthworms is (a)_____; the excretory system consists of paired (b)_____ in most segments; and the gas exchange surface is the (c)_____.

C. The leeches belong to class Hirudinea

V. ARTHROPODS HAVE JOINTED APPENDAGES AND AN EXOSKELETON OF CHITIN

■22 The word "Arthropoda" means (a)_____, which is one of the phylum's distinguishing characteristics. Another is the (b)_____, an armor-like body covering composed of (c)_____.

A. The arthropod body has three regions

■23 The three body segments in arthropods are the _____.

■24 Gas exchange in terrestrial arthropods occurs in a system of thin, branching tubes called _____.

B. Living arthropods can be assigned to three subphyla

■25 The three subphyla of contemporary arthropods are (a)_____, the arthropods such as horseshoe crabs and arachnids that lack antennae; the (b)_____, which includes lobsters, crabs, shrimp, and barnacles that all have biramous appendages and two pairs of antennae; and (c)_____, which have unbranched appendages and one pair of antennae.

C. Arthropods are closely related to annelids

■26 The most primitive arthropods are the (a)_____, which co-existed in Paleozoic waters with (b)_____, the subphylum that is thought to have arisen from these ancient arthropods.

D. Subphylum Chelicerata includes the horseshoe crabs and Arachnids

■ 27 Subphylum Chelicerata includes class (a)_____ (the horseshoe crabs) and class (b)_____ (spiders, mites, and their relatives).

■ 28 The arachnid body consists of a (a)_____ and abdomen; there are (b)_____(#?) pairs of jointed appendages, of which (c)_____(#?) pairs serve as legs.

■ 29 Gas exchange in arachnids occurs in tracheae and/or across thin, vascularized plates called _____.

■ 30 Silk glands in spiders secrete an elastic protein that is spun into web fibers by organs called _____.

E. Subphylum Crustacea includes the lobsters, crabs, shrimp, and their relatives

■ 31 Crustaceans are characterized by the (a)_____, the third pair of appendages used for biting food; (b)_____ appendages; and (c)_____(#?) pairs of antennae. In most crustaceans, (d)_____(#?) pairs of appendages are adapted for walking.

■ 32 _____ are the only sessile crustaceans.

■ 33 Lobsters, crayfish, crabs, and shrimp are members of the largest crustacean order, the _____.

■ 34 Among the many highly specialized appendages in decapods are the (a)_____, two pairs of feeding appendages just behind the mandibles; the (b)_____ that are used to chop food and pass it to the mouth; the (c)_____ or pinching claws on the fourth thoracic segments; and the pairs of (d)_____ on the last four thoracic segments.

■ 35 Appendages on the first abdominal segment of decapods are part of the (a)_____ system. (b)_____ on the next four abdominal segments are paddle-like structures adapted for swimming and holding eggs.

F. Subphylum Uniramia includes the insects, centipedes, and millipedes

■ 36 Insects are described as articulated, meaning (a)_____, tracheated, meaning (b)_____, hexapods, which means (c)_____.

■ 37 Adult insects typically have (a)_____(#?) pairs of legs, (b)_____(#?) pairs of wings, and (c)_____(#?) pair(s) of antennae. The excretory organs are called (d)_____.

■ 38 The members of the class (a)_____ or "centipedes" have (b)_____ pair(s) of legs per body segment, whereas the members of the class (c)_____ or millipedes have (d)_____(#?) pair(s) of legs per body segment.

KEY TERMS

Frequently Used Prefixes and Suffixes: Use combinations of these prefixes and suffixes to generate terms for the definitions below.

Prefixes	The Meaning		Suffixes	The Meaning
arthro-	joint, jointed		-pod	foot, footed
bi-	twice, two			
cephalo-	head			
endo-	within			
exo-	outside, outer, external			
hexa-	six			
tri-	three			
uni-	one			

Prefix	Suffix	Definitions
_____	-skeleton	1. Bony and cartilaginous supporting structures within the body that provide support from within.
_____	-skeleton	2. An external skeleton, such as the shell of arthropods.
_____	-valve	3. A mollusk that has two shells (valves) hinged together, as the oyster, clam, or mussel; a member of the class Bivalvia.
_____	-ramous	4. Consisting of or divided into two branches.
_____	-ramous	5. Unbranched; consisting of only one branch.
_____	-lobite	6. An extinct marine arthropod characterized by two dorsal grooves that divide the body into three longitudinal parts (or lobes).
_____	-thorax	7. Anterior part of the body in certain arthropods consisting of the fused head and thorax.
_____	_____	8. Having six feet; an insect.
_____	_____	9. An animal with paired, jointed legs.

Other Terms You Should Know

abdomen	centipede	excurrent siphon
adductor muscle	Cephalopoda	Gastropoda
Annelida	cerebral ganglion (-ia)	gizzard
antenna (-ae)	Chelicerata	head
antennal gland	cheliped	hemocoel
anus (-ses)	Chilopoda	hermaphroditic
aorta (-as)(-ae)	chiton	Hirudinea
appendage	clitellum (-la)	horseshoe crab
Arachnida	cocoon	incurrent siphon
Arthropoda	crop	ink sac
articulated	Crustacea	Insecta
Bivalvia	Decapoda	intestine
book lung	Diplopoda	leech
carapace	esophagus (-gi)	Malpighian tubule

mandible
mantle
maxilla (-ae)
maxilliped
Merostomata
metamerism
metanephridium (-ia)
millipede
Mollusca
mollusk
molting
nudibranch
Oligochaeta
Onychophora
open circulatory system
ostium (-ia)

palp
parapodium (-ia)
pedipalp
pharynx (-nges)(-xes)
Polychaeta
Polyplacophora
prostomium (-ia)
pulmonate
radula
reflex (-xes)
seminal receptacle
septum (-ta)
seta (-ae)
silk gland
statocyst
stomach

subpharyngeal ganglion (-ia)
swimmeret
tagma (-ata)
telson
thorax (-aces)(-axes)
torsion
trachea (-ae)
tracheated
trochophore larva (-ae)
typhlosole
Uniramia
uropod
veliger larva (-ae)
visceral mass
walking leg

SELF TEST

Multiple Choice: Some questions may have more than one correct answer.

1. An octopus is a
    a. vertebrate.
    b. mollusk.
    c. chilopod.
    d. protostome.
    e. cephalopod.

2. Butterflies, potato bugs, termites, and black widows all
    a. are insects.
    b. are arthropods.
    c. have paired, jointed appendages.
    d. have closed circulatory systems.
    e. have an abdomen.

3. The most primitive arthropods are
    a. crustaceans.
    b. trilobites.
    c. in the class Chilopoda.
    d. spiders.
    e. insects.

4. Animals with most of the organs in a visceral mass that is covered by a heavy fold of tissue are
    a. annelids.
    b. oligochaets.
    c. onychophorans.
    d. mollusks.
    e. cheliceratans.

5. The animal that has a shell consisting of eight separate, overlapping transverse plates is
    a. an insect.
    b. a chiton.
    c. in the same class as squids.
    d. in the class Polyplacophora.
    e. a mollusk.

6. Earthworms are
    a. annelids.
    b. oligochaets.
    c. in the phylum Polychaeta.
    d. protostomes.
    e. acoelomate animals.

7. Garden snails and abalone, considered by many to be gourmet delicacies, are
    a. in different phyla.
    b. in the same phylum.
    c. in different classes.
    d. in the same class.
    e. nudibranchs.

8. A marine animal with parapodia and a trochophore larva might very well be
   a. a tubeworm.
   b. an oligochaet.
   c. bilaterally symmetrical.
   d. a leech.
   e. a polychaet.

9. Animals thought by many zoologists to be a link between annelids and arthropods are
   a. wormlike.
   b. acoelomates.
   c. found in humid tropical rain forests.
   d. onychophorans.
   e. in the class Hirudinea.

10. Earthworms are held together in copulation by mucous secretions from the
   a. clitellum.
   b. typhlosole.
   c. seminal receptacles.
   d. epidermis.
   e. prostomium.

VISUAL FOUNDATIONS

Color the parts of the illustration below as indicated.

RED   ❑   digestive tract
GREEN   ❑   shell
YELLOW   ❑   foot

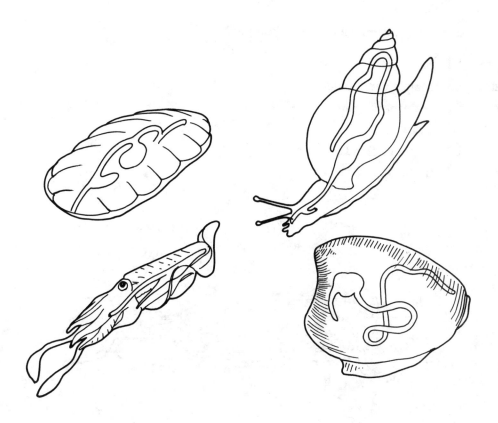

Color the parts of the illustration below as indicated.

RED	☐	dorsal blood vessel
GREEN	☐	nephridium
YELLOW	☐	ventral nerve cord
BLUE	☐	parapodium
ORANGE	☐	septum
BROWN	☐	pharynx
TAN	☐	coelom

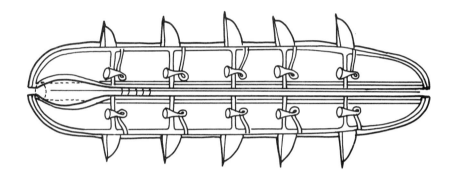

Color the parts of the illustration below as indicated.

RED	☐	heart
GREEN	☐	Malpighian tubules
YELLOW	☐	brain, nerve cord
BLUE	☐	ovary
ORANGE	☐	digestive gland
BROWN	☐	intestine

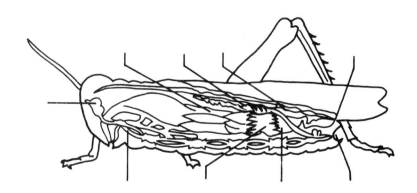

CHAPTER 30

The Animal Kingdom: The Deuterostomes

OUTLINE

Phylum Echinodermata: Spiny-skinned animals of the sea
Phylum Chordata: Animals with a notochord, dorsal
tubular nerve cord, and pharyngeal gill slits

Deuterostomes all have a true coelom. They owe their name to the fact that the second opening that develops in the embryo becomes the mouth, and the first becomes the anus. There are four groups of deuterostomes: the Echinodermata, Chaetognatha, Hemichordata, and Chordata. The echinoderms and the chordates are by far the more ubiquitous and dominant deuterostomes. The echinoderms all live in the sea. They have spiny skins, a water vascular system, and tube feet. There are five groups of echinoderms: the feather stars and sea lilies, the starfish, the basket stars and brittle stars, the sea urchins and sand dollars, and the sea cucumbers. The chordates include animals that live both in the sea and on land. They have a notochord, a dorsal tubular nerve cord, and pharyngeal gill slits, at least at some time in their life cycle. There are three groups of chordates: the tunicates, the lancelets, and the vertebrates. The vertebrates are characterized by a vertebral column, a cranium,. pronounced cephalization, a differentiated brain, muscles attached to an endoskeleton, and two pairs of appendages. There are seven major groups of vertebrates: jawless fishes, cartilaginous fishes, bony fishes, amphibians, reptiles, birds, and mammals. Mammals are classified first on the basis of the site of embryonic development. Those that complete embryonic development in an egg are the monotremes; the duck-billed platypus is a monotreme. Those that complete embryonic development in a maternal pouch are the marsupials; the kangaroo is a marsupial. Those that complete embryonic development within an organ of exchange, i.e., a placenta, are the placental mammals; the human is an example of a placental mammal.

CHAPTER OUTLINE AND CONCEPT REVIEW (Fill in the blanks)

I. INTRODUCTION

■ 1 Echinoderms and chordates are thought to be related because they are both deuterostomes and therefore share many developmental characteristics. Among other things, this means that the (a)_____ forms from the first embryonic opening, and the (b)_____ forms from a second opening.

II. ECHINODERMS ARE SPINY-SKINNED ANIMALS OF THE SEA

■ 2 Phylum echinodermata includes marine animals with (a)_____ skins, a water vascular system, and tube feet; larvae have (b)_____ symmetry; most adults have (c)_____ symmetry. Echinoderms also have a true (d)_____ that houses the internal organs.

A. Class Crinoidea includes the feather stars and sea lilies

■ 3 Class Crinoidea includes sea lilies and feather stars. Unlike other echinoderms, the oral surface of crinoids is located on the _____ surface.

B. Class Asteroidea includes the sea stars

■ 4 Sea stars have a (a)_____ from which radiate five or more arms or rays. Each ray contains hundreds of small (b)_____, each of which contains a canal that leads to a central canal.

C. Class Ophiuroidea includes basket stars and brittle stars

D. Class Echinoidea includes sea urchins and sand dollars

■ 5 Class echinoidea includes the sea urchins and sand dollars, animals that lack arms; they have a solid shell called a _____, and their body is covered with spines.

E. Class Holothuroidea includes sea cucumbers

■ 6 Class Holothuroidea consists of sea cucumbers, animals with flexible bodies. The mouth is surrounded by a circle of modified _____ that serve as tentacles.

III. CHORDATES HAVE A NOTOCHORD, DORSAL TUBULAR NERVE CORD, AND PHARYNGEAL GILL SLITS DURING SOME TIME IN THEIR LIFE CYCLE

■ 7 At some time in its life cycle, a chordate has a notochord, a dorsal tubular nerve cord, and pharyngeal gill slits. Chordates are all coelomates with _____ symmetry.

■ 8 The three characteristics that distinguish chordates from all other groups are the dorsal, flexible rod called the (a)_____; the single, dorsal, hollow (b)_____; and the embryonic (c)_____ that develop into functional respiratory structures in aquatic chordates and entirely different structures in terrestrial chordates.

A. Subphylum Urochordata includes the tunicates

■ 9 Subphylum Urochordata comprises the tunicates, which are sessile, filter-feeding marine animals that have tunics made of _____.

B. Subphylum Cephalochordata includes the lancelets

■ 10 Subphylum Cephalochordata consists of the lancelets, small segmented fishlike animals that exhibit all three chordate characteristics. A common genus, *Branchiostoma*, also known as _____, is often selected as a representative, typical chordate.

■ 11 The (a)_____ is a vestibule of the digestive tract just anterior to the mouth. Food particles swept into the mouth are trapped in mucus in the (b)_____, then passed back to the intestine.

C. The success of the vertebrates is linked to the evolution of key adaptations

■ 12 Subphylum Vertebrata includes animals with a backbone or (a)_____, a braincase or (b)_____, and concentration of nerve cells and sense organs in a definite head, a phenomenon known as (c)_____.

■ 13 The classes in the subphylum Vertebrata include the jawless fish or (a)_____; the cartilaginous fish or (b)_____; the bony fish or (c)_____; tetrapod frogs, toads, and salamanders in the class (d)_____; the lizards, snakes, turtles, and alligators in the class (e)_____; birds or (f)_____; and animals with hair and mammary glands in the class (g)_____.

D. The jawless fish are the most primitive vertebrates

■ 14 It is thought that the now extinct _____, that existed some 500 million years ago, gave rise to the present day lampreys and hagfishes.

E. The earliest jawed fishes are now extinct

F. Class Chondrichthyes includes the sharks, rays, and skates

■ 15 The skin of cartilaginous fish contains numerous _____, toothlike structures composed of layers of enamel and dentine.

■ 16 (a)_____ are sensory grooves along the sides of fish that contain cells capable of perceiving water movements. Similarly, the (b)_____ on the head can sense very weak electrical currents, like those generated by another animal's muscle contractions.

■ 17 Sharks may be (a)_____, that is, lay eggs; or (b)_____, which means their eggs hatch internally; or (c)_____, giving live birth to their young.

G. Bony fish belong to class Osteichthyes

■ 18 The lobe-finned (a)_____, generally considered to be the ancestors of land vertebrates, are survived by the genus *Latimeria* in the suborder of (b)_____.

■ 19 In the Devonian period, bony fishes diverged into two major groups: the (a)_____ fish or actinopterygians and the (b)_____ with their fleshy, lobed fins and lungs. The actinopterygians gave rise to the (c)_____, the modern bony fish in which lungs became modified as (d)_____.

H. Class Amphibia includes frogs, toads, and salamanders

■ 20 The three orders of modern amphibians are: (a)_____, the amphibians with long tails, including salamanders, mud puppies, and newts; the (b)_____, tailless frogs and toads; and (c)_____, the wormlike caecilians.

■ 21 Most amphibians return to water to reproduce; frog embryos develop into tadpoles, which undergo metamorphosis to become adults. Some salamanders, such as the mud puppy, retain several larval characteristics in adulthood, a process known as _____.

■ 22 Amphibians use lungs and their (a)_____ for gas exchange. They have a (b)_____(#?)-chambered heart with systemic and pulmonary circulations, and they have (c)_____ in the skin that help keep the body surface moist.

I. Class Reptilia includes turtles, lizards, snakes, and alligators

■ 23 Reptiles are true terrestrial animals; fertilization is internal; most secrete a protective shell around the egg; the embryo develops an amnion and other extraembryonic membranes. Most reptiles have a (a)_____(#?)-chambered heart, excrete nitrogenous wastes in the form of (b)_____, and are (c)_____, which means they cannot regulate body temperature.

■ 24 Living reptiles are in three orders: the (a)_____ includes turtles and tortoises; the lizards, snakes, iguanas, and geckos in the order (b)_____; and (c)_____ with its crocodiles, alligators, and caimans.

■ 25 Reptiles dominated the earth during the (a)_____ era; then, during the (b)_____ period, most of them, including all of the dinosaurs, became extinct.

J. The birds belong to class Aves

■ 26 Adaptations for flight in birds include feathers, wings, and light hollow bones. Birds also have a (a)_____(#?)-chambered heart, very efficient lungs, a high metabolic rate, and they are (b)_____, meaning they can maintain a constant body temperature. They excrete wastes as semi-solid (c)_____. Birds also have a well-developed nervous system and excellent vision and hearing. Of all these characteristics, birds are the only animals that have (d)_____.

■ 27 The saclike portion of a bird's digestive system that temporarily stores food is the (a)_____, whereas the (b)_____ portion of the stomach secretes powerful gastric juices, and the (c)_____ grinds the food.

■ 28 The earliest know bird, _____, was a medium sized bird with rather feeble wings. With its teeth, long tail, and clawed wing tips, it looked somewhat like a reptile.

K. Mammals have hair and mammary glands

■ 29 The distinguishing characteristics of mammals are _____ _____. They also maintain a constant body temperature, and they have a highly developed nervous system and a muscular diaphragm.

■ 30 Mammals probably evolved from the reptilian group (a)_____, about 200 million years ago in the (b)_____ period.

■ 31 _____ are mammals that lay eggs. They include the duck-billed platypus and the spiny anteater.

■ 32 (a)_____ are pouched mammals. The young are born immature and complete development in the (b)_____, where they are nourished from the mammary glands.

■ 33 The placental mammals are characterized by an organ of exchange (placenta) that develops between the embryo and the mother; this organ supplies _____ to the fetus, enabling it to complete development within the uterus.

KEY TERMS

Frequently Used Prefixes and Suffixes: Use combinations of these prefixes and suffixes to generate terms for the definitions below.

Prefixes	The Meaning	Suffixes	The Meaning
a-	without	-derm	skin
Chondr(o)-	cartilage	-gnatha	jaw
echino-	sea urchin, "spiny"	-ichthy(es)	fish
Oste(o)-	bone	-pod(al)	foot, footed
tetra-	four	-ur(o)(an)	tail
Uro-	tail		

Prefix	Suffix	Definitions
_____	_____	1. A fish without jaws (jawless fish); a member of the class of vertebrates including lampreys and hagfishes.
an-	_____	2. An amphibian with legs but no tail; a tailless frog or toad.
_____	_____	3. Pertains to those amphibians with no feet or legs, i.e., the wormlike caecilians.
_____	_____	4. The class comprising the cartilaginous fishes.
_____	_____	5. The class comprising the bony fishes.
_____	_____	6. A vertebrate with four legs; a member of the superclass Tetrapoda.
_____	-dela	7. The order which comprises the amphibians with long tails, i.e., the salamanders, mudpuppies, and newts.
_____	_____	8. A spiny-skinned animal.

Other Terms You Should Know

Acanthodii
Agnatha
amnion (-ns)(-ia)
amniotic fluid
Amphibia
ampulla (-ae)
Anura
Apoda
Asteroidea
atriopore
atrium (-ia)
Aves
branchial arch (-hes)
carnivore
cephalization
Cephalochordata
cerebral cortex
Chaetognatha
Chelonia

Chordata
cirrus (-ri)
clasper
cloaca (-ae)
coelacanth
cotylosaur
cranium (-ia)
Crinoidea
Crocodilia
crop
diaphragm
Diplodocus
dorsal tubular nerve cord
Echinodermata
Echinoidea
ectothermic
electroreceptor
endoskeleton
endostyle

endothermic
Eutheria
excurrent siphon
gill
gizzard
hair
herbivore
Holothuroidea
incurrent siphon
labyrinthodont
lancelet
lateral line organ
lobe-finned fish (-hes)
lungfish (-hes)
Mammalia
mammary gland
marsupial
marsupium (-ia)
Mesozoic era

Metatheria	Pisces	tadpole
monotreme	placenta (-as)(-ae)	teleost
neoteny	placental mammals	Tetrapoda
nerve cord	placoderm	therapsid
nerve ring	placoid scale	tube feet
notochord	postanal tail	tunicate
operculum	Prototheria	Urochordata
Ophiuroidea	proventriculus (-li)	velum (-la)
oral hood	pulmonary circulation	ventricle
ostracoderm	ray-finned fish (-hes)	vertebra (-ae)
oviparous	Reptilia	vertebral column
ovoviviparous	sensory pit	Vertebrata
pedicellaria	Squamata	viviparous
pharyngeal gill slit	swim bladder	water vascular system
pharyngeal pouch (-hes)	systemic circulation	

SELF TEST

Multiple Choice: Some questions may have more than one correct answer.

1. The group(s) of animals with a part of the stomach that secretes gastric juices and a separate part of the stomach that grinds food is/are
   a. Reptilia.
   b. Mammalia.
   c. Aves.
   d. Amphibia.
   e. Pisces.

2. Dolphins and porpoises are members of the taxon(a)
   a. Lagomorpha.
   b. Rodentia.
   c. that has chisel-like incisors that grow continually.
   d. that includes placental mammals.
   e. Cetacea.

3. The first successful land vertebrates were
   a. tetrapods.
   b. reptiles.
   c. labyrinthodonts.
   d. amphibians.
   e. in the superclass Pisces.

4. A frog is a member of the taxon(a)
   a. Reptilia.
   b. Chelonia.
   c. Urodela.
   d. Amphibia.
   e. Anura.

5. A turtle is a member of the taxon(a)
   a. Reptilia.
   b. Chelonia.
   c. Urodela.
   d. Amphibia.
   e. Anura.

6. Unique features of the echinoderms include
   a. radial larvae.
   b. ciliated larvae.
   c. endoskeleton composed of $CaCO_3$ plates.
   d. gas exchange by diffusion.
   e. water vascular system.

7. Mammals probably evolved from
   a. amphibians.
   b. a type of fish.
   c. reptiles.
   d. monotremes.
   e. therapsids.

8. The group(s) of animals that maintain a constant internal body temperature is/are
   a. Reptilia.
   b. Mammalia.
   c. Aves.
   d. Amphibia.
   e. Pisces.

9. The lateral line organ is found
   a. only in sharks.
   b. only in bony fish.
   c. only in agnathans.
   d. in all fish.
   e. in all fish except placoderms.

10. The acorn worm
    a. has a notochord.
    b. is a deuterostome.
    c. is in the same subphylum as *Amphioxus.*
    d. is in the phylum Chordata.
    e. is a hemichordate.

11. A sting ray is
    a. a bony fish.
    b. a jawless fish.
    c. in the same class as sharks.
    d. in the same superclass as a rainbow trout.
    e. in the class Chondrichthyes.

12. The lungs of the ancestors of most modern fish became modified as
    a. gills.
    b. a vascular pump.
    c. an organ that stores oxygen.
    d. a hydrostatic organ.
    e. a swim bladder.

13. The group(s) of animals with both four-chambered hearts and double circuit of blood flow is/are the
    a. reptiles.
    b. birds.
    c. mammals.
    d. class Aves.
    e. amphibians.

14. Members of the phylum Chordata all have
    a. a notochord.
    b. gill slits.
    c. a dorsal, tubular nerve cord.
    d. well-developed germ layers.
    e. bilateral symmetry.

15. The phylum(a) that many biologists believe had a common ancestry with our own phylum is/are
    a. mollusca.
    b. annelida.
    c. arthropoda.
    d. echinodermata.
    e. chordata.

16. Humans are members of the taxon(a)
    a. Lagomorpha.
    b. Rodentia.
    c. that has chisel-like incisors that grow continually.
    d. that includes placental mammals.
    e. Cetacea.

17. The "spiny-skinned" animals include
    a. sea stars.
    b. sand dollars.
    c. snails.
    d. lancets.
    e. marine worms.

18. A rabbit is a member of the taxon(a)
    a. Lagomorpha.
    b. Rodentia.
    c. that has chisel-like incisors that grow continually.
    d. that includes placental mammals.
    e. Cetacea.

19. A hermaphroditic animal that traps food in mucus secreted by cells of the endostyle is a/an
   a. holothuroidean.   d. sea cucumber.
   b. enchinoderm.   e. tunicate.
   c. urochordate.

20. Guinea pigs are members of the taxon(a)
   a. Lagomorpha.   d. that includes placental mammals.
   b. Rodentia.   e. Cetacea.
   c. that has chisel-like incisors that grow continually.

21. Cats and dogs are members of the taxon(a)
   a. Lagomorpha.   d. that includes placental mammals.
   b. Rodentia.   e. Cetacea.
   c. that has chisel-like incisors that grow continually.

22. An animal that may eviscerate under undesirable environmental conditions, as well as attack an enemy by shooting tubules out of its anus is a/an
   a. holothuroidean.   d. sea cucumber.
   b. enchinoderm.   e. animal with tube feet.
   c. urochordate.

VISUAL FOUNDATIONS

Color the parts of the illustration below as indicated.

RED	❑	heart
GREEN	❑	mouth
YELLOW	❑	brain and dorsal, hollow nerve tube
BLUE	❑	notochord
ORANGE	❑	pharyngeal gill slits
BROWN	❑	stomach, intestine
TAN	❑	muscular segments

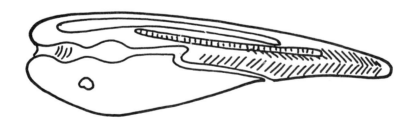

PART VI

Structure and Life Processes in Plants

CHAPTER 31

Plant Structure, Growth, and Differentiation

OUTLINE

Embryonic development in plants
Factors that affect seed germination
Localized growth in plants
Plant tissues
Roots, stems, leaves, flowers, and fruits
Genetic and environmental control of differentiation

Plant growth and development is precise and orderly, regulated by interacting genetic and environmental controls. Seed germination, for example, is affected by many factors, such as the seed's requirements for oxygen, water, temperature, and light, and by the embryo's maturity and the presence or absence within the seed of germination inhibitors. Plants grow by increasing both in girth and length. Unlike growth in animals, plant growth is localized in areas of unspecialized cells, and involves cell proliferation, elongation, and differentiation. The plant body is a complex structure, composed of cells that are organized into tissues, and tissues that are organized into organs. Each organ performs a single function or group of functions, and is dependent on other organs for its survival. Leaves, for example, photosynthesize, providing sugar molecules to the rest of the plant, and roots absorb water and minerals, making them available to the entire plant body. All vascular plants have three tissue systems: the dermal tissue system provides a covering for the plant body; the vascular tissue system conducts various substances throughout the plant body; and the ground tissue system has a variety of functions. The tissue systems are composed of various simple and complex tissues. The major plant tissues are parenchyma, collenchyma, sclerenchyma, xylem, phloem, epidermis, and periderm. The plant body is organized into a root system and a shoot system. The latter system consists of the stems and leaves.

CHAPTER OUTLINE AND CONCEPT REVIEW (Fill in the blanks)

I. INTRODUCTION

- **1** As is the case in animals, plant cells are organized in aggregations that perform specific functions. These groups of cells are (a)_____, which in turn are organized to form (b)_____ such as roots, stems, and leaves.

- **2** A (a)_____ is a fertilized egg that results from the fusion of haploid gametes. It multiplies by mitosis, forming the (b)_____, which eventually develops into a mature plant.

*II. EMBRYONIC DEVELOPMENT IN PLANTS FOLLOWS AN ORDERLY AND
PREDICTABLE PLAN*

■ 3 Plant embryos develop in the seed in an orderly, predictable fashion. After only one division, the zygote already exhibits polarity — one cell develops into a multicellular structure that anchors the embryo to the endosperm called the (a)_____, the other cell divides to form a chain of cells known as the (b)_____.

■ 4 The mature embryo consists of the embryonic root or (a)_____, a shoot, and one or two cotyledons. The shoot has two parts: the (b)_____ is part of the stem above the cotyledon(s) and the (c)_____ is continuous with the roots below the cotyledon(s).

III. A NUMBER OF EXTERNAL AND INTERNAL FACTORS AFFECT SEED GERMINATION

■ 5 External environmental factors that may affect seed germination include

_____.

■ 6 The presence of an immature embryo and inhibitors such as _____ are internal factors that prevent seed germination.

■ 7 The first part of the germinating embryo to emerge from the seed is the (a)_____, which is protected by a layer of cells on its growing edge called the (b)_____.

IV. PLANTS EXHIBIT LOCALIZED GROWTH AFTER GERMINATION

■ 8 Continuous growth, such as that exhibited by stems and roots, that continues throughout the life of the organism, is known as (a)_____, while growth that stops after attaining a certain size is called (b)_____.

■ 9 Plant growth is localized in regions called (a)_____, and involves three cell activities, namely, (b)_____.

A. Primary growth takes place at apical meristems

■ 10 Plants have two kinds of growth: (a)_____ growth, an increase in length; and (b)_____ growth, an increase in width of the plant.

■ 11 The root apical meristem consists of three zones, which, in order from the root cap inward, are the (a)_____, (b)_____, and (c)_____.

■ 12 The stem apical meristem contains embryonic leaves called (a)_____ and (b)_____ or embryonic buds.

B. Secondary growth takes place at lateral meristems

■ 13 Two lateral meristems are responsible for secondary growth (increase in girth), the (a)_____, which forms a cylinder of cells around the stem, and the (b)_____, consisting of patches of cells in the outer bark.

■ 14 Plants with secondary growth that live year-after-year are (a)_____, while those that survive for only one season and have only primary growth are (b)_____. Plants that complete their life cycles in two years are (c)_____.

V. CELLS AND TISSUES DIFFERENTIATE IN THE GROWING PLANT

A. Parenchyma cells have thin primary walls

■15 Parenchyma tissue is composed of relatively unspecialized, living cells with very thin cell walls. Their three primary functions include _____ _____.

B. Collenchyma cells have unevenly thickened primary walls

■16 Collenchyma tissue is composed of living cells that function primarily to _____ the plant.

C. Sclerenchyma cells have both a primary wall and a thick secondary wall

■17 Sclerenchyma tissue is composed of dead cells with very thick cell walls containing _____, a molecule that strengthens and hardens the walls.

D. Xylem has two kinds of conducting cells, tracheids and vessel elements

■18 The tracheid and vessel elements in xylem conduct (a)_____ _____ to stems and leaves. Xylem also contains (b)_____ for storage and (c)_____ for support.

E. Sieve tube members are the conducting cells of phloem

■19 Phloem is a complex tissue that functions to conduct _____ throughout the plant.

■20 Highly specialized cells in phloem called (a)_____ conduct food in solution. They are stacked to form tubes with continuous cytoplasm running through holes, the (b)_____, between the cells.

F. Epidermis is the outermost layer of cells on a herbaceous plant

■21 The epidermis covers the plant body and functions primarily for (a)_____. They secrete a layer called the (b)_____ that restricts water loss. (c)_____ are openings in this layer through which gases diffuse.

G. Periderm is the outermost layer of cells on a woody plant

■22 The periderm covers the plant body in plants with secondary growth. Its primary function is for _____.

VI. ROOTS, STEMS, LEAVES, FLOWERS, AND FRUITS MAKE UP THE PLANT BODY

■23 The plant body is organized into a root system and a shoot system. Typical dicot root systems, like those in dandelions, consisting of one primary root, are called (a)_____ systems. Grasses and most other monocots have multiple, branching (b)_____ root systems.

■24 The shoot system is composed of the stem and leaves. The point on the stem where a leaf attaches is the (a)_____. Stems also have undeveloped embryonic shoots or (b)_____ of two types: the (c)_____ at the tip of the stem, and the (d)_____ in leaf axils.

■25 Leaves typically have two parts: the broad, expanded (a)_____, and the stalk or (b)_____ that attaches to the stem.

VII. DIFFERENTIATION IN PLANTS IS UNDER BOTH GENETIC AND ENVIRONMENTAL CONTROLS

■ 26 Plant development is controlled by genetic factors and by nongenetic external factors. Other plant tissues and organs, whose controls are generally mediated by _____, also exert a profound influence on plant development.

■ 27 Growth and differentiation are studied by observing *in vitro* cell cultures. One significant development in this field that preceded the generation of whole plants from a single cell was the production of a _____, which is a mass of undifferentiated cells.

KEY TERMS

Frequently Used Prefixes and Suffixes: Use combinations of these prefixes and suffixes to generate terms for the definitions below.

Prefixes	The Meaning
inter-	between, among
pro-	"before"
trich-	hair

Prefix	Suffix	Definitions
_____	-ome	1. A hair or other appendage growing out from the epidermis of plants.
_____	-node	2. The section of a stem between nodes.
_____	-embryo	3. Early stage in plant embryology before development of the mature embryo.

Other Terms You Should Know

alternate leaf arrangement
annual
apical meristem
area of cell division
area of cell elongation
area of cell maturation
axil
biennial
blade
bud
bud primordium (-ia)
bud scale
bud scale scar
bundle scar
callus (-ses)
coleoptile
collenchyma
companion cell
complex tissue

compound leaf
cork cambium
cuticle
determinate growth
differentiate
dormant
epidermis
fiber
fibrous root system
germinate
globular embryo
guard cell
heart stage
herbaceous
hormone
indeterminate growth
lateral bud
lateral meristem
leaf primordium (-ia)

leaf scar
leaflet
lenticel
lignin
meristem
netted venation
node
opposite leaf arrangement
organ
palmate venation
parallel venation
parenchyma
perennial
perforation
periderm
petiole
phloem
pinnate venation
pit

primary cell wall
primary growth
radicle
root cap
sclereid
sclerenchyma
secondary cell wall
secondary growth
sieve plate

sieve tube member
simple leaf
simple tissue
stoma (-ata)
stomate
storage root
suspensor
taproot system
terminal bud

tissue
torpedo stage
tracheid
vascular cambium
vessel
vessel element
whorled leaf arrangement
xylem

SELF TEST

Multiple Choice: Some questions may have more than one correct answer.

1. Localized areas of plant growth are called
   - a. primordia.
   - b. mitotic zones.
   - c. meristems.
   - d. primary growth centers.
   - e. secondary growth centers.

2. The types of cells in phloem include
   - a. tracheids.
   - b. vessel elements.
   - c. parenchyma.
   - d. sieve tube members.
   - e. fiber cells.

3. The types of cells in xylem include
   - a. tracheids.
   - b. vessel elements.
   - c. parenchyma.
   - d. sieve tube members.
   - e. fiber cells.

4. A mature seed will not germinate until it has
   - a. imbibed water.
   - b. enough light.
   - c. been exposed to the appropriate temperature.
   - d. an adequate supply of oxygen.
   - e. developed past the globular embryo stage.

5. A leaf can be distinguished from a leaflet by the presence or absence of a _____ at their bases.
   - a. lateral bud
   - b. bundle scar
   - c. bud scale
   - d. axil
   - e. callus

6. The kind of growth that results in an increase in the width of the plant is known as
   - a. differentiation.
   - b. elongation.
   - c. apical meristem growth.
   - d. primary growth.
   - e. secondary growth.

7. Hairlike outgrowths of plant epidermis are called
   - a. periderm.
   - b. fiber elements.
   - c. lateral buds.
   - d. trichomes.
   - e. companion cells.

8. A root system with several main roots developing from the end of the stem is called a _____ system.
   - a. taproot
   - b. storage root
   - c. root hair
   - d. secondary root
   - e. fibrous root

9. The cell type found throughout the plant body that often functions in photosynthesis and storage is called
   a. sclerenchyma.
   b. collenchyma.
   c. parenchyma.
   d. companion cells.
   e. a tracheid.

10. All plant cells have
   a. cell walls.
   b. secondary growth.
   c. the capacity to form a complete plant.
   d. primary cell walls.
   e. secondary cell walls.

11. The seed plant embryo passes through several stages of development. The elongated stage that leads to the mature embryo is the
   a. proembryo.
   b. torpedo stage.
   c. globular stage.
   d. zygote.
   e. heart stage.

12. The first part of the plant to emerge during germination is the
   a. radicle.
   b. embryonic root.
   c. coleoptile.
   d. proembryo.
   e. shoot.

13. Plant differentiation in a given tissue is affected by
   a. genes.
   b. hormones.
   c. its location in the embryo.
   d. other tissues and organs.
   e. the external environment.

14. Indeterminate growth typically occurs in
   a. roots.
   b. flowers.
   c. leaves.
   d. stems.
   e. all plant parts.

15. Increase in the girth of a plant is due to growth of the
   a. vascular cambium.
   b. cork cambium.
   c. area of cell maturation.
   d. area of cell elongation.
   e. lateral meristems.

VISUAL FOUNDATIONS

Color the parts of the illustration below as indicated.

RED   ❑   embryonic shoot
GREEN   ❑   cotyledons
YELLOW   ❑   embryonic root
BLUE   ❑   hook
ORANGE   ❑   primary root
BROWN   ❑   seed coat
TAN   ❑   shriveled cotyledon

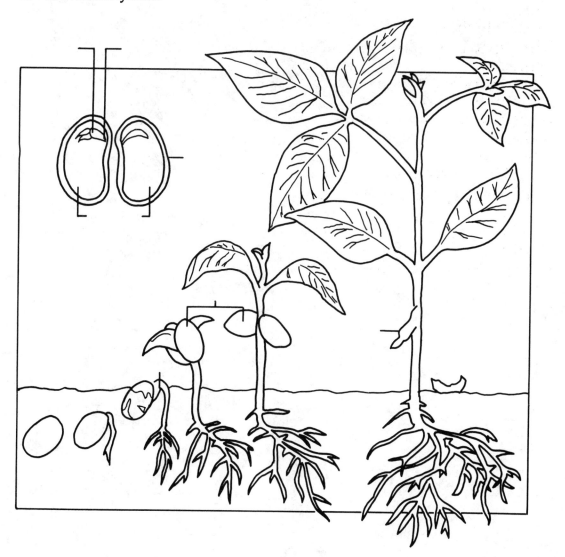

Color the parts of the illustration below as indicated.

RED ☐ tracheid
GREEN ☐ parenchyma
YELLOW ☐ vessel element
BLUE ☐ sclerenchyma
ORANGE ☐ companion cell
BROWN ☐ sieve tube member
TAN ☐ collenchyma

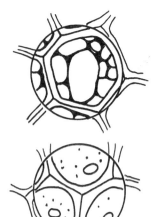

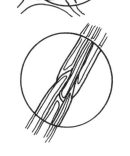

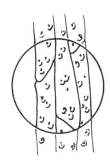

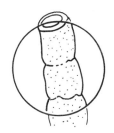

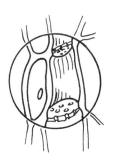

Color the parts of the illustration below as indicated.

RED	❑	flower
GREEN	❑	blade
YELLOW	❑	internode
BLUE	❑	petiole
ORANGE	❑	node
BROWN	❑	taproot
TAN	❑	branch root

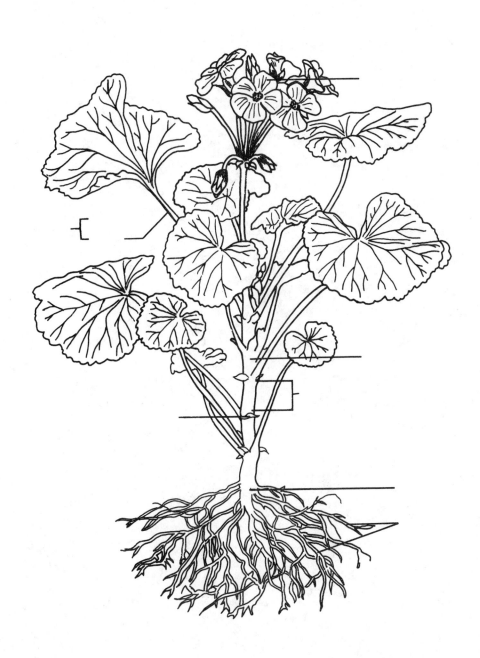

CHAPTER 32

□

Leaf Structure and Function

OUTLINE

Major tissues of the leaf
Structure-function relationships in leaves
The potassium ion mechanism
Transpiration and guttation
Leaf abscission
Modifications in leaf structure

Leaves, as the principal photosynthetic organs in plants, are highly adapted to collect radiant energy, convert it into the chemical bonds of carbohydrates, and transport the carbohydrates to the rest of the plant. They are also adapted to permit gas exchange, and to receive water and minerals transported into them by the plant vascular system. Water loss from leaves is controlled in part by the cell walls of the epidermal cells, by presence of a surface waxy layer secreted by epidermal cells, and by tiny pores in the leaf epidermis that open and close in response to various environmental factors. Most of the water absorbed by land plants is lost, either in the form of water vapor or as liquid water. The rate of water loss is affected by various environmental factors. Survival at low temperatures is facilitated in some plants by the loss of leaves. The process is complex, involving both physiological and anatomical changes. The leaves of many plants are modified for functions other than photosynthesis. Some, for example, have modifications for protection, others for grasping and holding onto structures, others for storing water or food, and still others for trapping animals.

CHAPTER OUTLINE AND CONCEPT REVIEW (Fill in the blanks)

I. INTRODUCTION

II. EPIDERMIS, MESOPHYLL, XYLEM, AND PHLOEM ARE THE MAJOR TISSUES OF THE LEAF

■ 1 Leaves are highly adapted for photosynthesis. The transparent epidermis allows light to penetrate into the photosynthetic tissue, the (a)_____. When this tissue is divided into two regions, the upper layer, nearest to the upper epidermis, is called the (b)_____ layer, and the lower portion is called the (c)_____ layer.

■ 2 (a)_____ in veins of a leaf conduct water and essential minerals to the leaf, while (b)_____ in veins conducts sugar produced by photosynthesis to the rest of the plant. Veins may be surrounded by a (c)_____, consisting of one or more layers of parenchyma or sclerenchyma cells.

■ 3 Monocot and dicot leaves can be distinguished based on their external morphology and their internal anatomy. Dicot leaves usually have (a)_____ venation and monocots usually have (b)_____ veins. Some monocot leaves have (c)_____ cells along the midvein that may by involved in folding the leaf during drought. In dicots, epidermal cells surrounding guard cells are like other epidermal cells, whereas in some monocots the guard cells are associated with special epidermal cells called (d)_____ cells.

III. STRUCTURE IS RELATED TO FUNCTION IN LEAVES

IV. THE POTASSIUM ION MECHANISM CONTROLS STOMATAL OPENING AND CLOSING

■ 4 Stomates open during the day and close at night. Factors that affect opening and closing of stomates include _____
_____

■ 5 The potassium ion (K^+) mechanism explains the opening and closing of stomates. Light triggers an influx of K^+ ions into the (a)_____ of (b)_____ cells, thereby increasing internal solute concentration. Water then moves into these cells changing their shape causing the pores to (c)_____(open or close?). In relatively low light, K^+ ions are pumped out of the cells, water leaves, the cells collapse, and the pores (d)_____(open or close?). The two photoreceptors that initiate stomatal opening and closing are thought to be (e)_____
_____.

V. LEAVES LOSE WATER BY TRANSPIRATION AND GUTTATION

■ 6 Loss of liquid water from leaves is (a)_____ and water vapor loss is (b)_____. Environmental factors that affect the rate of water loss include (c)_____.

VI. LEAF ABSCISSION ALLOWS PLANTS IN TEMPERATE CLIMATES TO SURVIVE WINTER

■ 7 Leaf abscission involves complex physiological and anatomical changes, all of them regulated by _____.

■ 8 The area where the petiole attaches to the stem is the (a)_____; it is a weak area because it contains relatively few strengthening (b)_____. In autumn, the waxy, impermeable substance (c)_____ permeates cork cells on the stem side of this petiole-stem interface, and enzymes dissolve the (d)_____, the intercellular cement that holds cells together. The leaf falls.

VII. LEAVES WITH FUNCTIONS OTHER THAN PHOTOSYNTHESIS EXHIBIT MODIFICATIONS IN STRUCTURE

■ 9 Leaves are variously modified for a plethora of specialized functions. For example, the hard, pointed (a)_____ on a cactus are leaves modified for protection, and (b)_____ are specialized leaves that anchor long, climbing vines to the supporting structures on which they are growing.

KEY TERMS

Frequently Used Prefixes and Suffixes: Use combinations of these prefixes and suffixes to generate terms for the definitions below.

Prefixes	The Meaning		Suffixes	The Meaning
epi-	upon, over, on		-phyte	plant
meso-	middle			
photo-	light			

Prefix	Suffix	Definitions
_____	-phyll	1. The photosynthetic tissue of the leaf sandwiched between (in the middle of) the upper and lower epidermis.
_____	-receptor	2. A light-sensitive pigment. (Also a sense organ specialized to detect light.)
_____	_____	3. A plant that is self-nourishing but that grows upon another plant, using it for position and support.

Other Terms You Should Know

abscise	epiphyte	spongy layer
abscission zone	guard cell	stomate
bundle sheath extension	guttation	suberin
bud scale	lower epidermis	subsidiary cell
bulb	mesophyll	tendril
bulliform cell	palisades layer	transpiration
bundle sheath	phloem	turgor
circadian rhythm	potassium ion mechanism	upper epidermis
cuticle	spine	xylem

SELF TEST

Multiple Choice: Some questions may have more than one correct answer.

1. The abscission zone of plants
   a. secretes suberin.
   b. is characteristic of desert plants.
   c. is composed primarily of thin-walled cells.
   d. is largely parenchymal cells.
   e. is an active area of photosynthesis.

2. Factors that tend to affect the amount of transpiration include
   a. the cuticle.
   b. wind velocity.
   c. ambient temperature.
   d. relative humidity.
   e. amount of light.

3. Trichomes are found on/in the
   a. palisade layer.
   b. spongy layer.
   c. entire mesophyll.
   d. epidermis.
   e. guard cells.

4. In general, leaf epidermal cells
   a. are living.
   b. lack chloroplasts.
   c. protect mesophyll cells from sunlight.
   d. are absent on the lower leaf surface.
   e. have a cuticle.

5. Guard cells generally
   a. have chloroplasts.
   b. form a pore.
   c. are found in the epidermis.
   d. are found only in monocots.
   e. are found only in dicots.

6. Active transport of potassium ions into guard cells
   a. requires ATP.
   b. closes stomates.
   c. causes guard cells to shrink and collapse.
   d. occurs more in daylight than at night.
   e. indirectly causes pores to open.

7. The portion of mesophyll that is usually composed of loosely and irregularly arranged cells is
   a. the palisade layer.
   b. the spongy layer.
   c. toward the leaf's underside.
   d. toward the leaf's upperside.
   e. an area of photosynthesis.

8. The process by which plants secrete water as a liquid is
   a. evaporation.
   b. transpiration.
   c. guttation.
   d. the potassium ion mechanism.
   e. found only in monocots.

9. The leaves of dicots usually have
   a. bulliform cells.
   b. netted venation.
   c. differentiated palisade and spongy tissues.
   d. special subsidiary cells.
   e. bean-shaped guard cells.

10. The receptor in plants that responds to light with the subsequent opening of stomates
   a. is chlorophyll a.
   b. is a pigment.
   c. is located in bulliform cells.
   d. is part of the abscission zone.
   e. doubles as a garage door opener.

VISUAL FOUNDATIONS

Color the parts of the illustration below as indicated.

RED	☐	xylem
GREEN	☐	mesophyll
YELLOW	☐	stoma
BLUE	☐	phloem
ORANGE	☐	bundle sheath
BROWN	☐	cuticle
TAN	☐	epidermis

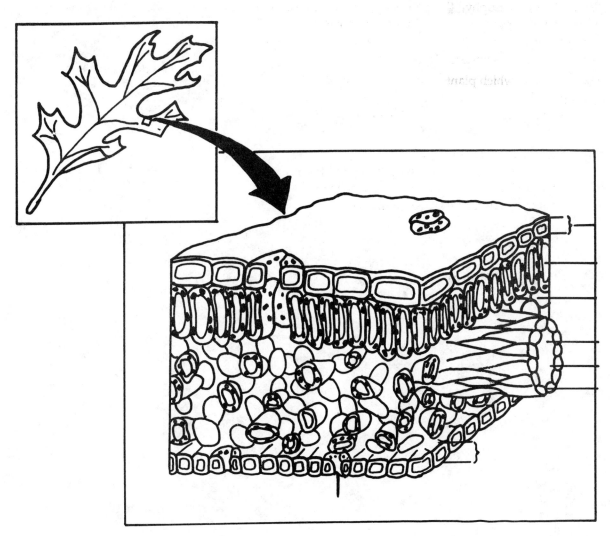

□

CHAPTER 33

Stems and Plant Transport

OUTLINE

Herbaceous dicot and monocot stems
Stems of gymnosperms and woody dicots
Transport in plants

Stems function to support leaves and reproductive structures, conduct materials absorbed by the roots and produced in the leaves to other parts of the plant, and produce new tissue throughout the life of the plant. Specialized stems have other functions as well. The stems of some plants increase only in length. These are the herbaceous, or nonwoody, plants. Although all herbaceous stems have the same basic tissues, the arrangement of the tissues in the monocot stem differs from that in the dicot stem. The stems of other plants increase in both length and girth. These include all gymnosperms and some dicots. Increases in the length of plants result from mitotic activity in apical meristems, and increases in the girth of plants result from mitotic activity in lateral meristems. Simple physical forces are responsible for the movement of food, water, and minerals in multicellular plants.

CHAPTER OUTLINE AND CONCEPT REVIEW (Fill in the blanks)

I. INTRODUCTION

■ 1 The primary functions of stems are to _____
_____.

II. HERBACEOUS DICOT AND MONOCOT STEMS CAN BE DISTINGUISHED BY THE ARRANGEMENT OF VASCULAR BUNDLES

A. The vascular bundles of herbaceous dicot stems are arranged in a circle in cross section

■ 2 Patches of vascular tissues in herbaceous dicots are called (a)_____.
Each patch contains two vascular tissues, the (b)_____, and a single layer of cells, the (c)_____, sandwiched between them.
Vascular tissue patches in dicot stems are arranged in a circle between the central (d)_____ and the peripheral tissue layer called the (e)_____.

B. Vascular bundles are scattered throughout monocot stems

■ 3 The parenchyma tissue in monocot stems that contains vascular bundles is called _____.

C. Each tissue in a herbaceous stem has a specific function

■ 4 Stem tissues provide a variety of functions. Primary stems are protected from mechanical wear and desiccation by the (a)_____. In green stems, photosynthesis occurs in the (b)_____ of dicots and the (c)_____ of monocots, and these tissues may also store essential materials.

III. GYMNOSPERMS AND WOODY DICOTS HAVE STEMS WITH SECONDARY GROWTH

■ 5 Secondary growth occurs in some dicots and all gymnosperms as a result of cell divisions in two lateral meristems, the (a)_____, which gives rise to secondary xylem and phloem, and the (b)_____, which gives rise to a periderm.

A. Vascular cambium gives rise to secondary xylem and secondary phloem

■ 6 The vascular cambium produces secondary xylem (wood) to the inside and secondary phloem (inner bark) to the outside, which provide vertical transport. Lateral transport takes place in (a)_____, which are chains of (b)_____ cells.

B. Cork cambium produces periderm, the functional replacement for the epidermis

■ 7 Cork cambium cells form tissues toward the inside and the outside. The layer toward the outside consists of (a)_____, heavily suberized cells that protect the plant. To the inside, cork cambium forms the (b)_____, which functions for storage.

C. Common terms associated with wood can be explained on the basis of plant structure

■ 8 The older, brownish wood in the center of a tree is called (a)_____, and the newer, more peripheral wood is (b)_____. Hardwood comes from the plant group (c)_____ and softwood comes from (d)_____.

■ 9 Annual rings are composed of two types of cells arranged in alternating concentric circles, each layer appropriately named for the season in which it developed. The (a)_____ has thin-walled, large vessels and tracheids and few fibers, whereas (b)_____ has thicker-walled, narrower vessels and tracheids and numerous fibers.

IV. TRANSPORT IN PLANTS OCCURS IN XYLEM AND PHLOEM

A. Water and minerals are translocated in xylem

■ 10 Water and dissolved minerals move upward in the xylem from the root to the stem to the leaves. One of the principal forces behind water movements through plants is a function of a cell's ability to absorb water by osmosis, also know as the "free energy of water" or the (a)_____. Water containing solutes has (b)_____(more or less?) free energy than pure water. Water moves from regions containing (c)_____(more or less?) free energy of water to regions containing (d)_____(more or less?) free energy of water.

■ 11 Root pressure is caused by the differences in water potential between _____.

■ 12 The (a)_____ "pulls" water up the plant, and the evaporation-pull of (b)_____ causes a tension at the top of the plant due to a gradient in water potentials from soil up through the plant to the atmosphere.

B. Sugar is translocated in phloem

■ 13 Dissolved food, predominantly sucrose, is transported up or down in the phloem. Movement of materials in the phloem is explained by the (a)_____ hypothesis. Sugar is actively loaded into the sieve tubes at the (b)_____, causing water to move into sieve tubes by osmosis, then sugar is actively unloaded from the sieve tubes at the (c)_____, causing water to leave sieve tubes by osmosis.

KEY TERMS

Terms You Should Know

annual ring
cork cambium
cortex (-tices)
epidermis
ground parenchyma
ground tissue
hardwood
heartwood
knot

late summerwood
periderm
phloem
phloem fiber cap
pith
pith ray
pressure flow
  hypothesis (-ses)
ray

root pressure mechanism
sapwood
softwood
springwood
tension-cohesion mechanism
translocation
vascular cambium
water potential
xylem

SELF TEST

Multiple Choice: Some questions may have more than one correct answer.

1. One daughter cell from a mother cell in the vascular cambium remains as part of the vascular cambium, the other divides to form
   a. wood or secondary phloem.
   b. wood or inner bark.
   c. outer bark.
   d. secondary tissue.
   e. primary xylem and phloem.

2. If soil water contains 0.1% dissolved materials and root water contains 0.2% dissolved materials, one would expect
   a. a negative water pressure in soil water.
   b. a negative water pressure in root water.
   c. less water pressure in roots than in soil.
   d. water to flow from soil into root.
   e. water to flow from root into soil.

3. Which of the following is/are true of dicot stems?
   a. stem covered with epidermis
   b. vascular tissues embedded in ground tissue
   c. stem has distinct cortex and pith
   d. vascular bundles scattered through the stem
   e. vascular bundles arranged in circles

4. Which of the following is/are true of monocot stems?
   a. stem covered with epidermis
   b. vascular tissues embedded in ground tissue
   c. stem has distinct cortex and pith
   d. vascular bundles scattered through the stem
   e. vascular bundles arranged in circles

5. When water is plentiful, wood formed by the vascular cambium is
   a. springwood.
   b. summerwood.
   c. late summerwood.
   d. composed of thin-walled vessels.
   e. composed of thick-walled vessels.

6. Newer wood closer to the bark is known as
   a. heartwood.
   b. sapwood.
   c. tracheids.
   d. hardwood.
   e. softwood.

7. If a plant is placed in a beaker of pure distilled water, then
   a. water will move out of the plant.
   b. water will move into the plant.
   c. water pressure in the plant is positive relative to water in the beaker.
   d. water pressure in the beaker is zero.
   e. water pressure in the plant is less than zero.

8. Secondary xylem and phloem are derived from
   a. cork parenchyma.
   b. cork cambium.
   c. vascular cambium.
   d. periderm.
   e. epidermis.

9. The outer portion of bark is formed primarily from
   a. cork parenchyma.
   b. cork cambium.
   c. vascular cambium.
   d. periderm.
   e. epidermis.

10. The pull of water up through a plant is due in part to
    a. cohesion.
    b. adhesion.
    c. the low water pressure in the atmosphere.
    d. transpiration.
    e. a water pressure gradient between soil and the atmosphere.

VISUAL FOUNDATIONS

Color the parts of the illustration below as indicated.

RED	☐	vascular cambium
GREEN	☐	secondary phloem
YELLOW	☐	pith
BLUE	☐	primary phloem
ORANGE	☐	secondary xylem
BROWN	☐	primary xylem
TAN	☐	periderm

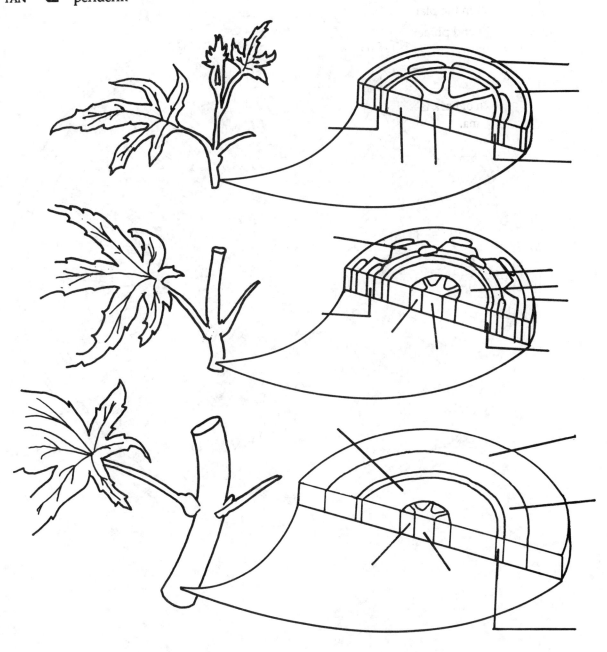

Color the parts of the illustration below as indicated.

RED ☐ vascular cambium
GREEN ☐ secondary phloem
YELLOW ☐ pith
BLUE ☐ primary phloem
ORANGE ☐ secondary xylem
BROWN ☐ primary xylem
TAN ☐ cork

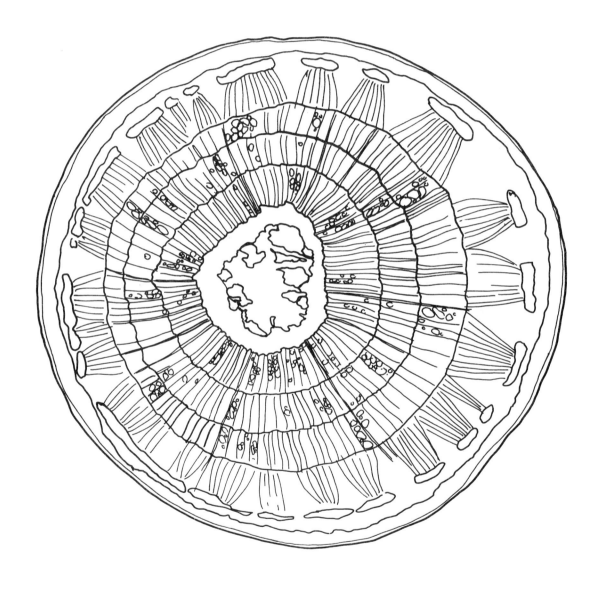

Color the parts of the illustration below as indicated.

RED   ❑   nucleus

GREEN   ❑   mesophyll cell

YELLOW   ❑   companion cell

BLUE   ❑   plasmodesmata

ORANGE   ❑   sieve plate

BROWN   ❑   sieve tube member

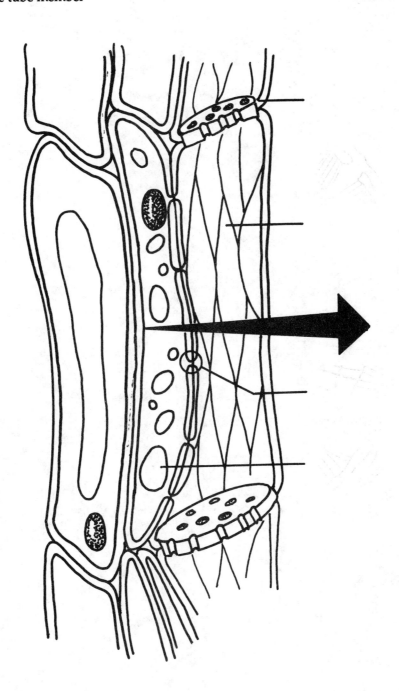

CHAPTER 34

□

Roots and Mineral Nutrition

OUTLINE

Primary roots
Roots with secondary growth
Soil
Essential minerals

The roots of most plants function to anchor plants to the ground, to absorb water and dissolved minerals from the soil, and, in some cases, to store food. Additionally, the roots of some plants are modified for support, aeration, and/or photosynthesis. Although herbaceous roots have the same tissues and structures found in stems, they also have other tissues and structures as well. There are structural differences between monocot and dicot roots. The gymnosperms and some dicots produce woody stems and roots that grow by increasing in both length and girth. Various factors influence soil formation. Soil is composed of inorganic minerals, organic matter, soil organisms, soil atmosphere, and soil water. Plants require at least 16 essential nutrients for normal growth, development, and reproduction. Some human practices, such as harvesting food crops, depletes the soil of certain essential elements, making it necessary to return these elements to the soil in the form of organic or inorganic fertilizer.

CHAPTER OUTLINE AND CONCEPT REVIEW (Fill in the blanks)

I. INTRODUCTION

■ 1 Roots function in _____.

■ 2 Roots that grow from unusual places on plants, such as aerial roots growing laterally from a stem, are known as (a)_____ roots. Two other types of such roots are (b)_____ roots that support the plant, like those in corn, and (c)_____ roots that pull bulbs deeper into the ground.

II. STRUCTURAL DIFFERENCES EXIST BETWEEN PRIMARY ROOTS AND PRIMARY STEMS

A. Herbaceous dicot and monocot roots can be distinguished by the arrangement of vascular tissues

■ 3 The inner layer of cortex in a dicot root is the (a)_____, the cells of which possess a bandlike region on their radial and transverse walls. This "band," or (b)_____ as it's called, contains suberin.

■ 4 Just inside the cortex is a layer of parenchyma cells with meristematic characteristics called the (a)_____. The central portion of dicot roots is occupied by (b)_____ tissue.

B. Structure is related to function in primary roots

■ 5 Most of the water that enters roots first moves along the (a)_____ rather than moving through intracellular spaces. In part, this is due to the powerful absorptive qualities of (b)_____. Later, when water reaches the waterproof walls of the (c)_____ cells, it does permeate cell membranes. Lateral movement of water through root tissues can be summarized as: root hair —> (d)_____ —> cortex —> (e)_____ —> pericycle —> (f)_____, where it is then transported upward.

■ 6 The primary function of the root cortex is (a)_____. Multicellular branch roots originate in the (b)_____.

III. GYMNOSPERMS AND CERTAIN DICOTS HAVE ROOTS WITH SECONDARY GROWTH

IV. SOIL SUPPORTS TERRESTRIAL PLANTS

A. Soil is composed of inorganic minerals, organic matter, soil organisms, soil atmosphere, and soil water

■ 7 The five major components of soil are _____
_____.

■ 8 Good, loamy agricultural soil contains about 40% of (a)_____ and about 20% of (b)_____. The organic portion of soil, or (c)_____, is also an important component. The spaces between soil particles are filled with (d)_____.

B. Soil erosion is a serious threat to cultivated soils in many regions of the world

V. ROOTS OBTAIN MOST OF THE MINERALS FOUND IN PLANTS

■ 9 In order to identify elements that are essential for plant growth, investigators must reduce the number of variables to a minimum. To do this, since soil is very complex, plants are grown in aerated water containing known quantities of elements, a process known as _____.

A. Sixteen elements are essential for plant growth

■ 10 The 16 elements that plants require for normal growth, development, and reproduction are called (a)_____. Of these, nine are needed in relatively large amounts. These macronutrients include: (b)_____
_____
_____.

The remaining seven elements are needed in trace amounts; these micronutrients include: (c)_____
_____.

B. Fertilizers replace essential elements if they are missing from the soil

■ 11 Some essential elements may be added to soil as organic or inorganic fertilizer. Plant growth is usually restricted by the essential factor in shortest supply, a virtually self evident assertion known as the (a)_____
_____. Three essential factors for plant growth are water, light, and essential elements. The three elements that most often limit plant growth are (b)_____.

KEY TERMS

Frequently Used Prefixes and Suffixes: Use combinations of these prefixes and suffixes to generate terms for the definitions below.

Prefixes	The Meaning
hydro-	water
macro-	large, long, great, excessive
micro-	small

Prefix	Suffix	Definitions
_____	-ponics	1. Growing plants in water (not soil) containing dissolved inorganic minerals.
_____	-nutrient	2. An essential element that is required in fairly large amounts for normal plant growth.
_____	-nutrient	3. An essential element that is required in trace (small) amounts for normal plant growth.

Other Terms You Should Know

adventitious root	epidermis	root cap
Casparian strip	humus	sand
clay	pericycle	silt
concept of limiting factors	phloem	topography (-hies)
contractile root	pith	vascular cambium
cortex (-tices)	prop root	xylem
endodermis	root hair	

SELF TEST

Multiple Choice: Some questions may have more than one correct answer.

1. The origin of multicellular branch roots is the
   a. cambium.
   b. Casparian strip.
   c. pericycle.
   d. parenchyma.
   e. cortex.

2. The principal ingredients in inorganic fertilizers are
   a. phosphorus.
   b. potassium.
   c. nitrogen.
   d. iron.
   e. hydrogen.

3. When water first enters a root, it usually
   a. enters parenchymal cells.
   b. is absorbed by cellulose.
   c. moves from a water negative potential to a positive water potential.
   d. moves along cell walls.
   e. enters cells in the Casparian strip.

4. Essential macronutrients include
   a. phosphorus.
   b. potassium.
   c. iron.
   d. hydrogen.
   e. magnesium.

5. The function(s) performed by all roots is/are
   a. absorption of water.
   b. absorption of minerals.
   c. food storage.
   d. anchorage.
   e. aeration.

6. Structures found in primary roots that are also found in stems include the
   a. cortex.
   b. cuticle.
   c. conducting vessels.
   d. apical meristem cap.
   e. epidermis.

7. Which of the following statements is/are accurate with respect to root hairs?
   a. They are short-lived.
   b. Occur in monocots, but not dicots.
   c. Occur in dicots, but not monocots.
   d. Occur in both monocots and dicots.
   e. Some develop into root branches.

8. Roots produced in unusual places on the plants, often as aerial roots, are called _____ roots.
   a. secondary
   b. enhancement
   c. adventitious
   d. contractile
   e. aerial water-absorbing

9. Vascular tissues in monocot roots in general
   a. form a solid cylinder.
   b. are absent.
   c. are in patches arranged in a circle.
   d. contain a vascular cambium.
   e. continue into root hairs.

10. The principal function(s) of the root cortex is/are
    a. conduction.
    b. storage.
    c. production of root hairs.
    d. water absorption.
    e. mineral absorption.

11. Essential micronutrients include
    a. phosphorus.
    b. potassium.
    c. iron.
    d. hydrogen.
    e. magnesium.

12. The inorganic materials in soil come from
    a. fertilizers.
    b. water runoff.
    c. weathered rock.
    d. the atmosphere.
    e. percolating water.

13. The two groups of organisms in soil that are the most important in decomposition and nutrient cycles are
    a. insects.
    b. fungi.
    c. worms.
    d. algae.
    e. bacteria.

VISUAL FOUNDATIONS

Color the parts of the illustration below as indicated.

RED   ❑   proton
GREEN   ❑   membrane channel
YELLOW   ❑   plasma membrane
BLUE   ❑   negatively charged ion
ORANGE   ❑   ATP
BROWN   ❑   ADP + *Pi*
TAN   ❑   cytoplasm
PINK   ❑   extracellular fluid
VIOLET   ❑   positively charged ion

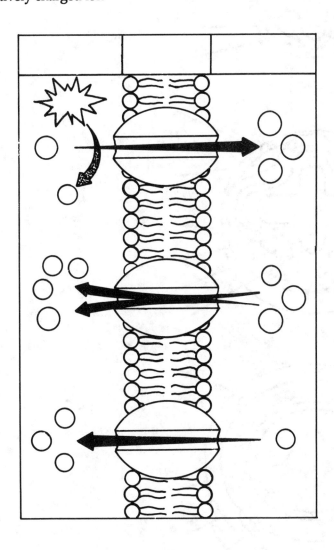

Color the parts of the illustration below as indicated.

RED	☐	vascular cambium
GREEN	☐	secondary phloem
YELLOW	☐	periderm
BLUE	☐	primary phloem
ORANGE	☐	secondary xylem
BROWN	☐	primary xylem
TAN	☐	pericycle

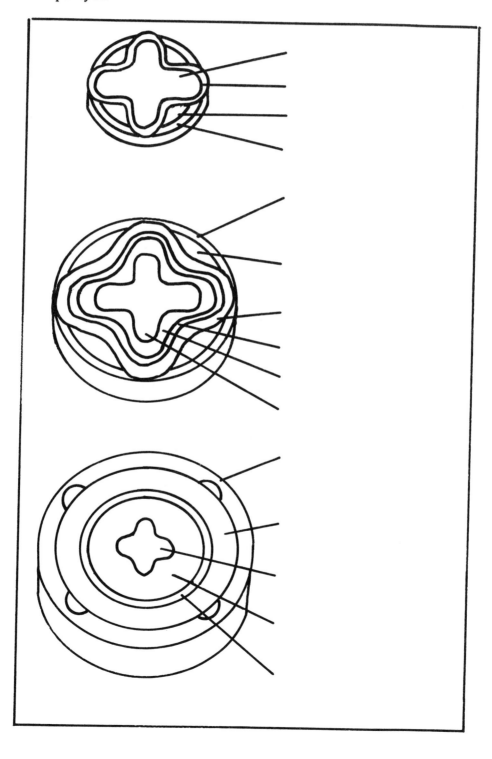

CHAPTER 35

□

Reproduction in Flowering Plants

OUTLINE

Asexual reproduction
Sexual reproduction
Environmental cues that induce flowering

All flowering plants reproduce sexually; some also reproduce asexually. Sexual reproduction involves flower formation, fertilization within the flower ovary, and seed and fruit formation. The resulting offspring exhibit a great deal of individual variation, due to gene recombination and the union of dissimilar gametes. Sexual reproduction offers the advantage of new combinations of genes that might make an individual plant better suited to its environment. Asexual reproduction involves only one parent; the offspring are genetically identical to the parent and each other. Asexual reproduction, therefore, is advantageous when the parent is well adapted to its environment. The stems, leaves, and roots of many flowering plants are modified for asexual reproduction. Also, in some plants, seeds and fruits are produced asexually without meiosis or the fusion of gametes. Initiation of sexual reproduction is often under environmental control. In some plants, it is induced by changes in duration of daylight, and in others, by changes in temperature.

CHAPTER OUTLINE AND CONCEPT REVIEW (Fill in the blanks)

I. INTRODUCTION

■ 1 The sexual reproductive structure in angiosperms is the (a)_____, in which, following fusion of haploid gametes, development of the (b)_____ takes place.

■ 2 Asexual reproduction, also called _____, produces genetically identical offspring as the result of mitosis of cells from a single parent.

II. ASEXUAL REPRODUCTION IN FLOWERING PLANTS MAY INVOLVE MODIFIED STEMS, LEAVES, OR ROOTS

■ 3 Various vegetative structures may be involved in asexual reproduction. Some of them, for example, are (a)_____, like those in bamboo and grasses that have horizontal underground stems; (b)_____, underground stems greatly enlarged for food storage, like those in potatoes; (c)_____, short underground stems with fleshy storage leaves, as in onions and tulips; (d)_____, thick stems covered with papery scales; and (e)_____ or runners that are above ground stems with long internodes, like those in strawberries.

■ 4 Roots may develop adventitious buds that form _____. These stems develop roots and may give rise to new plants.

■ 5 _____ is the production of seeds and fruits without sexual reproduction.

III. SEXUAL REPRODUCTION IN FLOWERING PLANTS INVOLVES FLOWERS, FRUITS, AND SEEDS

A. Fruits are mature, ripened ovaries

■ 6 The four basic types of fruits are _____
_____.

■ 7 One type of fruit, the (a)_____ fruit, develops from a single ovary. It may be fleshy or dry. Fruits that are fleshy throughout, like those in tomatoes and grapes, are called (b)_____, while fleshy fruits that have a hard pit are known as (c)_____. Dry fruits that release their seeds by splitting open at maturity are classified as (d)_____ fruits, and (e)_____ fruits, like corn and wheat kernels, do not split open. Of those that do split, some, like the milkweed (f)_____, split along *one* seam, while (g)_____ split along *two* seams, and (h)_____ split along *multiple* seams.

■ 8 (a)_____ fruits result from the fusion of several developing ovaries in a *single* flower. The raspberry is an example. Similarly, (b)_____ fruits form from the fusion of several ovaries of many flowers that are in close proximity. The pineapple is an example.

■ 9 (a)_____ fruits contain other plant tissues in addition to ovaries. When one eats a strawberry, for example, he or she is eating the fleshy (b)_____ of the flower, and when eating apples and pears she or he is consuming the (c)_____ that surrounds the ovary.

B. Fruit and seed dispersal is highly varied in flowering plants

■ 10 Angiosperm seeds and fruits are adapted for various means of dispersal, among which four common means are _____
_____.

IV. ENVIRONMENTAL CUES MAY INDUCE FLOWERING IN PLANTS

A. Flowering may be initiated by changes in light and dark periods

■ 11 _____ is the response of plants to the duration and timing of light and dark. Flowering is a photoperiodic response.

■ 12 The photoreversible pigment in plants that responds to light is _____.

B. Temperature may also affect reproduction

■ 13 _____ is the promotion of flowering by exposure to cold temperatures.

KEY TERMS

Frequently Used Prefixes and Suffixes: Use combinations of these prefixes and suffixes to generate terms for the definitions below.

Prefixes	The Meaning		Suffixes	The Meaning
phyto-	plant		-chrom(e)	color

Prefix	Suffix	Definitions
_____	_____	1. A blue-green, proteinaceous pigment involved in photoperiodism and a number of other light-initiated physiological responses of plants.

Other Terms You Should Know

accessory fruit	drupe	plantlet
achene	floral tube	receptacle
aggregate fruit	follicle	rhizome
apomixis	fruit	short-day plant
berry	grain	simple fruit
bulb	indehiscent	stolon
capsule	legume	sucker
corm	long-day plant	tuber
day-neutral plant	multiple fruit	vernalization
dehiscent	photoreceptor	

SELF TEST

Multiple Choice: Some questions may have more than one correct answer.

1. A fruit that forms from many separate carpels in a single flower is a/an
   a. follicle.
   b. achene.
   c. drupe.
   d. aggregate fruit.
   e. accessory fruit.

2. The process by which an embryo develops from a diploid cell in the ovary without fusion of haploid gametes is called
   a. lateral bud generation.
   b. dehiscence.
   c. asexual reproduction by means of suckers.
   d. apomixis.
   e. spontaneous generation.

3. The tomato is actually a
   a. drupe.
   b. berry.
   c. fleshy lateral bud.
   d. fruit.
   e. mature ovary.

4. Which of the following are somehow related to dry dehiscent fruits?
   a. drupe
   b. legume
   c. grain
   d. capsule
   e. green bean

5. The "eyes" of a potato are
   a. diploid gametes.
   b. parts of a stem.
   c. rhizomes.
   d. lateral buds.
   e. capable of producing complete plants.

6. A fruit composed of ovary tissue and other plant parts is a
   a. follicle.
   b. achene.
   c. drupe.
   d. aggregate fruit.
   e. accessory fruit.

7. Horizontal, asexually reproducing stems that run above ground are
   a. diploid gametes.
   b. rhizomes.
   c. tubers.
   d. corms.
   e. stolons.

8. Raspberries and blackberries are examples of
   a. follicles.
   b. achenes.
   c. drupes.
   d. aggregate fruits.
   e. multiple fruits.

9. Short-day plants typically flower when
   a. nights are long.
   b. days are long.
   c. exposed exclusively to 730 nm wavelengths of light.
   d. P_R concentration is high.
   e. P_{FR} concentration is high.

10. In some plants, vernalization affects
    a. hormone levels.
    b. the stolons.
    c. growth of tubers.
    d. vegetative reproduction.
    e. sexual reproduction.

11. After a long night, plants typically contain
    a. more P_R than P_{FR}.
    b. more P_{FR} than P_R.
    c. about equal amounts of P_R and P_{FR}.
    d. more active phytochrome than inactive phytochrome.
    e. more inactive phytochrome than active phytochrome.

VISUAL FOUNDATIONS

Color the parts of the illustration below as indicated.

RED   ☐   ovule
GREEN   ☐   sepal
YELLOW   ☐   ovary (and derived from ovary)
BLUE   ☐   stamen
ORANGE   ☐   style
BROWN   ☐   stigma
TAN   ☐   seed
PINK   ☐   floral tube (and derived from floral tube)
VIOLET   ☐   petal

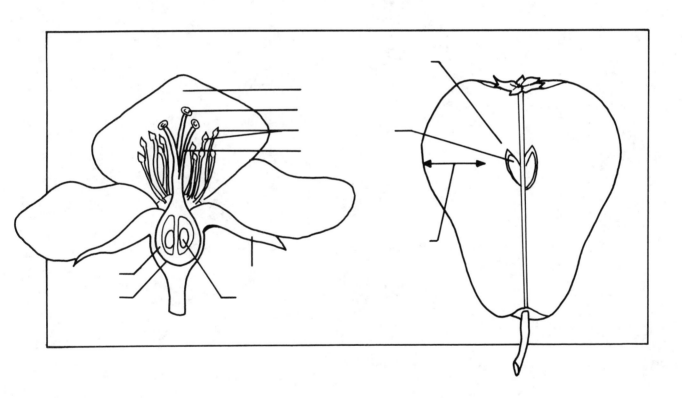

CHAPTER 36

❑

Plant Hormones and Responses

OUTLINE

Temporary plant movements
Plant responses timed by a biological clock
Tropisms
Plant hormones

Plants display a variety of growth movements and responses, varying from gradual to rapid. The more gradual movements and responses include the tropisms and circadian rhythms. The more rapid include the turgor movements. All aspects of plant growth and development are affected by hormones, chemical messengers produced in one part of the plant and transported to another part where they have their effect. Five different hormones interact in complex ways with one another to produce a variety of responses in plants. These are the auxins which are involved in cell elongation, phototropism, gravitropism, apical dominance, and fruit development; the giberellins which are involved in stem elongation, flowering, and seed germination; the cytokinins which promote cell division and differentiation, delay senescence, and interact with auxins in apical dominance; ethylene which has a role in the ripening of fruits and leaf abscission; and abscissic acid which is involved in stomatal closure due to water stress, and bud and seed dormancy.

CHAPTER OUTLINE AND CONCEPT REVIEW (Fill in the blanks)

I. INTRODUCTION

■ **1** The five major hormones that regulate responses in plants are _____
_____.

II. TURGOR MOVEMENTS CAN INDUCE TEMPORARY PLANT MOVEMENTS

■ **2** Some plants respond to external stimuli by changes in turgor in special cells in an organ at the base of a petiole called the (a)_____. A mechanical stimulus may initiate an electrical "impulse" that causes changes in cell membrane permeability. When this occurs, (b)_____ ions flow out of the affected cells, resulting in intracellular hypotonicity and the net movement of water out of these cells. This sudden change in water volume causes the leaf to move. Such movements, known as (c)_____, are temporary.

■ **3** The ability of plant parts to align with the sun as it moves across the sky is known as _____.

III. A BIOLOGICAL CLOCK INFLUENCES MANY PLANT RESPONSES

■ **4** _____ are regular daily rhythms in growth or activities of the plant.

IV. A TROPISM IS PLANT GROWTH IN RESPONSE TO AN EXTERNAL STIMULUS

■ **5** Tropisms are categorized according to the stimulus that causes them to occur. For example, growth or movement initiated by light is (a)_____, response to gravity is (b)_____, and movement caused by a mechanical stimulus is (c)_____.

V. HORMONES REGULATE PLANT GROWTH AND DEVELOPMENT

A. Charles Darwin first provided evidence for the existence of auxins, which cause a variety of physiological effects

■ **6** By definition, auxins are compounds that stimulate phototropic (a)_____ in oat coleoptiles, the principal one in plants generally being (b)_____ _____. These hormones always move in one direction, specifically from the (c)_____ toward the roots. Such unidirectional movement is called (d)_____. Auxins are also involved in phototropism, gravitropism, apical dominance, and fruit development.

■ **7** According to the (a)_____ Hypothesis, auxins influence cell elongation by "turning on" a proton pump in the plasma membranes of cells in the (b)_____ region of the plant.

B. Gibberellins, first discovered in a fungus, cause a variety of physiological effects

■ **8** In addition to influencing stem elongation, gibberellins are also involved in _____.

C. Cytokinins promote cell division and cause other physiological responses

■ **9** Cytokinins mainly promote (a)_____. They also delay (b)_____(aging), and they interact with auxins in apical dominance.

D. Ethylene, the only gaseous plant hormone, causes a variety of physiological effects

■ **10** Ethylene is involved in many aspects of aging. Among them two principal influences are on _____.

E. Abscisic acid promotes bud and seed dormancy as well as causing other physiological responses

■ **11** Abscisic acid is known as the stress hormone. Among other things, it is involved in stomatal closure due to _____ stress.

KEY TERMS

Frequently Used Prefixes and Suffixes: Use combinations of these prefixes and suffixes to generate terms for the definitions below.

Prefixes	The Meaning	Suffixes	The Meaning
organo-	organ, organic	-gen(esis)	production of
photo-	light	-tropism	turn, turning

Prefix	Suffix	Definitions
_____	_____	1. The growth response of an organism to light; usually the turning toward or away from the light source.
gravi-	_____	2. The growth response of an organism to gravity; usually the turning toward or away from the direction of gravity.
_____	_____	3. The origin and development (production) of an organ.

Other Terms You Should Know

abscisic acid	ethylene	senescence
acid-growth hypothesis	gibberellin	solar tracking
apical dominance	hormone	thigmotropism
auxin	hybrid vigor	tropism
bolting	indoleacetic acid	turgor movement
circadian rhythms	polar transport	zeatin
cytokinin	pulvinus (-ni)	

SELF TEST

Multiple Choice: Some questions may have more than one correct answer.

1. Which of the following is/are correct about hormones?
   a. They are effective in very small amounts.   d. The effects of different hormones overlap.
   b. They are organic compounds.   e. Each plant hormone has multiple effects.
   c. They are produced in one part of the plant and transported to other parts.

2. The hormone(s) that trigger(s) changes in plants that are exposed to unfavorable environmental conditions is/are
   a. auxin.   d. ethylene.
   b. gibberellin.   e. abscisic acid.
   c. cytokinin.

3. The hormone(s) principally responsible for cell division and differentiation is/are
   a. auxin.   d. ethylene.
   b. gibberellin.   e. abscisic acid.
   c. cytokinin.

4. If a plant that touches your house continues to grow toward and attach itself to the house, the plant is exhibiting a
   a. tropism.   d. thigmotropism.
   b. phototropism.   e. turgor movement.
   c. gravitropism.

5. If potassium ions and tannins leave special cells at the base of a petiole when the plant's leaf is touched, the leaf may move. This is an example of a
   a. tropism.   d. thigmotropism.
   b. phototropism.   e. turgor movement.
   c. gravitropism.

6. Which of the following applies to the acid-growth hypothesis?
   a. Auxin triggers a proton pump.
   b. Cell walls are acidified.
   c. Activated enzymes break bonds between cell wall molecules.
   d. Water is lost from vacuoles.
   e. H^+ ions flow from cell walls into cytoplasm.

7. The plant hormone(s) about which Charles Darwin gathered information is/are
   a. auxin.
   b. gibberellin.
   c. cytokinin.
   d. ethylene.
   e. abscisic acid.

8. The hormone(s) with a five ring structure that is/are involved in rapid stem elongation just prior to flowering is/are
   a. auxin.
   b. gibberellin.
   c. cytokinin.
   d. ethylene.
   e. abscisic acid.

9. The hormone(s) principally responsible for the growth of a coleoptile toward light is/are
   a. auxin.
   b. gibberellin.
   c. cytokinin.
   d. ethylene.
   e. abscisic acid.

10. If a plant that normally opens its stomates in daylight and closes its stomates during the night is placed in total darkness for a week, one would expect that it would
    a. open its stomates all week.
    b. close its stomates all week.
    c. assume a different periodicity for opening and closing stomates.
    d. continue opening and closing stomates as usual.
    e. soon exhibit a hormone imbalance.

VISUAL FOUNDATIONS

Color the parts of the illustration below as indicated.

RED ❑ proton
GREEN ❑ membrane channel
YELLOW ❑ plasma membrane
BLUE ❑ cell wall
ORANGE ❑ ATP
BROWN ❑ ADP + Pi
TAN ❑ cytoplasm

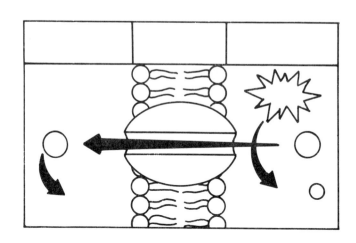

PART VII

Structures and Life Processes in Animals

□

The Animal Body: Tissues, Organs, and Organ Systems

O U T L I N E

Epithelial tissue
Connective tissue
Muscle tissue
Nervous tissue
Organs and organ systems

In most animals, cells are organized into tissues, tissues into organs, and organs into organ systems. The principal animal tissues are epithelial, connective, muscular, and nervous. Epithelial tissues are characterized by tight-fitting cells and the presence of a basement membrane. Covering body surfaces and lining cavities, they function in protection, absorption, secretion, and sensation. Epithelial tissues are classified on the basis of cell shape and arrangement of cell layers. Connective tissue joins together other tissues, supports the body and its organs, and protects underlying organs. There are many different types of connective tissues, consisting of a variety of cell types. For example, there is loose and dense connective tissue, elastic connective tissue, reticular connective tissue, adipose tissue, cartilage, bone, and blood. Muscle tissue is composed of cells that are specialized to contract. There are three major types of muscle tissue -- cardiac, smooth, and skeletal. Nervous tissue is composed of cells that are specialized for conducting impulses and those that support and nourish the conducting cells. Organs are comprised of two or more kinds of tissues. Complex animals have many organs and ten organ systems.

CHAPTER OUTLINE AND CONCEPT REVIEW (Fill in the blanks)

I. INTRODUCTION

■ **1** New cells formed by cell division remain associated in multicellular animals. The size of an animal is determined by the _____ of cells that make up its body, not the size of the individual cells.

■ **2** A group of cells that carry out a specific function is called a (a)_____, and these groups associate to form (b)_____, which in turn are grouped into the (c)_____ of the body.

II. EPITHELIAL TISSUES COVER THE BODY AND LINE ITS CAVITIES

■ **3** Epithelial tissues cover body surfaces and cavities, forming (a)_____ of tightly joined cells that are attached to underlying tissues by a fibrous noncellular (b)_____.

■ 4 Four of the major functions of epithelial cells include _____

_____ .

■ 5 Epithelial tissues vary in the number of cell layers and the shapes of their cells. For example, (a)_____ epithelium is made up of only one layer of cells and (b)_____ epithelium has two or more cell layers. (c)_____ cells are flat, (d)_____ cells resemble dice, and (e)_____ cells are tall, slender cells shaped like cylinders.

III. CONNECTIVE TISSUES JOIN AND SUPPORT OTHER BODY STRUCTURES

■ 6 Connective tissues join, support, and protect other tissues. There are many kinds of connective tissues, the main types of which are _____

_____

_____ .

■ 7 Characteristically, connective tissues contain very few cells, an (a)_____ in which cells and (b)_____ are embedded, and a (c)_____ that is secreted by the cells.

■ 8 The _____ associated with connective tissue is largely responsible for the nature and function of the tissue.

A. Connective tissue contains collagen, reticular, and elastic fibers

■ 9 There are three types of fibers in connective tissues. Collagen fibers are numerous, strong fibers composed of the protein (a)_____. Elastic fibers are composed of (b)_____ and can stretch. Reticular fibers, composed of (c)_____, form delicate networks of connective tissue.

B. Connective tissue contains specialized cells

■ 10 Fibroblast cells produce (a)_____ in connective tissues. (b)_____ are phagocytic scavengers.

■ 11 The (a)_____ in connective tissue release histamines, and (b)_____ secrete antibodies.

C. Loose connective tissue is widely distributed

■ 12 The most abundant tissue in the body is called loose connective tissue or (a)_____ tissue. It is found in spaces between body parts, around muscles, nerves, and blood vessels, and under the skin where it is called the (b)_____ .

D. Dense connective tissue consists mainly of fibers

■ 13 Dense connective tissues are predominantly composed of (a)_____ fibers. If these fibers are arranged in a specific pattern, the tissue is referred to as (b)_____ connective tissue. These strong tissues form "cablelike" cords called (c)_____ . (d)_____ _____ connective tissue consists of fibers arranged in various directions.

E. Elastic tissue is found in structures that must expand

■ 14 The walls of large _____ contain elastic connective tissue.

F. Reticular connective tissue provides support

■ **15** Reticular connective tissue forms an internal structural support network called (a)_____ for soft organs such as the (b)_____ _____.

G. Adipose tissue stores energy

■ **16** Two of the most important functions of adipose tissue are _____ _____.

H. Cartilage and bone provide support

■ **17** All vertebrates have an internal supporting structure called the _____ that is composed of cartilage and/or bone.

■ **18** The (a)_____ that provides support in the vertebrate embryo is largely replaced by (b)_____ in all vertebrate adults except (c)_____.

■ **19** Cartilage cells called (a)_____ are surrounded by a rubbery (b)_____ in which collagen fibers are embedded. "Imprisoned" by their own secretions, the cells are isolated in holes called (b)_____. Nutrients diffuse through the matrix to the cells.

■ **20** Bone is similar to cartilage in that it too has a (a)_____ dotted with lacunae in which cells called (b)_____ are "imprisoned." Unlike cartilage however, bone is highly (c)_____, a term that refers to their abundant supply of blood vessels.

■ **21** The matrix of bone contains (a)_____, a material that inhibits the diffusion of nutrients through the matrix. Consequently, osteocytes communicate with one another through small channels called (b)_____.

■ **22** The structural unit of bone is called an (a)_____. At its center is a (b)_____ containing blood vessels and nerves. Concentric layers of osteocytes, known as (c)_____, surround the canal.

I. Blood and lymph are circulating tissues

■ **23** The noncellular component of blood is the _____.

■ **24** Red blood cells, or (a)_____ as they are called, contain the respiratory pigment (b)_____. Their shape is generally described as a (c)_____.

■ **25** The white blood cells, collectively known as the _____, have important roles in defending the body against disease-causing organisms.

■ **26** Platelets are small cell fragments that originate in _____. In some vertebrates they assist in blood clotting.

IV. MUSCLE TISSUE IS SPECIALIZED TO CONTRACT

■ **27** Skeletal muscle cells are called _____.

■ **28** Contractile proteins called (a)_____, or "thin filaments," and (b)_____, or "thick filaments," are contained within the elongated (c)_____ of muscle cells.

■29 Vertebrates have three kinds of muscles: (a)_____ muscles that attach to bones and cause body movements, (b)_____ muscle found in internal organs and the walls of blood vessels, and heart or (c)_____ muscle. Both (d)_____ muscles are under involuntary control, while (e)_____ muscle control is voluntary.

V. NERVOUS TISSUE CONTROLS MUSCLES, GLANDS, AND OTHER ORGANS

■30 Cells in nervous tissue that conduct impulses are (a)_____, and support cells in this tissue are called (b)_____ cells.

■31 Neurons communicate with one another at cellular junctions called _____.

■32 A _____ is a collection of neurons bound together by connective tissue.

■33 Neurons generally contain three functionally and anatomically distinct regions: the (a)_____ which contains the nucleus, the (b)_____ that receive incoming impulses, and the (c)_____ that carry away outgoing impulses.

VI. COMPLEX ANIMALS HAVE ORGANS AND ORGAN SYSTEMS

■34 The ten major organ systems of complex animals are the _____ _____ _____ systems.

KEY TERMS

Frequently Used Prefixes and Suffixes: Use combinations of these prefixes and suffixes to generate terms for the definitions below.

Prefixes	The Meaning	Suffixes	The Meaning
chondro-	cartilage	-blast	embryo
fibro-	fiber	-cyte	cell
inter-	between, among	-phage	eat, devour
macro-	large, long, great, excessive		
multi-	many		
myo-	muscle		
osteo-	bone		
pseudo-	false		

Prefix	Suffix	Definitions
_____	-cellular	1. Composed of many cells.
_____	-stratified	2. An arrangement of epithelial cells in which the cells falsely appear to be stratified.
_____	_____	3. A cell, especially active in developing ("embryonic") tissue and healing wounds, that produces connective tissue fibers.
_____	-cellular	4. Situated between or among cells.
_____	_____	5. A large cell, common in connective tissues, that phagocytizes ("eats") foreign matter including bacteria.

_____ _____ 6. A cartilage cell.

_____ _____ 7. A bone cell.

_____ -fibril 8. A small longitudinal contractile fiber inside a muscle cell.

Other Terms You Should Know

actin	erythrocyte	organ
adipose tissue	fiber	organ system
areolar tissue	fibril	osteoclast
axon	gland	osteocyte
basement membrane	glial cell	osteon
blood	goblet cell	plasma
bone	haversian canal	platelet
canaliculus (-li)	hemoglobin	red blood cell
cardiac muscle	intercellular substance	regular dense CT
cartilage	irregular dense CT	reticular CT
cell body	lacuna (-ae)	reticular fiber
chondrocyte	lamella (-ae)	simple epithelium (-ia)
collagen fiber	loose CT	skeletal muscle
columnar epithelial cell	marrow cavity	smooth muscle
connective tissue (CT)	matrix (-trices)	squamous epithelial cell
cuboidal epithelial cell	mesenchyme	stratified epithelium (-ia)
dendrite	muscle tissue	striation
dense CT	myofibril	stroma
elastic CT	myosin	synapsis (-ses)
elastic fiber	nerve	tissue
endothelium (-ia)	nervous tissue	white blood cell
epithelium (-ia)	neuron	

SELF TEST

Multiple Choice: Some questions may have more than one correct answer.

1. Chondrocytes are
   a. part of cartilage.
   b. found in bone.
   c. cells that produce the matrix in cartilage.
   d. eventually found in the lacunae of a matrix.
   e. embryonic osteocytes.

2. Cartilage
   a. supports vertebrate embryos.
   b. comprises the shark's skeleton.
   c. is largely replaced by bone in most vertebrates.
   d. is derived from embryonic adipose tissue.
   e. is produced by chondrocytes.

3. Blood is one type of
   a. endothelium.
   b. collagen.
   c. mesenchyme.
   d. connective tissue.
   e. cardiac tissue.

4. Large muscles attached to bones are
   a. skeletal muscles.
   b. smooth muscles.
   c. composed largely of actin and myosin.
   d. nonstriated.
   e. multinucleated.

5. Cells that nourish and support nerve cells are known as
   a. neurons.
   b. glial cells.
   c. neuroblasts.
   d. synaptic cells.
   e. neurocytes.

6. A group of closely associated cells that carries out a specific function is a/an
   a. organ.
   b. system.
   c. organ system.
   d. tissue.
   e. clone.

7. Cells lining internal cavities that secrete a lubricating mucus are
   a. goblet-cells.
   b. pseudostratified.
   c. part of glandular tissue.
   d. epithelial.
   e. part of connective tissue.

8. Platelets are
   a. bone marrow cells.
   b. fragments of cells.
   c. a type of white blood cells.
   d. a type of RBCs.
   e. derived from bone marrow cells.

9. Epithelium located in the ducts of glands would most likely be
   a. simple columnar.
   b. simple cuboidal.
   c. simple squamous.
   d. stratified squamous.
   e. stratified cuboidal.

10. The portion of the neuron that is specialized to receive a nerve impulse is the
    a. cell body.
    b. synapse.
    c. dendrite.
    d. axon.
    e. glial body.

11. The connective tissue matrix is
    a. noncellular.
    b. mostly lipids.
    c. a gel.
    d. a polysaccharide.
    e. fibrous.

12. Collagen is
    a. part of blood.
    b. part of bone.
    c. part of connective tissue.
    d. composed of fibroblasts.
    e. fibrous.

13. Haversian canals
    a. contain nerves.
    b. run through cartilage.
    c. run through bone.
    d. are matrix lacunae.
    e. are synonymous with canaliculi.

14. Adipose tissue is a type of
    a. cartilage.
    b. epithelium.
    c. connective tissue.
    d. modified blood tissue.
    e. marrow.

15. The nucleus of a neuron is typically found in the
    a. cell body.
    b. synapse.
    c. dendrite.
    d. axon.
    e. glial body.

16. Endothelium
    a. lines ducts.
    b. lines blood vessels.
    c. lines lymph vessels.
    d. is derived from mesenchyme.
    e. is a form of sensory epithelium.

17. Myosin and actin are the main components of
    a. cartilage.
    b. blood.
    c. collagen.
    d. muscle.
    e. adipose tissue.

18. The basement membrane
    a. lies beneath epithelium.
    b. contains polysaccharides.
    c. has the typical fluid mosaic membrane structure.
    d. is synonymous with plasma membrane.
    e. is noncellular.

19. The major classes of animal tissues include
    a. epithelial.
    b. cardiovascular.
    c. muscular.
    d. skeletal.
    e. connective.

20. Epithelium cells that are flattened and thin are called
    a. columnar.
    b. cuboidal.
    c. squamous.
    d. stratified.
    e. endothelium.

Visual Foundations on next page ☞

VISUAL FOUNDATIONS

Color the parts of the illustration below as indicated.

RED	☐	blood vessel
GREEN	☐	lacuna
YELLOW	☐	compact bone
BLUE	☐	osteocyte
ORANGE	☐	haversian canal
BROWN	☐	osteon
TAN	☐	spongy bone

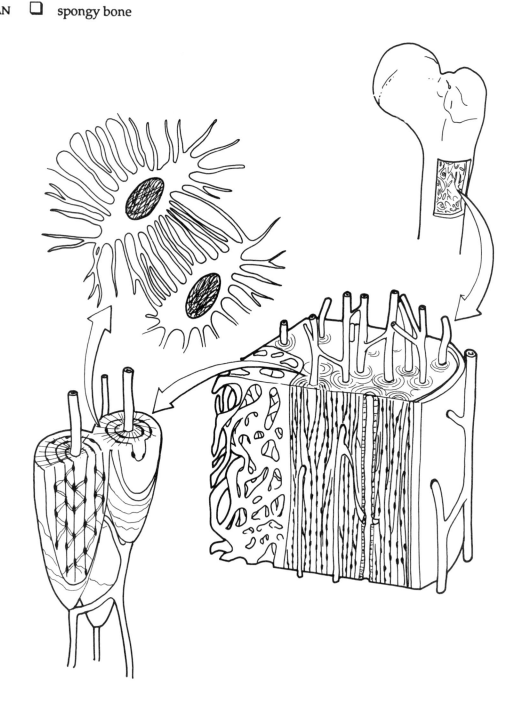

CHAPTER 38

Protection, Support, and Movement: Skin, Skeleton, and Muscle

O U T L I N E

Outer coverings: Protection
Skeletons: Locomotion, protection, and support
Muscles: Movement

Three organ systems — the integumentary, the skeletal, and the muscular — are closely interrelated in function and significance. Epithelium covers all external and internal body surfaces. In invertebrates, the external epithelium may contain secretory cells that produce a protective cuticle, secrete lubricants or adhesives, produce odorous or poisonous substances, or produce threads for nests or webs. In vertebrates, specifically humans, the external epithelium (skin) includes nails, hair, sweat glands, oil glands, and sensory receptors. In other vertebrates, it may include feathers, scales, mucous, and pigmentation. The skeleton supports and protects the body and transmits mechanical forces generated by contractile cells. Among the invertebrates are found hydrostatic skeletons (cnidarians and annelids) and exoskeletons (mollusks and arthropods). Exoskeletons, composed of nonliving material above the epidermis, prevent growth, necessitating periodic molting. Endoskeletons on the other hand, extensive in echinoderms and chordates, are composed of living tissue that can grow. Whereas some bones in the body develop from a noncartilage connective tissue model, most bones of the body develop from cartilage replicas. All animals have the ability to move. Muscle tissue is found in most invertebrates and all vertebrates. As muscle contracts, it moves body parts by pulling on them. Muscle contraction, fueled by ATP, involves a complex sequence of events. Muscle action is antagonistic -- the movement produced by one muscle can be reversed by another.

CHAPTER OUTLINE AND CONCEPT REVIEW (Fill in the blanks)

I. INTRODUCTION

■ 1 Muscle can contract. When it is anchored to the _____ system, muscle contraction causes movement (locomotion).

II. OUTER COVERINGS PROTECT THE BODY

■ 2 The external and internal surfaces of the animal body are covered by _____ tissue.

A. The protective epithelial covering of invertebrates may function in secretion or gas exchange

■ **3** Epithelial tissue in invertebrates may contain _____ cells that produce a protective cuticle, secrete lubricants or adhesives, produce odorous or poisonous substances, or produce threads for nests or webs.

B. The vertebrate skin functions in protection and temperature regulation

■ **4** Human _____ includes nails, hair, sweat glands, oil glands, and sensory receptors.

■ **5** Feathers, scales, mucus coverings, and poison glands are all structures associated with the _____ of vertebrates.

C. The epidermis is a waterproof protective barrier

■ **6** The outer layer of skin is called the (a)_____. It consists of several sublayers, or (b)_____.

■ **7** Cells in the _____ continuously divide. As they are pushed upward, these cells mature, produce keratin, and eventually die.

■ **8** Epithelial cells produce an insoluble, elaborately coiled protein called (a)_____, which functions in the skin for _____.

D. The dermis contains blood vessels and other structures

■ **9** The dermis consists of dense, fibrous (a)_____ resting on a layer of (b)_____ tissue composed largely of fat.

III. SKELETONS ARE IMPORTANT IN LOCOMOTION, PROTECTION, AND SUPPORT

■ **10** The skeleton transmits mechanical forces generated by _____ and also supports and protects the body.

■ **11** The support system in some organisms is an _____ consisting of a non-living substance overlying the epidermis.

A. In hydrostatic skeletons, body fluids are used to transmit force

■ **12** Many invertebrates (e.g., hydra) have a (a)_____ in which fluid is used to transmit forces generated by contractile cells or muscle. In these animals, contractile cells are arranged in two layers, an outer (b)_____ oriented layer and an inner (c)_____ arranged layer. When the _____ (outer or inner?) layer contracts, the animal becomes shorter and thicker, and when the _____ (outer or inner?) contracts, the animal becomes longer and thinner.

■ **13** Annelid worms have sophisticated hydrostatic skeletons. The body cavity is divided by transverse partitions called _____, creating isolated, fluid-filled segments that can operate independently.

B. Mollusks and Arthropods have nonliving exoskeletons

■ **14** The main function of the mollusk exoskeleton or (a)_____ is for (b)_____.

■ **15** Arthropod exoskeletons, composed mainly of (a)_____, are jointed for flexibility. This nonliving skeleton prevents growth, necessitating periodic (b)_____.

C. Internal skeletons are living tissues capable of growth

■ **16** Internal skeletons, called (a)_____, are extensively developed only in the echinoderms and the (b)_____.

■ **17** The endoskeletons of _____ are internal "shells" formed from spicules and plates composed of calcium salts.

■ **18** The two main divisions of the vertebrate endoskeleton are the (a)_____ skeleton along the long axis of the body and the (b)_____ skeleton comprised of the bones of the limbs and girdles.

■ **19** Components of the axial skeleton include the (a)_____, which consists of cranial and facial bones; the (b)_____, made up of a series of vertebrae; and the (c)_____, consisting of the sternum and ribs. Vertebrae have a bony, weight-bearing (d)_____ and a dorsal (e)_____ around the spinal cord.

■ **20** Components of the appendicular skeleton include the (a)_____, consisting of clavicles and scapulas; the (b)_____ which consists of large, fused hipbones; and the (c)_____, each of which terminates in digits. Digits can be specialized, as, for example, with the (d)_____ of humans and other primates that enables them to grasp and manipulate objects.

■ **21** The long bones of the limbs, such as the radius, have characteristic structures. They are covered by a connective tissue layer called the (a)_____. The ends of these bones are called the (b)_____, and the shaft is the (c)_____. A cartilaginous "growth center" in children called the (d)_____ becomes an (e)_____ in adults. The hollow core of a long bone, the (f)"_____," is lined with a thin membrane called the (g)_____. (h)_____, made up of osteons, is a thin covering over bone. The epiphyseal ends of long bones are filled with spongy, or (i)_____, bone tissue.

■ **22** Long bones develop from cartilage replicas; this is (a)_____ bone formation. Other bones develop from noncartilage connective tissue replica; this is (b)_____ bone development.

■ **23** Junctions between bones are called (a)_____. They are classified according to the degree of their movement: (b)_____, such as sutures, are tightly bound by fibers; (c)_____, like those between vertebrae; and the most common type of joint, the (d)_____ _____.

IV. MUSCLE IS THE CONTRACTILE TISSUE THAT PERMITS MOVEMENT IN COMPLEX ANIMALS

■ **24** (a)_____ is a ubiquitous contractile protein in eukaryotic cells. It makes up the thin microfilaments that function in such cellular processes as skeletal muscle contraction, ameboid movement, and cell division. In many cells, as, for example, in skeletal muscle myofibrils, it functions in association with the contractile protein (b)_____.

■ 25 Most animals move. Most invertebrates have specialized muscles for movements. For example, bivalve mollusks have both smooth and striated muscle. The (a)_____ muscle act to hold their shells closed for long periods of time, whereas the (b)_____ muscle enables them to shut their shells rapidly.

A. A vertebrate muscle may consist of thousands of muscle fibers

■ 26 Each skeletal muscle of a vertebrate consists of bundles of muscle fibers called _____.

■ 27 Striations of skeletal muscle fibers reflect the interdigitations of (a)_____ filaments. A unit of actin and myosin filaments makes up a (b)_____.

B. Muscle contraction occurs when actin and myosin filaments slide past each other

■ 28 During muscle contractions, actin filaments move _____(inward or outward?) between the myosin filaments.

■ 29 A motor neuron releases (a)_____ into the myoneural cleft, where it combines with receptors on the surface of the muscle fiber, depolarizing the sarcolemma and initiating an (b)_____.

■ 30 An impulse spreads through t tubules and stimulates calcium release. (a)_____ initiates a process that uncovers the binding sites of the (b)_____ filaments. Cross bridges of the (c)_____ filaments attach to the binding sites.

■ 31 Cross bridges flex and reattach to new binding sites so that the filaments are pulled past one another, _____ the muscle.

C. ATP powers muscle contraction

■ 32 (a)_____ is the immediate energy source for muscle contraction. (b)_____ is an energy storage compound. (c)_____ is the fuel stored in muscle fibers.

D. Skeletal muscle action depends on muscle pairs that work antagonistically

■ 33 Skeletal muscles pull on cords of connective tissue called (a)_____, which then pull on bones. Usually, muscles cross a joint and are attached to opposite sides of the joint. At most joints, muscles act (b)_____, which means that the action of one muscle (the agonist) is opposed by the action of the other muscle, called the (c)_____.

E. Smooth, cardiac, and skeletal muscle are specialized for particular types of response

■ 34 (a)_____ muscle is capable of slow, sustained contractions and (b)_____ muscle contracts rhythmically. (c)_____ muscle contains two sets of specialized fibers: the (d)_____ fibers that are specialized for slow responses and the (e)_____ fibers specialized for fast responses.

KEY TERMS

Frequently Used Prefixes and Suffixes: Use combinations of these prefixes and suffixes to generate terms for the definitions below.

Prefixes	The Meaning		Suffixes	The Meaning
endo-	within		-blast	embryo, "formative cell"
epi-	upon, over, on		-chondr(al)	cartilage
intra-	within		-dermis	skin
myo-	muscle		-fibr(il)	fiber
osteo-	bone		-oste(um)	bone
peri-	about, around, beyond			

Prefix	Suffix	Definitions
_____	_____	1. The outermost layer of skin, resting on the dermis.
_____	-cuticle	2. The outer portion of the cuticle.
_____	_____	3. Connective tissue membrane on the surface of bone that is capable of forming bone.
_____	_____	4. Pertains to the occurrence or formation of "something" within cartilage.
_____	-membranous	5. Pertains to the occurrence or formation of "something" within a membrane.
_____	_____	6. A bone-forming cell.
_____	-clast	7. Cells that break down bone.
_____	_____	8. Threadlike structure running lengthwise through muscle fiber.
_____	-filament	9. Subunit of myofibril consisting of either actin or myosin.

Other Terms You Should Know

acetylcholine
acne
actin
actin filament
action potential
agonist
antagonist
antagonistic
appendicular skeleton
atlas
axial skeleton
canaliculus (-li)
cancellous bone
centrum
cervical
chitin
clavicle

coccygeal
coccyx (-yges)
compact bone
creatine phosphate
cross bridge
cuticle
depolarization
dermis
diaphysis (-yses)
digit
endochondral bone
endosteum (-ea)]
epiphyseal line
epiphysis (-yses)
exoskeleton
fascicle
fast-twitch fiber

fiber
freely movable joint
girdle
haversian canal
haversian system
immovable joint
impulse
intramembranous bone
keratin
lacuna (-ae)
lumbar
marrow cavity (-ties)
metaphysis (-yses)
molt
muscle tone
myoglobin
myosin

myosin filament
neural arch (-hes)
osteon
oxygen debt
pectoral girdle
pelvic girdle
rib cage
sacral
sacrum (-ra)
sarcolemma
sarcomere

sarcoplasm
sarcoplasmic reticulum
scapula
sebum
septum (-ta)
simple twitch
skull
slightly movable joint
slow-twitch fiber
spongy bone
sternum (-na)

stratum basale
stratum corneum
striation
synovial fluid
T tubule
tetanus
thoracic
tropomyosin
troponin
vertebra (-ae)
vertebral column

SELF TEST

Multiple Choice: Some questions may have more than one correct answer.

1. The human skull is part of the
   a. cervical complex.
   b. girdle.
   c. axial skeleton.
   d. appendicular skeleton.
   e. atlas.

2. Most of the mechanical strength of bone is due to
   a. spongy bone.
   b. cancellous bone.
   c. intramembranous bone.
   d. endochondral bone.
   e. marrow.

3. You perceive touch through sense organs in your
   a. epidermis.
   b. stratum basale.
   c. stratum corneum.
   d. epithelium.
   e. dermis.

4. Myofilaments are composed of
   a. myofibrils.
   b. fibers.
   c. actin.
   d. myosin.
   e. sarcoplasmic reticulum.

5. The outer layer of a vertebrate's skin is the
   a. epidermis.
   b. stratum basale.
   c. stratum corneum.
   d. epithelium.
   e. dermis.

6. An exoskeleton would not work for an organism the size of a walrus because
   a. muscle attachments can't carry the load.
   b. their eggs are too small.
   c. the number of ecdyses is limited.
   d. exoskeletons weigh too much.
   e. invertebrate muscles are not strong enough.

7. The tissue around bones that lays down new layers of bone is the
   a. metaphysis.
   b. epiphysis.
   c. marrow.
   d. periosteum.
   e. endosteum.

8. The human axial skeleton includes the
   a. ulna.
   b. shoulder blades.
   c. centrum.
   d. femur.
   e. breastbone.

9. Cartilaginous growth centers in children are called
   a. metaphysis.
   b. epiphysis.
   c. marrow.
   d. periosteum.
   e. endosteum.

10. The human appendicular skeleton includes the
    a. ulna.
    b. shoulder blades.
    c. centrum.
    d. femur.
    e. breastbone.

11. Vertebrate appendages are connected to
    a. the cervical complex.
    b. girdles.
    c. the axial skeleton.
    d. the appendicular skeleton.
    e. the atlas.

12. Skeletons that operate entirely by hydrostatic pressure are found in
    a. annelids.
    b. echinoderms.
    c. hydra.
    d. lobsters and crayfish.
    e. vertebrates.

13. An opposable digit is found in
    a. lobsters.
    b. great apes.
    c. crayfish.
    d. some insects.
    e. humans.

14. A snail's slime track, the sticky substance on a fly's feet, and a spider's web are all produced by
    a. the stratum basale.
    b. epithelial cells.
    c. epicuticle.
    d. specialized dermal cells.
    e. stratum corneum.

15. Internal skeletons are found in
    a. annelids.
    b. echinoderms.
    c. hydra.
    d. lobsters and crayfish.
    e. vertebrates.

Visual Foundations on next page ☞

VISUAL FOUNDATIONS

Color the parts of the illustration below as indicated.

RED	❑	artery
GREEN	❑	sweat gland
YELLOW	❑	nerve ending
BLUE	❑	vein
ORANGE	❑	hair erector muscle
BROWN	❑	hair shaft, hair follicle
TAN	❑	dermis
PINK	❑	epidermis
VIOLET	❑	sebaceous gland

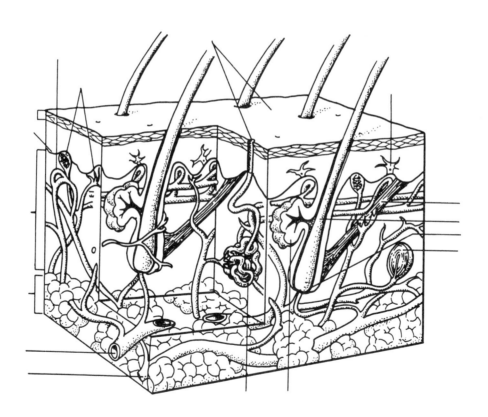

Color the parts of the illustration below as indicated. Label a sarcomere and the Z-lines.

RED ❑ nucleus
GREEN ❑ T tubule
YELLOW ❑ sarcoplasmic reticulum
BLUE ❑ mitochondria
ORANGE ❑ myofibrils
BROWN ❑ sarcolemma
TAN ❑ H zone
PINK ❑ I band
VIOLET ❑ A band and M band

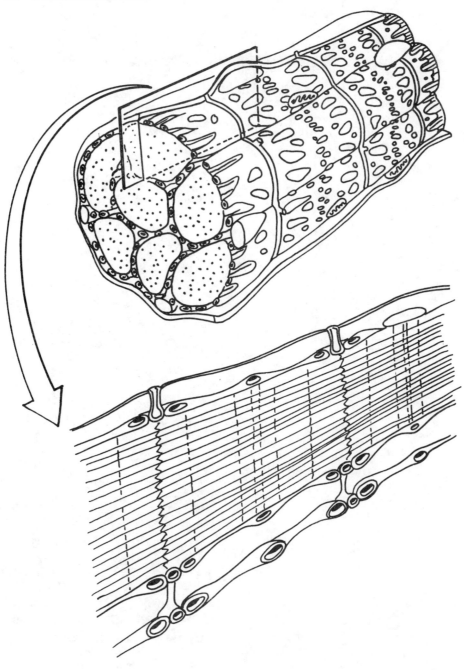

CHAPTER 39

Neural Control: Neurons

OUTLINE

Neurons and glial cells
Neurons transmit information
Synaptic transmission
Neural integration
Neural circuits

Two organ systems regulate behavior and physiological processes in animals -- the endocrine system and the nervous system. Endocrine regulation is slow and long lasting, whereas nervous regulation is rapid. Changes within an animal's body or in the outside world are detected by receptors and transmitted in the form of rapidly moving electrical signals or impulses to the central nervous system (CNS). The CNS sorts and interprets the information and determines an appropriate response. Impulses are then transmitted to muscle or glands where the response occurs. The structural and functional unit of the nervous system that carries the impulses is the neuron. The typical neuron consists of a cell body, dendrites, and an axon. The cell body houses the nucleus and has branched cytoplasmic projections called dendrites that receive stimuli. The axon is a single, long structure that transmits impulses from the cell body to an adjacent neuron, muscle, or gland. Transmission along a neuron is an electrochemical process involving changes in ion distribution between the inside and outside of the neuron. Transmission from one neuron to the next generally involves the release of chemicals — neurotransmitters — into the space between neurons. The neurotransmitters diffuse across the space and bring about a response in the adjacent neuron (or muscle or gland). Many types of neurotransmitters have been identified. Because every neuron networks with hundreds of other neurons, hundreds of messages may arrive at a single neuron at the same time. These messages must be integrated before the neuron can respond. After the neuron has completed its integration, it may or may not initiate an impulse along its axon. Most neural integration occurs in the CNS. Neurons are organized into complex neural pathways or circuits.

CHAPTER OUTLINE AND CONCEPT REVIEW (Fill in the blanks)

I. INTRODUCTION

■ 1 In complex animals, two systems are responsible for the regulation of physiological processes and behavior. Regulation by the (a)_____ system is slow and long-lasting, whereas regulation by the (b)_____ system is rapid.

■ 2 The ability of an organism to survive is largely dependent on its ability to detect and respond appropriately to _____, which are defined as changes in either the internal or external environments.

■ 3 Information flow through the nervous system begins with the detection of a stimulus by a "sensory device," a process called (a)_____. The "sensory device" is associated with sensory or (b)_____ neurons that (c)_____ or relay the information to the central nervous system (CNS). Association neurons (also called (d)_____) in the CNS "manage" the information, sorting and interpreting the information in a process called (e)_____. Once the appropriate response is selected, the information is transmitted from the CNS by motor or (f)_____ neurons to muscles and/or glands (the (g)_____) that carry out the response.

II. THE CELL TYPES OF THE NERVOUS SYSTEM ARE NEURONS AND GLIAL CELLS

■ 4 The _____ is the structural and functional unit of the nervous system.

A. Glial cells support and protect neurons

■ 5 _____, meaning "nerve glue," are unique cells that protect, insulate, and support neurons.

B. A typical neuron consists of a cell body, dendrites, and an axon

■ 6 Neurons are highly specialized cells with a distinctive structure consisting of a cell body and two types of cytoplasmic extensions. Numerous, short (a)_____ receive impulses and conduct them to the (b)_____ which integrates impulses. Once integrated, the impulse is conducted away by the (c)_____, a long process that terminates at another neuron or effector.

■ 7 Branches at the ends of axons, called (a)_____, form tiny (b)_____ that release a transmitter chemical called a (c)_____.

■ 8 Many axons outside of the CNS are covered by sheaths produced by cells called (a)_____ that wrap around these axons. The outer (b)_____, or neurilemma, is important in the regeneration of neurons. The inner (c)_____ acts as an insulator, due largely to the lipid-rich material called (d)_____ in the plasma membranes of sheath-producing cells.

■ 9 A (a)_____ is a bundle of axons outside the CNS, whereas a (b)_____ is a bundle of axons within the CNS.

■ 10 Masses of cell bodies outside the CNS are called (a)_____, whereas masses of cell bodies within the CNS are called (b)_____.

III. NEURONS CONVEY INFORMATION BY TRANSMITTING RAPIDLY MOVING ELECTRICAL IMPULSES

A. The resting potential is the difference in electrical charge across the plasma membrane

■ 11 The steady potential difference that exists across the plasma membrane of a nonconducting neuron is the (a)_____. Its value is expressed as (b)_____.

■ 12 When a neuron is not conducting an electrical impulse, the _____(inner or outer?) surface of the plasma membrane is negatively charged compared to the interstitial fluid.

■ **13** In most cells, potassium ion concentration is highest (a)_____ (inside or outside?) the cell and sodium ion concentration is highest)_____ (inside or outside?) the cell.

■ **14** The ionic balance across the plasma membrane of neurons is a result of several factors. The (a)_____ is an active transport system that moves sodium out of the cell and potassium into the cell. (b)_____ _____ are part of a facilitated diffusion system that allows sodium and potassium ions to follow their gradients. Finally, the cytoplasm of the neuron is negatively charged due to the presence of large (c)_____.

B. The nerve impulse is an action potential

■ **15** If a stimulus is strong enough, it will provoke a response from the neuron called an (a)_____, which is an alteration of the resting potential caused by an increase in membrane permeability to (b)_____. The potential value for this response is expressed as (c)_____.

■ **16** (a)_____ in the plasma membrane of the neuron open when they detect a critical level of voltage change in the membrane potential. This value, known as the (b)_____, is generally given as about (c)_____.

■ **17** A (a)_____ is a sharp rise and fall of an action potential. The sudden rise is largely caused by the opening of _____.

■ **18** When depolarization reaches threshold level, an action potential may be generated. An action potential is a _____ that spreads along the axon.

■ **19** As the action potential moves down the axon, _____ occurs behind it.

C. Saltatory conduction is rapid

■ **20** In saltatory conduction in myelinated neurons, depolarization skips along the axon from one _____ to the next.

D. The neuron obeys an all-or-none law

■ **21** Once a stimulus depolarizes a neuron to the threshold, the neuron will respond fully with an action potential. All action potentials have the same strength. A response that always occurs at a given value, or it won't occur at all, is referred to as an _____ response.

E. Certain substances affect excitability

IV. SYNAPTIC TRANSMISSION OCCURS BETWEEN NEURONS

■ **22** A junction between two neurons, or between a neuron and an effector, is called a (a)_____. There are two kinds of neuroeffector junctions:
(b)_____ junctions between neurons and gland cells and
(c)_____ between neurons and muscle cells.

■ **23** The neuron that terminates at a synapse is known as the (a)_____ neuron and the neuron that begins at that synapse is called the
(b)_____ neuron.

■ 24 (a)_____ are gap junctions that allow rapid communication between neurons. Most synapses are (b)_____ that involve (c)_____ that cross the physical space or (d)_____ between neurons.

■ 25 _____ generally depends upon release of a neurotransmitter from vesicles in the synaptic knobs of the presynaptic neuron.

■ 26 The neurotransmitter diffuses across the synaptic cleft and combines with specific (a)_____ on the postsynaptic neuron. This opens (b)_____ that permit either depolarization or hyperpolarization.

A. Signals may be excitatory of inhibitory

■ 27 A depolarization of the postsynaptic membrane that brings the neuron closer to firing is called an (a)_____. A hyperpolarization of the postsynaptic membrane that reduces the probability that the neuron will fire is called an (b)_____ _____.

B. Graded potentials vary in magnitude

■ 28 Local responses in the postsynaptic membrane that vary in magnitude, fade over distance, and can be summated are called _____. EPSPs and IPSPs are examples.

■ 29 Graded potentials can be added together in a process called (a)_____. When a second EPSP occurs before the depolarization caused by the first EPSP has decayed, (b)_____ occurs. In (c)_____ _____, several synapses in the same area generate EPSPs simultaneously. Adding EPSPs together can bring the neuron to threshold.

C. Many types of neurotransmitters are known

■ 30 Cells that release (a)_____ are called cholinergic neurons. This neurotransmitter functions at (b)_____ junctions and some synapses in the autonomic nervous system.

■ 31 (a)_____ release norepinephrine, a neurotransmitter that, along with epinephrine and dopamine, belongs to the (b)_____ class.

D. Nerve fibers may be classified in terms of speed of conduction

■ 32 Large, heavily myelinated neurons conduct impulses fastest. In general, the _____ (further apart or closer?) the nodes of Ranvier, the faster the axon conducts.

V. NEURAL IMPULSES MUST BE INTEGRATED

■ 33 Integration of EPSPs and IPSPs occurs in the _____ of the postsynaptic neuron.

VI. NEURONS ARE ORGANIZED INTO CIRCUITS

■ 34 Complex neural pathways are possible because of such neuronal associations as _____.

KEY TERMS

Frequently Used Prefixes and Suffixes: Use combinations of these prefixes and suffixes to generate terms for the definitions below.

Prefixes	The Meaning	Suffixes	The Meaning
inter-	between, among	-neuro(n)	nerve
multi-	many, much, multiple		
neur(i)(o)-	nerve		
post-	behind, after		
pre-	before, prior to, in advance of		

Prefix	Suffix	Definitions
_____	_____	1. A nerve cell that carries impulses to other nerves, and is between an effector and sense receptor but not directly associated with either.
_____	-glia	2. Connecting and supporting cells in the central nervous system surrounding the neurons.
_____	-lemma	3. The cellular sheath formed by Schwann cells surrounding the axons of certain nerve cells.
_____	-transmitter	4. Substance used by neurons to transmit impulses across a synapse.
_____	-polar	5. Pertains to a neuron with more than two (often many) processes or projections.
_____	-synaptic	6. Pertains to a neuron that begins after a specific synapse.
_____	-synaptic	7. Pertains to a neuron that ends before a specific synapse.

Other Terms You Should Know

absolute refractory period
acetylcholine
action potential
adenylate cyclase
adrenergic neuron
afferent neuron
all-or-none law
association neuron
autonomic system
axon
axon terminal
biogenic amine
cAMP
catecholamine
cell body (-dies)
cellular sheath
central nervous system
chemical synapse

chemically-activated ion
  channel
cholinergic neuron
cholinesterase
circuit
CNS
collateral
continuous conduction
convergence
cyclic AMP
dendrite
depolarization
divergence
dopamine
effector
efferent neuron
electrical synapse
epinephrine
EPSP

excitatory postsynaptic
  potential
excitatory stimulus (-li)
facilitation
ganglion (-ia)
graded potential
inhibitory postsynaptic
  potential
inhibitory stimulus (-li)
integration
IPSP
low-calcium tetany
membrane potential
monoamine oxidase
motor end plate
motor neuron
multiple sclerosis
myelin
myelin sheath

nerve
nerve fiber
nerve impulse
neural impulse
neuroglandular junction
neuromuscular junction
neuron
neuronal pool
nodes of Ranvier
norepinephrine
nucleus (-ei)
pathway
polarized
postsynaptic neuron
presynaptic neuron

reception
receptor
reflex action
relative refractory period
repolarize
response
resting neuron
resting potential
reverberating circuit
saltatory conduction
Schwann cell
sensory neuron
somatic system
spatial summation
spike

stimulus (-li)
summation
synapse
synaptic cleft
synaptic knob
synaptic vesicle
temporal summation
threshold level
tract
transmission
transmitter substance
voltage-activated ion channel
wave of depolarization

SELF TEST

Multiple Choice: Some questions may have more than one correct answer.

1. The inner surface of a resting neuron is _____ compared with the outside.
   a. positively charged
   b. negatively charged
   c. polarized
   d. 70 mV
   e. −70 mV

2. The part of the neuron that transmits an impulse from the cell body to an effector cell is the
   a. dendrite.
   b. axon.
   c. Schwann body.
   d. collateral.
   e. hillock.

3. The structural and functional unit of the nervous system is the
   a. cell body.
   b. nerve cell.
   c. neuron.
   d. nerve.
   e. synapse.

4. Saltatory conduction
   a. occurs between nodes of Ranvier.
   b. occurs only in the CNS.
   c. is more rapid than the continuous type.
   d. requires less energy than continuous conduction.
   e. involves depolarization at nodes of Ranvier.

5. The nodes of Ranvier are
   a. on the cell body.
   b. on the dendrites.
   c. on the axon.
   d. gaps between adjacent Schwann cells.
   e. insulated with myelin.

6. A neuron that begins at a synaptic cleft is called a
   a. presynaptic neuron.
   b. postsynaptic neuron.
   c. synapsing neuron.
   d. neurotransmitter.
   e. acetylcholine releaser.

7. Which of the following correctly expresses the movement of ions by the sodium-potassium pump?
   a. sodium out, potassium in
   b. sodium in, potassium out
   c. sodium and potassium both in and out, but in different amounts
   d. more sodium out than potassium in
   e. less sodium out than potassium in

8. The neurilemma is
   a. a cellular sheath.
   b. the neuron membrane.
   c. composed of Schwann cells.
   d. found primarily in neurons of the CNS.
   e. associated with nodes of Ranvier.

9. A nerve pathway in the CNS is
   a. one neuron.
   b. a group of nerves.
   c. a bundle of axons.
   d. a ganglion.
   e. a bundle of cell bodies.

10. An axon cannot transmit an action potential no matter how great a stimulus is applied when it is
    a. hyperpolarized.
    b. depolarized.
    c. in the absolute refractory period.
    d. in the relative refractory period.
    e. in the resting state.

11. If a neurotransmitter hyperpolarizes a postsynaptic membrane, the change in potential is referred to as
    a. spatial summation.
    b. temporal summation.
    c. a threshold level impulse.
    d. IPSP.
    e. EPSP.

12. Aside from the sodium-potassium pump, the resting potential is due mainly to
    a. inward diffusion of chloride ions.
    b. inward diffusion of sodium ions.
    c. outward diffusion of potassium ions.
    d. open sodium channels in the membrane.
    e. large protein anions inside the cell.

13. Branchlike extensions of the cell body involved in receiving stimuli are
    a. dendrites.
    b. axons.
    c. Schwann bodies.
    d. collaterals.
    e. hillocks.

VISUAL FOUNDATIONS

Color the parts of the illustration below as indicated.

RED ☐ synaptic knob
GREEN ☐ axon terminal
YELLOW ☐ myelin sheath
BLUE ☐ axon
ORANGE ☐ cellular sheath
PINK ☐ cell body and dendrite
VIOLET ☐ nucleus

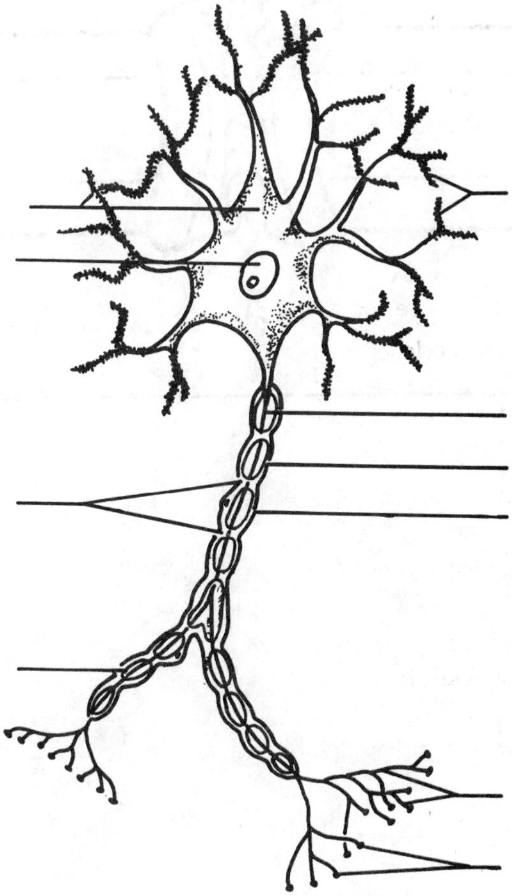

Color the parts of the illustration below as indicated.

RED	❑	sodium ion
GREEN	❑	potassium ion
YELLOW	❑	plasma membrane
BLUE	❑	arrows indicating diffusion into the cell
ORANGE	❑	sodium channel
BROWN	❑	potassium channel
TAN	❑	cytoplasm
PINK	❑	extracellular fluid
VIOLET	❑	arrows indicating diffusion out of the cell

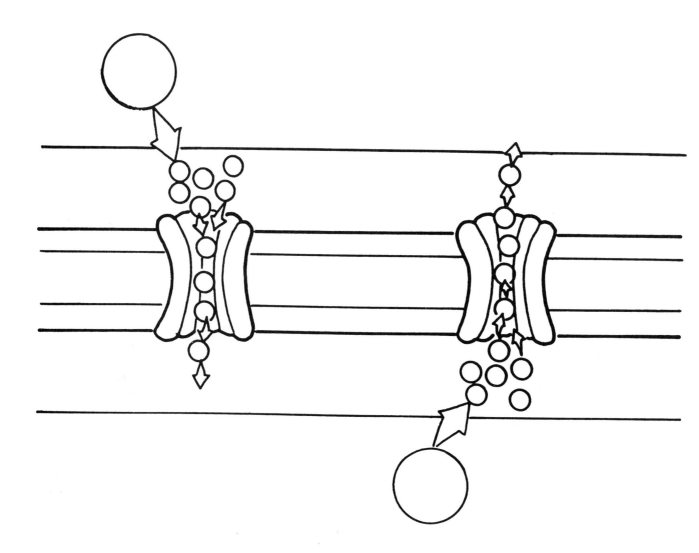

Neural Regulation: Nervous Systems

OUTLINE

Invertebrate nervous systems
The vertebrate nervous system
Evolution of the vertebrate brain
The human central nervous system: Structure and function
The human peripheral nervous system: Somatic and
autonomic divisions
Effects of drugs on the nervous system

The simplest organized nervous system is the nerve net found in cnidarians. A nerve net consists of nerve cells scattered throughout the body; there is no central control organ and no definite nervous pathways. The nervous system of echinoderms is more complex, with a nerve ring and nerves that extend into various parts of the body. Bilaterally symmetric animals have more complex nervous systems. The vertebrate nervous system has two main divisions: a central nervous system (CNS) consisting of a complex tubular brain that is continuous with a tubular nerve cord, and a peripheral nervous system (PNS). The CNS provides centralized control, integrating incoming information and determining appropriate responses. The PNS consists of sensory receptors and cranial and spinal nerves. The spinal cord transmits impulses to and from the brain and controls many reflex activities. The outer layer of the cerebrum, the largest subdivision of the brain, is functionally divided into sensory areas that receive incoming signals from the sense organs, motor areas which control voluntary movement, and association areas that link the sensory and motor areas and are responsible for thought, learning, language, memory, judgment, and personality. The reticular activating system is responsible for maintaining wakefulness and for arousal from deep sleep. The limbic system influences the emotional aspects of behavior, sexual behavior, biological rhythms, autonomic responses, and motivation. Brain waves, electrical potentials given off by the brain, reflect a person's state of relaxation and mental activity. Sleep is a state of unconsciousness marked by decreased electrical activity. Learning is a function of the brain. Just how memory is stored is not known, however, in at least some types of learning, physical or chemical changes take place in brain structure. Memory appears to be located throughout the association areas of the cerebrum. Many drugs affect the nervous system, some by changing the levels of neurotransmitters within the brain.

CHAPTER OUTLINE AND CONCEPT REVIEW (Fill in the blanks)

I. INTRODUCTION

■ 1 In general, the lifestyle of an animal is closely related to the design and complexity of its nervous system. For example, the simplest nervous system, called a (a)_____ _____, is found in (b)_____, which are sessile animals like *Hydra*.

II. MANY INVERTEBRATES HAVE COMPLEX NERVOUS SYSTEMS

■ 2 Echinoderms have a modified nerve net. In this system, a _____ surrounds the mouth, and large, radial nerves extend into each arm. Branches from these nerves coordinate the animal's movements.

■ 3 Planarian worms have a (a)_____ nervous system with a (b)_____ in the head region that serves as a simple brain.

■ 4 Each body segment in annelids contains a pair of (a)_____ which are connected by ventrally located (b)_____.

III. KEY FEATURES OF THE VERTEBRATE NERVOUS SYSTEM ARE THE HOLLOW DORSAL NERVE CORD AND THE WELL-DEVELOPED BRAIN

■ 5 The vertebrate nervous system is divided into a (a)_____ system consisting of the brain and spinal cord, and a (b)_____ system consisting of sense receptors, nerves, and ganglia. This latter system is further divided into the somatic or voluntary portion and the autonomic portion that regulates the internal environment. The autonomic system has two efferent pathways made up of (c)_____ nerves.

IV. THE EVOLUTION OF THE VERTEBRATE BRAIN IS MARKED BY INCREASING COMPLEXITY, ESPECIALLY OF THE CEREBRUM AND CEREBELLUM

■ 6 The (a)_____ of the vertebrate embryo differentiates anteriorly into the brain and posteriorly into the spinal cord. Three bulges in the anterior end develop into the (b)_____.

A. The hindbrain develops into the medulla, pons, and cerebellum

■ 7 The hindbrain, or _____, divides to form the metencephalon and myelencephalon.

■ 8 The (a)_____, which coordinates muscle activity, and the (b)_____, which connects the spinal cord and medulla with higher brain centers, are formed from the (c)_____. The (d)_____, made up of nerve tracks and nuclei, is formed from the (e)_____.

B. The midbrain is most prominent in fish and amphibians

■ 9 The _____ is made up of the medulla, pons, and midbrain.

■ 10 The midbrain is the largest part of the brain in fish and amphibians. It is their main _____ area, linking sensory input and motor output.

■ 11 In reptiles, birds, and mammals, the midbrain consists of the (a)_____ which regulate visual reflexes and the (b)_____ which regulate some auditory reflexes. It also contains the (c)_____ that integrates information about muscle tone and posture.

C. The forebrain gives rise to the thalamus, hypothalamus, and cerebrum

■ 12 The forebrain, or (a)_____, differentiates into the diencephalon and the telencephalon, which in turn develop into the (b)_____ and (c)_____ respectively.

■ **13** Cavities called (a)_____ are located within the cerebrum. They communicate with the (b)_____, a cavity in the diencephalon.

■ **14** The (a)_____ is a relay center for motor and sensory signals in all vertebrate classes. Located below this center, the (b)_____ regulates autonomic responses and links the nervous and endocrine systems. The (c)_____ gland is connected to the hypothalamus.

■ **15** Olfactory bulbs develop from the telencephalon, as does the _____, a structure in birds associated with innate behavioral patterns.

■ **16** The right and left cerebral (a)_____ are predominantly comprised of axons, which are collectively referred to as (b)_____. Mammals and most reptiles have a layer of grey matter called the (c)_____ that covers the cerebrum. In complex mammals the surface area of the cortex is increased by ridges called (d)_____, shallow folds called (e)_____, and deep fissures.

V. THE HUMAN CENTRAL NERVOUS SYSTEM IS THE MOST COMPLEX BIOLOGICAL MECHANISM KNOWN

■ **17** The human central nervous system consists of the brain and spinal cord. Both are protected by bone and by three meninges, the (a)_____ _____, and both are bathed by cerebrospinal fluid produced in a specialized capillary bed called the (b)_____.

A. The spinal cord transmits impulses to and from the brain and controls many reflex activities

■ **18** The spinal cord extends from the base of the brain to the _____ vertebra.

B. The spinal cord consists of gray and white matter

■ **19** Grey matter in the spinal cord is shaped like the letter "H." It is surrounded by (a)_____ that contains nerve pathways or (b)_____.

■ **20** The spinal cord consists of (a)_____ that transmit information to the brain, and (b)_____ that transmit information from the brain.

C. Withdrawal reflexes are protective

■ **21** A relatively fixed reaction pattern in response to a simple stimulus is called a (a)_____. A simple reflex, such as the withdrawal reflex, involves three neurons: a (b)_____ afferent neuron that receives information and transmits it to the CNS, an (c)_____ neuron in the CNS, and a (d)_____ efferent neuron that transmits the information to an effector.

■ **22** The human cerebral cortex consists of three functional areas. (a)_____ areas receive incoming sensory information, (b)_____ areas control voluntary movement, and (c)_____ areas link the other two areas.

D. The largest, most prominent part of the human brain is the cerebrum

■ 23 Each hemisphere of the human cerebrum is divided into lobes.
(a)_____ contain primary motor areas and are separated from the primary sensory areas in the (b)_____ lobes by a groove called the (c)_____. (d)_____ lobes contain centers for hearing and the (e)_____ lobes are associated with vision.

E. The reticular activating system is an arousal system

■ 24 The reticular activating system (RAS) is responsible for maintaining consciousness. It is located within the _____.

F. The limbic system influences emotional aspects of behavior

■ 25 The limbic system affects the emotional aspects of behavior, motivation, sexual behavior, autonomic responses, and biological rhythms. It consists of parts of the

_____.

G. The brain exhibits electrical activity

■ 26 Electrical activity in the form of brain waves can be detected by a device that produces brain wave tracings called an (a)_____.
The slow (b)_____ waves are associated with sleep, while (c)_____ waves are associated with relaxation, and (d)_____ waves are generated by an active, thinking brain.

H. Sleep may occur when signals from the RAS slow

■ 27 There are two main stages of sleep: (a)_____ sleep which is associated with dreaming, and (b)_____ sleep which is characterized by delta waves.

I. Learning involves the storage of information and its retrieval

■ 28 According to current theory, there are three levels of memory. (a)_____ memory is very short, intense, and requires attention; (b)_____ memory stores information for a few seconds, and this information may get processed into more-or-less permanent (c)_____ memory.

■ 29 An altered circuit, created when a memory is stored, is called a

_____.

J. Experience affects the brain

■ 30 Research shows that early stimulation can enhance the development of motor areas, result in a thicker cerebral _____ and stimulate memory.

VI. THE PERIPHERAL NERVOUS SYSTEM INCLUDES SOMATIC AND AUTONOMIC SUBDIVISIONS

A. The somatic system helps the body adjust to the external environment

■ 31 The somatic nervous system consists of (a)_____ that detect stimuli, (b)_____ that transmit information to the CNS, and (c)_____ that carry information from the CNS to effectors.

■ **32** Neurons are organized into nerves. The 12 pairs of (a)_____ nerves connect to the brain and are largely involved with sense receptors. Thirty-one pairs of (b)_____ nerves connect to the spinal cord.

■ **33** Each spinal nerve splits into two roots: a (a)_____ root that carries only sensory afferent neurons and a (b)_____ root that carries only motor efferent neurons.

B. The autonomic system helps maintain homeostasis in response to changes in the internal environment

■ **34** The efferent component of the autonomic nervous system is divided into two systems. The (a)_____ system enables the body to respond to stressful situations. The (b)_____ system influences organs to conserve and restore energy.

VII. MANY DRUGS AFFECT THE NERVOUS SYSTEM

KEY TERMS

Frequently Used Prefixes and Suffixes: Use combinations of these prefixes and suffixes to generate terms for the definitions below.

Prefixes	The Meaning
circum-	around, about
hypo-	under
para-	beside, near
post-	behind, after
pre-	before, prior to, in advance of

Prefix	Suffix	Definitions
_____	-thalamus	1. Part of the brain located below the thalamus; principal integration center for the regulation of the viscera.
_____	-ganglionic	2. Pertains to a neuron located distal to (after) a ganglion.
_____	-ganglionic	3. Pertains to a neuron located proximal to (before) a ganglion.
_____	-vertebral	4. Pertains to structures located beside the vertebral column.
_____	-esophageal	5. Pertains to a ring of ganglia located around the esophagus in cephalopods.

Other Terms You Should Know

addiction	brain stem	cerebrospinal fluid
alpha wave	central canal	cerebrum
arachnoid	central nervous system	choroid plexus (-ses)
ascending tract	central sulcus	CNS
association area	cerebellum	collateral ganglion (-ia)
autonomic system	cerebral aqueduct	column
beta wave	cerebral cortex	convolution
biofeedback	cerebral ganglion (-ia)	corpus striatum

cranial nerve
delta wave
diencephalon
dorsal root
dura mater
EEG
electroencephalograph
engram
euphoria
first ventricle
fissure
forebrain
fourth ventricle
frontal lobe
ganglion (-ia)
general interpretative area
gray matter
gyrus (-ri)
habituation
hemisphere
hindbrain
horn
inferior colliculus (-li)
interventricular foramen
irritable
ladder-type nervous system
lateral ventricle
limbic system
long-term memory
medulla
memory trace
meningitis

meninx (-nges)
metencephalon
midbrain
motor area
myelencephalon
neopallium
nerve net
neural tube
non-REM
occipital lobe
olfactory bulb
optic lobe
parasympathetic system
paravertebral sympathetic
    ganglion (-ia)
parietal lobe
pathway
peripheral nervous system
physical addiction
pia mater
plexus (-ses)
PNS
pons
primary motor area
primary sensory area
psychological dependence
pyramidal tract
RAS
red nucleus
reflex action
releasing hormone
REM

reticular activating system
second ventricle
sensitization
sensory area
sensory memory (-ries)
serotonin
short-term memory
sleep
somatic
somatic nervous system
spinal cord
spinal ganglion (-ia)
spinal nerve
spinothalamic tract
subarachnoid space
sulcus (-ci)
superior colliculus (-li)
sympathetic system
telencephalon
temporal lobe
terminal ganglion (-ia)
thalamus
third ventricle
tolerance
tract
ventral root
vital center
Wernicke's area
white matter
withdrawal reflex (-xes)
withdrawal symptom

SELF TEST

Multiple Choice: Some questions may have more than one correct answer.

1. Regulation of body temperature under ordinary circumstances is under the control of the
   a. autonomic n.s.
   b. sympathetic n.s.
   c. parasympathetic n.s.
   d. cranial nerve VI.
   e. dorsal root ganglion.

2. The principal integration center for the regulation of viscera is the
   a. cerebellum.
   b. cerebrum.
   c. thalamus.
   d. hypothalamus.
   e. red nucleus.

3. In non-REM sleep, as compared to REM sleep, there is/are
   a. more delta waves.
   b. faster breathing.
   c. lower blood pressure.
   d. more dream consciousness.
   e. more norepinephrine released.

4. When you are eating one of your favorite foods, the cranial nerve(s) directly involved in your perception of this pleasurable experience is/are
   a. facial.
   b. VII.
   c. X.
   d. glossopharyngeal.
   e. XI.

5. In higher mammals that engage in complex association functions, there would likely be
   a. a convoluted cerebral cortex.
   b. a small amount of gray matter.
   c. an expanded cortical surface area.
   d. a well-developed neopallium.
   e. fissures.

6. Cerebrospinal fluid
   a. is produced by choroid plexuses.
   b. circulates through ventricles.
   c. is between the arachnoid and pia mater.
   d. is within layers of the meninges.
   e. is in the subarachnoid space.

7. The brain and spinal cord are wrapped in connective tissue called
   a. gray matter.
   b. meninges.
   c. gyri.
   d. sulcus.
   e. neopallium.

8. The part of the brain that coordinates muscular activity is the
   a. cerebrum.
   b. cerebellum.
   c. myelencephalon.
   d. medulla oblongata.
   e. midbrain.

9. If you are having strong sexual feelings one minute and enraged over little or nothing the next minute, chances are very good that you have just had a very active
   a. frontal lobe.
   b. limbic system.
   c. thalamus.
   d. reticular activating system (RAS).
   e. cerebellum.

10. The part of the mammalian brain that integrates information about posture and muscle tone is the
    a. cerebrum.
    b. cerebellum.
    c. myelencephalon.
    d. medulla oblongata.
    e. midbrain.

11. When you are very sleepy, there's a good chance that
    a. the meninges are damaged.
    b. beta waves are diminishing.
    c. you have convolutions, but not gyri.
    d. your reticular activating system is not very active.
    e. your brain lacks gray matter.

12. If only one lateral temporal lobe is damaged, one might expect
    a. blindness in one eye.
    b. total blindness.
    c. loss of hearing in one ear.
    d. partial loss of hearing in both ears.
    e. inability to detect odors.

13. Skeletal muscles are controlled by the
    a. parietal lobes.
    b. frontal lobes.
    c. central sulcus.
    d. temporal lobes.
    e. cerebral ganglion.

14. In vertebrates, the cerebrum is typically
    a. divided into two hemispheres.
    b. mostly gray matter.
    c. mainly cell bodies and some sensory neurons.
    d. mostly white matter.
    e. mainly axons connecting parts of the brain.

VISUAL FOUNDATIONS

Color the parts of the illustration below as indicated.

RED ☐ cerebral hemisphere
GREEN ☐ cerebellum
YELLOW ☐ medulla
BLUE ☐ olfactory bulb, olfactory tract, olfactory lobe
ORANGE ☐ optic lobe
BROWN ☐ epiphysis
TAN ☐ corpus striatum

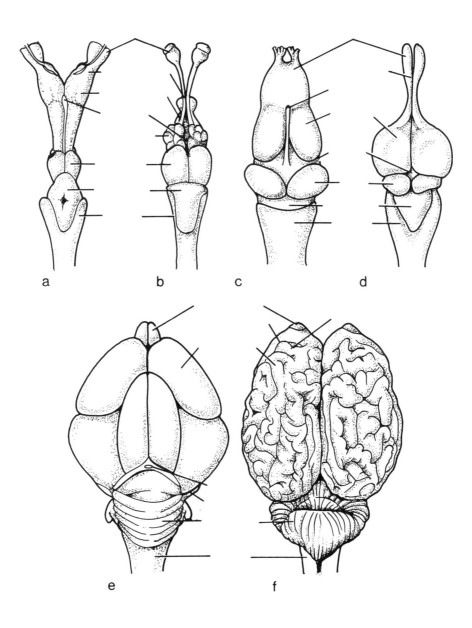

a   b   c   d

e   f

Color the parts of the illustration below as indicated.

RED   ❑   sensory neuron

GREEN   ❑   association neuron

YELLOW   ❑   receptor

BLUE   ❑   motor neuron

ORANGE   ❑   nerve cell body of sensory neuron

BROWN   ❑   muscle

TAN   ❑   central nervous system

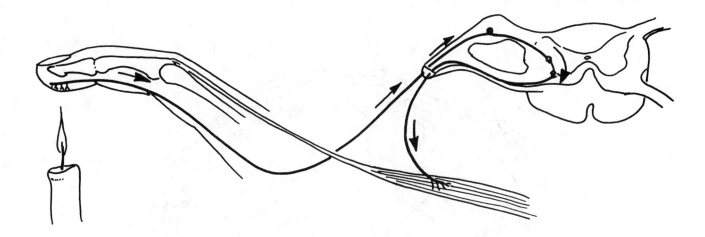

Color the parts of the illustration below as indicated.

RED ❑ diencephalon
GREEN ❑ pituitary
YELLOW ❑ spinal cord
BLUE ❑ ventricle
ORANGE ❑ medulla and pons
BROWN ❑ midbrain
TAN ❑ cerebellum
PINK ❑ corpus callosum and fornix
VIOLET ❑ cerebrum

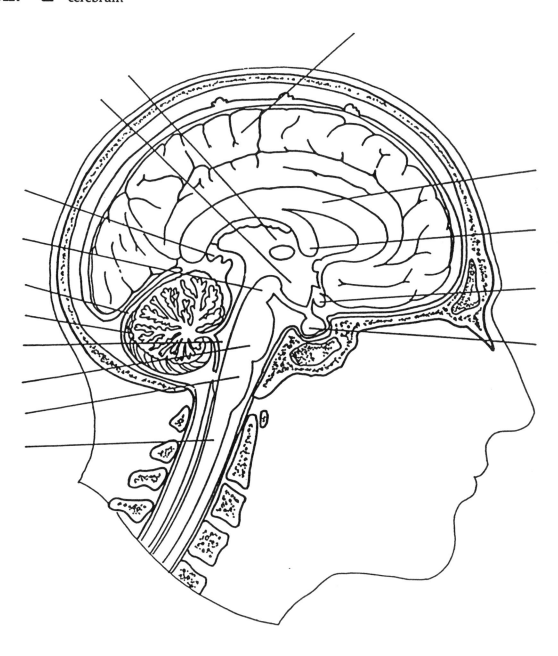

CHAPTER 41

□

Sense Organs

Types of sense organs
How sense organs work
Mechanoreceptors
Chemoreceptors
Thermoreceptors
Electroreceptors
Photoreceptors

Sense organs are specialized structures with receptor cells that detect changes in the internal or external environment and transmit the information to the nervous system. Receptor cells may be neuron endings or specialized cells in contact with neurons. Sense organs are classified according to the location of the stimuli and according to the types of energy to which they respond. They work by absorbing energy and converting it into electrical energy that depolarizes the receptor cell. When the state of depolarization reaches a threshold level, an impulse is generated in the axon, thereby transmitting the information to the CNS. Many receptors do not continue to respond at the initial rate even if the stimulus maintains its intensity. Such adaptation permits an animal to ignore persistent unpleasant or unimportant stimuli. Sense organs that respond to mechanical energy include the various tactile receptors in the skin, the lateral line organs in fish, the receptors that respond continuously to tension and movement in muscles and joints, the receptors that enable organisms to maintain their orientation, and the receptors responsible for hearing. Sense organs that respond to chemical energy are responsible for the sense of taste and smell. Those that respond to heat and cold provide important cues about body temperature, and help some animals locate a warm-blooded host. Those that detect electrical energy are found in some fish; they can detect electric currents in water, and in some species can generate a shock for defense or to stun prey. Sense organs that respond to light energy are responsible for vision.

CHAPTER OUTLINE AND CONCEPT REVIEW (Fill in the blanks)

I. INTRODUCTION

■ 1 (a)_____ in a sense organ detect changes in the environment.
These cells may be specialized for the task or merely (b)_____ endings.

■ 2 The five human senses that have been recognized traditionally are (a)_____
_____. (b)_____
_____ are now also recognized as
human senses.

II. SENSE ORGANS CAN BE CLASSIFIED ACCORDING TO THE STIMULI TO WHICH THEY RESPOND

■ 3 Sense receptors can be classified by their location. Those that detect stimuli in the external environment are called (a)_____; (b)_____ are located within joints, muscles, and tendons; and (c)_____ detect change in the internal environment.

■ 4 Classification of sense organs can also be based on the type of stimuli detected. For example, (a)_____ respond to gravity, pressure, and stretching; (b)_____ respond to electrical energy; heat and cold are detected by (c)_____; and chemicals provoke responses in (d)_____.

III. SENSE ORGANS WORK BY PRODUCING RECEPTOR POTENTIALS

■ 5 Various forms of energy are transduced into (a)_____ energy by receptor cells that then produce a (b)_____ potential.

■ 6 A receptor potential is a kind of (a)_____ potential. If it can (b)_____ the axon hillock to threshold, it can initiate an (c)_____ that will travel along the (d)_____ afferent neuron to the central nervous system.

■ 7 In general, the stronger the stimulus and the longer it lasts, the (a)_____ (greater or lower?) the amplitude and duration of the (b)_____ potential.

■ 8 A strong stimulus will generate _____(more or less?) action potentials than a weak stimulus.

A. Sensation depends on transmission of a "coded" message

■ 9 Sensation is determined by the part of the _____ that receives the message from the sense organ.

■ 10 Impulses from sense organs are encoded into messages based upon the (a)_____ _____ of neurons (or fibers) carrying the message and the (b)_____ of action potentials carried by the neuron (fiber).

B. Receptors adapt to stimuli

■ 11 When a receptor continues to receive a constant stimulus, yet the sensory neuron becomes less responsive to that stimulus, the receptor is demonstrating _____ _____.

IV. MECHANORECEPTORS RESPOND TO TOUCH, PRESSURE, GRAVITY, STRETCH, OR MOVEMENT

■ 12 Mechanoreceptors are activated when they are mechanically _____.

A. Touch receptors are located in the skin

■ 13 The tactile receptors in the skin are _____ that respond to mechanical displacement of hairs or of the receptor cells themselves.

■ 14 In human skin, (a)_____ are responsible for the perception of deep pressure, (b)_____ are responsible for touch, and free nerve endings perceive (c)_____.

B. Many invertebrates have gravity receptors called statocysts

■ 15 (a)_____ are tiny granules that stimulate (b)_____ when pulled down by gravity. They occur in sense organs called statocysts.

C. Lateral line organs supplement vision in fish

■ 16 Lateral line organs consist of a (a)_____ lined with (b)_____ that runs the length of the animal's lateral surface. Water disturbances move the (c)_____ on the end of hairs in the receptor cells, generating an electrical response.

D. Proprioceptors help coordinate muscle movement

■ 17 _____ are proprioceptors that respond continuously to tension and movement in the muscles and joints.

E. The vestibular apparatus of the vertebrate ear functions to maintain equilibrium

■ 18 Gravity detectors in the form of ear stones called (a)_____ are housed in the (b)_____ of the vestibular apparatus.

■ 19 Turning movements, referred to as (a)_____, cause movement of a fluid called (b)_____ in the semicircular canals, which in turn stimulate the receptor cells of (c)_____.

F. Auditory receptors are located in the cochlea

■ 20 Auditory receptors in an inner ear structure of birds and mammals, called the (a)_____, contain (b)_____ that can detect pressure waves.

■ 21 In terrestrial vertebrates, sound waves first initiate vibrations in the eardrum, or (a)_____. Three tiny earbones, the (b)_____ _____, transmit the vibration to fluids in the inner ear through an opening called the (c)_____.

■ 22 Fluids in the cochlea, in response to vibrations from the oval window, initiate vibrations in the (a)_____ which in turn cause stimulation of hair cells in the (b)_____. The hair cells initiate impulses in the (c)_____.

V. CHEMORECEPTORS ARE ASSOCIATED WITH THE SENSES OF TASTE AND SMELL

■ 23 (a)_____ refers to the sense of taste, and (b)_____ applies to the sense of smell, both of which are highly sensitive (c)_____ systems.

A. Taste buds are the organs of taste in mammals

■ 24 The four basic tastes are _____.

B. The olfactory epithelium is responsible for the sense of smell

■ 25 The (a)_____ of terrestrial vertebrates is located on the roof of the nasal cavity. It reacts to as many as (b)_____ different odors.

VI. THERMORECEPTORS ARE SENSITIVE TO HEAT

■ 26 In two types of snakes, the _____, thermoreceptors are used to locate prey.

■ 27 Thermoreceptors that detect internal temperature changes in mammals are found in the _____ of the brain.

VII. ELECTRORECEPTORS DETECT ELECTRICAL CURRENTS IN WATER

VIII. PHOTORECEPTORS USE PIGMENTS TO ABSORB LIGHT

■ 28 Cephalopods, arthropods, and vertebrates all have photosensitive pigments called _____ in their eyes.

A. Eyespots, simple eyes, and compound eyes are found among invertebrates

■ 29 Planaria have _____, which are photoreceptive organs capable of differentiating the intensity of light.

■ 30 Photoreceptors in simple eyes detect (a)_____, but these eyes do not form images effectively. Effective image formation and interpretation is called (b)_____, and it requires a (c)_____ capable of focusing.

■ 31 The compound eye in insects and crustaceans consists of _____, which collectively form a mosaic image.

B. Vertebrate eyes form sharp images

■ 32 The tough outer coat of the mammalian eye, called the (a)_____, helps maintain the (b)_____ of the eyeball. The anterior, transparent part of this coat, which allows the entry of light, is called the (c)_____.

■ 33 Light is focused by the (a)_____ on the (b)_____, which contains the photoreceptive cells. (c)_____, containing the pigment (d)_____, are concentrated in the periphery of the retina. They function best in dim light and perceive black and white.

■ 34 (a)_____ are responsible for color vision and are densest in the (b)_____ in the center of the retina.

KEY TERMS

Frequently Used Prefixes and Suffixes: Use combinations of these prefixes and suffixes to generate terms for the definitions below.

Prefixes	The Meaning	Suffixes	The Meaning
endo-	within	-lith	stone
oto-	ear		
proprio-	one's own		

Prefix	Suffix	Definitions
_____	-ceptor	1. Sense organs within muscles, tendons, and joints that enable the animal to perceive the position of its own body parts.
_____ _____		2. Calcium carbonate "stone" in the inner ear of vertebrates.
_____	-lymph	3. Fluid within the semicircular canals of the vertebrate ear.

Other Terms You Should Know

accessory cell
action potential
adaptation
ampulla (-ae)
angular acceleration
anterior cavity (-ties)
aqueous fluid
basilar membrane
binocular vision
chemoreceptor
choroid
ciliary body (-dies)
ciliary muscle
cochlea
cochlear duct
compound eye
cone
cornea
crista (-ae)
cupula
depolarize
eardrum
electroreceptor
external auditory
   meatus (-ses)
exteroceptor
eyespot
facet
flicker
fovea (-ae)
Golgi tendon organ

gravity receptor
harmonics
incus (-udes)
interoceptor
iris (-ses)
joint receptor
labyrinth
lateral line organ
linear acceleration
malleus (-ei)
mechanoreceptor
metarhodopsin II
mosaic picture
muscle spindle
ocellus (-li)
olfactory epithelium (-ia)
ommatidium (-ia)
optic nerve
organ of Corti
oval window
overtones
perilymph
phasic receptor
photon
photoreceptor
plane of polarization
posterior cavity (-ties)
pupil
quantum (-ta)
receptor cell
receptor potential

retina (-as)(-ae)
retinal
rhabdome
rhodopsin
rod
round window
saccule
sclera
semicircular canal
sense organ
sensory adaptation
smell
stapes
statocyst
statolith
tactile receptor
taste
taste bud
tectorial membrane
thermoreceptor
tonic sense organ
transduce
tympanic canal
tympanic membrane
utricle
vestibular apparatus
vestibular canal
vision
vitreous body (-dies)

SELF TEST

Multiple Choice: Some questions may have more than one correct answer.

1. The increasing inability with age to focus light from near objects is due to
   a. deformity of the retina.
   b. deterioration of rhodopsin.
   c. inability of the lens to become roundish.
   d. loss of lens elasticity.
   e. decreasing sensitivity of rods and/or cones.

2. Movement in ligaments is detected by
   a. muscle spindles.
   b. joint receptors.
   c. Golgi tendon organs.
   d. proprioceptors.
   e. tonic sense organs.

3. Variations in the quality of sound are recognized by the
   a. number of hair cells stimulated.
   b. pattern of hair cells stimulated.
   c. frequency of nerve impulses.
   d. intensity of stimulation.
   e. amplitude of response.

4. Receptors within muscles, tendons, and joints that perceive position and body orientation are
   a. exteroceptors.
   b. proprioceptors.
   c. interoceptors.
   d. mechanoreceptors.
   e. electroreceptors.

5. One means of coding sensory perceptions may be related to the frequency of the action potentials in a neuron. This mechanism of coding is known as
   a. temporal patterning.
   b. spatial localization.
   c. cross-fiber patterning.
   d. fiber specificity.
   e. adaptation.

6. Sense organs that detect changes in pH, osmotic pressure, and temperature within body organs are
   a. exteroceptors.
   b. proprioceptors.
   c. interoceptors.
   d. mechanoreceptors.
   e. electroreceptors.

7. The state of depolarization in a receptor neuron that is caused by a stimulus is the
   a. depolarization potential.
   b. resting potential.
   c. receptor potential.
   d. action potential.
   e. threshold potential.

8. Receptors that respond only when stimulated by motion are known as
   a. mechanoreceptors.
   b. tonic receptors.
   c. labyrinths.
   d. phasic receptors.
   e. proprioceptors.

9. The membrane at the opening of the inner ear that is in contact with the stirrup is in the
   a. tectorial membrane.
   b. basilar membrane.
   c. oval window.
   d. eardrum.
   e. round window.

VISUAL FOUNDATIONS

Color the parts of the illustrations below as indicated.

RED ☐ hair cell

GREEN ☐ cochlear duct

YELLOW ☐ cochlear nerve

BLUE ☐ tectorial membrane

ORANGE ☐ cilia

BROWN ☐ tympanic canal and vestibular canal

TAN ☐ bone

PINK ☐ basilar membrane

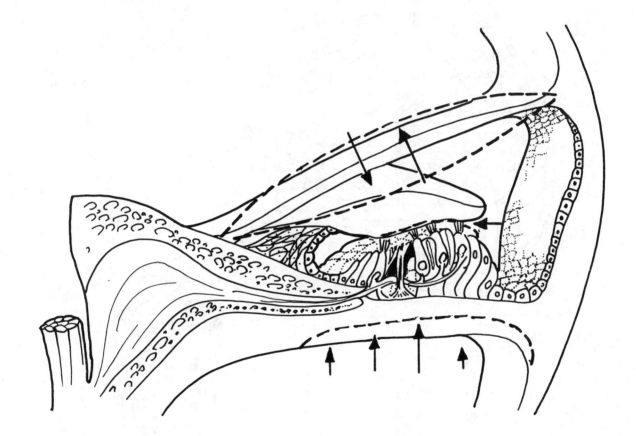

Color the parts of the illustrations below as indicated.

RED	☐	ommatidium
GREEN	☐	facets
YELLOW	☐	optic nerve
BLUE	☐	retinular cell
ORANGE	☐	optic ganglion
BROWN	☐	cornea
TAN	☐	lens
PINK	☐	crystalline cone
VIOLET	☐	rhabdome

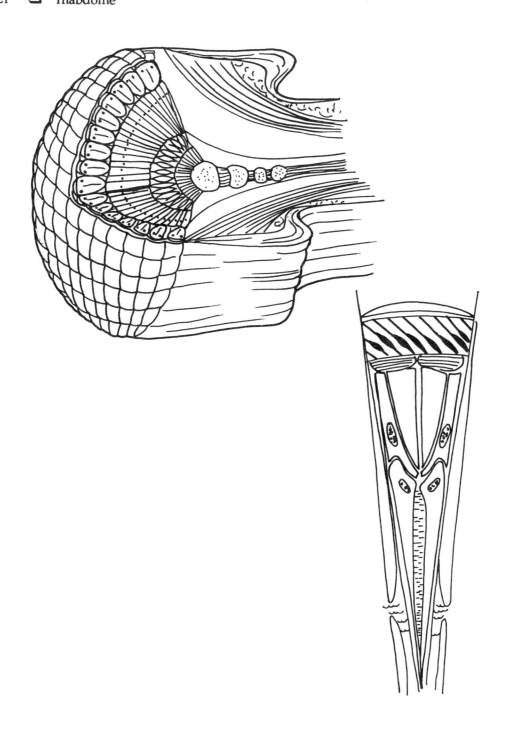

Color the parts of the illustrations below as indicated.

RED   ❑   light ray
GREEN   ❑   bipolar cell
YELLOW   ❑   optic nerve fiber
BLUE   ❑   cone cell
ORANGE   ❑   horizontal cell and amacrine cell
BROWN   ❑   pigmented epithelium
TAN   ❑   choroid layer and sclera
PINK   ❑   rod cell
VIOLET   ❑   ganglion cell

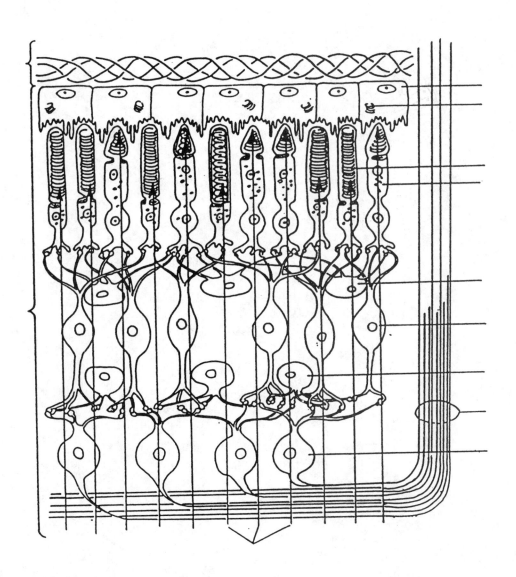

CHAPTER 42

Internal Transport

OUTLINE

Invertebrates with no circulatory system
Open circulatory systems
Closed circulatory systems
Functions of the vertebrate circulatory system
The blood of vertebrates
Types of blood vessels
Evolution of the vertebrate heart
The human heart
Blood pressure
The circulation of blood in mammals
The lymphatic system

Some animals are so small that diffusion alone is effective at transporting materials to and from their cells. Larger animals, however, require a circulatory system, not only to transport materials, but also to help maintain fluid balance, to defend the body against invading microorganisms, and, in some animals, to regulate body temperature. Some invertebrates have an open circulatory system in which blood is pumped from a heart into vessels that have open ends. Blood spills out of the vessels into the body cavity and baths the tissues directly. The blood then passes back into the heart, either directly through openings in the heart (arthropods) or indirectly, passing first through open vessels that lead to the heart (some mollusks). Other animals have a closed circulatory system in which blood flows through a continuous circuit of blood vessels. The walls of the smallest vessels are thin enough to permit exchange of materials between the vessels and the extracellular fluid that baths the tissue cells. The vertebrate circulatory system consists of a muscular heart that pumps blood into a system of blood vessels: the arteries that carry blood away from the heart to the capillaries, the capillaries where the exchange of materials between the blood and tissues occurs, and the veins that carry the blood back to the heart. Vertebrate blood consists of liquid plasma in which red blood cells, white blood cells, and platelets are suspended. The red blood cells transport oxygen, the white blood cells defend the body against disease organisms, and the platelets function in blood clotting. The vertebrate heart consists of one or two chambers that receive blood, and one or two that pump blood into the arteries. In birds and mammals, there are two circuits of blood vessels. One pumps blood from the heart to the lungs, and the other, from the heart to every tissue in the body. The lymphatic system, a subsystem of the circulatory system, collects extracellular fluid and returns it to the blood. It also functions to defend the body against disease organisms and to absorb lipids from the digestive tract.

CHAPTER OUTLINE AND CONCEPT REVIEW (Fill in the blanks)

I. INTRODUCTION

■ 1 _____ are animal groups that have no specialized circulatory structures.

■ 2 The bodies of complex animals need a (a)_____ system to transport materials because simple (b)_____, which distributes materials adequately in thin-bodied, small animals, cannot supply all cells in larger animals.

■ 3 Components of a circulatory system include a fluid, the (a)_____, that is usually pumped by a (b)_____ through a system of spaces or blood (c)_____.

■ 4 An (a)_____ system consists of a blood cavity, or (b)_____, a heart, and open-ended vessels. In a (c)_____ _____ system, blood is contained within the heart and a continuous system of blood vessels.

II. SOME INVERTEBRATES HAVE NO CIRCULATORY SYSTEM

■ 5 The _____ of cnidarians serves as both a circulatory organ and a digestive organ.

III. MANY INVERTEBRATES HAVE AN OPEN CIRCULATORY SYSTEM

■ 6 Blood is called (a)_____ in animals with open circulatory systems because blood is indistinguishable from (b)_____ fluid.

■ 7 (a)_____ have an open circulatory system in which hemolymph flows into large spaces, called (b)_____, of the hemocoel.

IV. SOME INVERTEBRATES HAVE A CLOSED CIRCULATORY SYSTEM

■ 8 (a)_____, a blood pigment that imparts a bluish color to hemolymph in some invertebrates, contains the metal (b)_____.

■ 9 Annelids have a (a)_____ blood vessel that conducts blood anteriorly, a (b)_____ blood vessel that conducts blood posteriorly, and (c)_____(#?) "hearts" that connect these two vessels.

■ 10 Hemoglobin is found in the _____ of earthworm blood.

V. THE CLOSED CIRCULATORY SYSTEM OF VERTEBRATES IS ADAPTED TO CARRY OUT A VARIETY OF FUNCTIONS

■ 11 The circulatory system of all vertebrates is a/an (a)_____(open or closed?) system with a muscular (b)_____ that pumps (c)_____ through blood vessels.

■ 12 The vertebrate circulatory system transports (a)_____ _____. Other functions include (b)_____, as well as thermoregulation in (c)_____.

VI. VERTEBRATE BLOOD CONSISTS OF PLASMA, BLOOD CELLS, AND PLATELETS

■ 13 Human blood consists of (a)_____ suspended in (b)_____.

A. Plasma is the fluid component of blood

■ 14 Plasma is in dynamic equilibrium with (a)_____ fluid, which bathes cells, and with (b)_____ fluid within cells.

■ 15 (a)_____, a protein involved in blood clotting, (b)_____, which are involved in immunity, and albumin are all (c)_____. (d)_____ in plasma transport triglycerides and cholesterol.

B. Red blood cells transport oxygen

■ 16 Red blood cells (erythrocytes) are specialized to transport (a)_____. They are produced in (b)_____, contain the respiratory pigment (c)_____, and live about (d)_____ days.

C. White blood cells defend the body against disease organisms

■ 17 White blood cells (leukocytes) called (a)_____ are the principle phagocytic cells in blood. These cells, together with (b)_____ _____ are called (c)_____ because of the distinctive granules in their cytoplasm.

■ 18 Agranular leukocytes are (a)_____ which secrete antibodies, and (b)_____ which develop into phagocytic (c)_____.

D. Platelets function in blood clotting

■ 19 Platelets, also known as (a)_____, play an important role in the control of bleeding, or (b)_____.

VII. VERTEBRATES HAVE THREE MAIN TYPES OF BLOOD VESSELS

■ 20 (a)_____ carry blood away from the heart; (b)_____ return blood to the heart.

■ 21 The wall of blood vessels is made up of the (a)_____, which contains endothelium; the (b)_____ with smooth muscle; and the (c)_____, which contains many elastic and collagen fibers.

■ 22 The exchange of nutrients and waste products takes place across the thin wall of _____.

VIII. THE EVOLUTION OF THE VERTEBRATE HEART CULMINATED IN A FOUR-CHAMBERED HEART AND CIRCULATION THROUGH A DOUBLE CIRCUIT OF VESSELS

■ 23 Chambers in the heart that pump blood into arteries are called (a)_____, and chambers that receive blood from veins are called (b)_____.

A. The fish heart has a single atrium and single ventricle

■ 24 Fish have a (a)_____(#?)-chambered heart. In this structure, veins first empty into the (b)_____, an accessory chamber that then pumps blood into the atrium. Blood leaving the ventricle empties into the (c)_____, from which it passes into the ventral aorta.

B. Amphibians have a three-chambered heart

■25 In the (a)_____(#?)-chambered amphibian heart, the (b)_____ pumps venous blood into the right atrium, and the (c)_____ helps to separate oxygen-rich blood from oxygen-poor blood.

C. In the reptilian heart the wall between the ventricles is incomplete

D. Birds and mammals have a four-chambered heart

■26 Because oxygen-rich blood can be isolated from oxygen-poor blood in a four chambered heart, tissues receive (a)_____(more or less?) oxygen, and can maintain a (b)_____(higher or lower?) metabolic rate. Animals with such a heart can maintain a constant body temperature, and are therefore called (c)_____.

IX. THE HUMAN HEART IS MARVELOUSLY ADAPTED FOR PUMPING BLOOD

■27 The human heart is enclosed by a (a)_____, creating a (b)_____ filled with fluid that serves to reduce friction.

■28 The (a)_____ septum separates the two ventricles, whereas the (b)_____ separates the two atria in a four-chambered heart.

■29 (a)_____(#?) valves control the flow of blood in the human heart. The (b)_____ prevent backflow into the atria during ventricular contractions. The valve on the right is also known as the (c)_____, and the valve on the left is known as the (d)_____. The (e)_____ "guard" the exits from the heart.

A. Each heartbeat is initiated by a pacemaker

■30 The conduction system of the heart contains a pacemaker called the (a)_____, an (b)_____ which links the atria to the (c)_____, which divides, sending a branch into each ventricle.

■31 The portion of the (a)_____ cycle in which contraction occurs is called (b)_____, and the portion in which relaxation occurs is called (c)_____.

B. Two main heart sounds can be distinguished

■32 Of the two main heart sounds, the (a)_____ occurs first and is associated with closure of the (b)_____, and the (c)_____ is produced by the closure of the (d)_____, which marks the beginning of ventricular (e)_____.

C. The electrical activity of the heart can be recorded

■33 The _____ is a written record of the electrical activity of the heart.

D. Cardiac output varies with the body's need

■34 (a)_____ is the volume of blood pumped by one ventricle in one minute. It is determined by multiplying the number of ventricular beats per minute, called the (b)_____, times the amount of blood pumped by one ventricle per beat, a value referred to as the (c)_____. The normal volume per minute value is about (d)_____, although it can vary dramatically.

E. Stroke volume depends on venous return

■35 Stroke volume is determined primarily by (a)_____ return, although (b)_____ mechanisms also have an effect.

F. Heart rate is regulated by the nervous system

■36 Although the heart can beat independently, it is regulated by _____ in the medulla.

X. BLOOD PRESSURE DEPENDS ON BLOOD FLOW AND RESISTANCE TO BLOOD FLOW

■37 High blood pressure is called (a)_____. It can be caused by an (b)_____(increase or decrease?) in blood volume such as frequently occurs with a high dietary intake of (c)_____.

■38 The most important factor in determining peripheral resistance to blood flow is the _____.

A. Blood pressure is highest in arteries

■39 Veins are low-pressure vessels that contain _____ to prevent blood backflow.

B. Blood pressure is carefully regulated

■40 _____ are sense organs in the walls of some arteries that detect (b)_____ and send the perceived information to (c)_____ in the medulla.

XI. IN BIRDS AND MAMMALS BLOOD IS PUMPED THROUGH A PULMONARY AND A SYSTEMIC CIRCUIT

■41 A double circulatory system consists of a (a)_____ between the heart and lungs, and (b)_____ between the heart and body.

A. The pulmonary circulation oxygenates the blood

■42 Pulmonary veins carry oxygen-(a)_____(rich or poor?) blood to the (b)_____ atrium of the heart.

B. The systemic circulation delivers blood to all of the tissues

■43 Arteries in the systemic circuit branch from the (a)_____, the largest artery in the body, and serve major body areas. For example, the carotid arteries feed the (b)_____, subclavian arteries supply the (c)_____, and iliac arteries feed the (d)_____.

■44 Veins returning blood from the head and neck empty into the large (a)_____ _____ vein, while venous return from the lower body empties into the (b)_____ vein.

C. The coronary circulation delivers blood to the heart

■ 45 Atherosclerosis leads to (a)_____ in which the heart muscle does not receive sufficient blood. (b)_____ is a very serious form of ischemic heart disease.

D. Four arteries deliver blood to the brain

E. The hepatic portal system delivers nutrients to the liver

XII. THE LYMPHATIC SYSTEM IS AN ACCESSORY CIRCULATORY SYSTEM

■ 46 Considered an accessory circulatory system in vertebrates, the lymphatic system returns (a)_____ to the blood. It is also involved with immune mechanisms and absorbs (b)_____ from the digestive tract.

A. The lymphatic system consists of lymphatic vessels and lymph tissue

■ 47 (a)_____, the fluid contained within lymphatic vessels, is formed from (b)_____, filtered by (c)_____, and emptied into (d)_____ veins by the (e)_____ duct on the left side and the (f)_____ duct on the right.

B. The lymphatic system plays an important role in fluid homeostasis

■ 48 _____, the excessive accumulation of interstitial fluid, can result if lymph vessels are obstructed.

KEY TERMS

Frequently Used Prefixes and Suffixes: Use combinations of these prefixes and suffixes to generate terms for the definitions below.

Prefixes	The Meaning	Suffixes	The Meaning
baro-	pressure	-cardium	heart
erythro-	red	-coel	cavity
hemo-	blood	-cyte	cell
leuk(o)-	white (without color)	-lunar	moon
neutro-	neutral	-phil	loving, friendly, lover
peri-	about, around, beyond		
semi-	half		
vaso-	vessel		

Prefix	Suffix	Definitions
_____	_____	1. The blood cavity that comprises the open circulatory system of arthropods and some mollusks.
_____	-cyanin	2. The copper-containing blood pigment in some mollusks and arthropods.
_____	_____	3. Red blood cell.
_____	_____	4. General term for all of the body's white blood cells.

_____ _____   5. The principal phagocytic cell in the blood that has an affinity for neutral dyes.

eosino-   _____   6. WBC with granules that have an affinity for eosin.

baso-   _____   7. WBC with granules that have an affinity for basic dyes.

_____ -emia   8. A form of cancer in which WBCs multiply rapidly within the bone marrow.

_____ -constriction   9. Constriction of a blood vessel.

_____ -dilatation   10. Relaxation of a blood vessel.

_____ _____   11. The tough connective tissue sac around the heart.

_____ _____   12. Shaped like a half-moon.

_____ -receptor   13. Pressure receptor.

Other Terms You Should Know

acetylcholine
adenoid
agranular leukocyte
albumin
anemia
angina pectoris
angiotensin
aorta
aortic valve
arteriole
artery (-ries)
atherosclerosis
atrioventricular node
atrioventricular valve
atrium (-ia)
auricle
AV bundle
AV node
AV valve
blood
blood flow
blood pressure
capillary (-ries)
cardiac center
cardiac cycle
cardiac output
carotid artery
cerebrovascular accident
chordae tendineae
circle of Willis
circulatory system
closed circulatory system
conus arteriosus
coronary artery
coronary sinus
cusp

CVA
diastole
differential WBC count
ECG
edema
EKG
electrocardiogram
elephantiasis
endothelium
epinephrine
fibrin
fibrinogen
filariasis
foramen ovale
fossa ovalis
functional syncytium
gamma globulin
globulin
granular leukocyte
HDL
heart
heart murmur
hemoglobin
hemolymph
hemolytic anemia
hemophiliac
hemostasis
heparin
hepatic portal system
hepatic portal vein
hepatic vein
high-density lipoprotein
histamine
hypertension
iliac artery
iliac vein

infarct
inferior vena cava
interatrial septum
intercalated disk
interstitial fluid
interventricular septum
ischemic
ischemic heart disease
jugular vein
LDL
lipoprotein
low-density lipoprotein
lymph
lymph node
lymph nodule
lymph tissue
lymphatic
lymphatic system
lymphatic vessel
lymphocyte
macrophage
mesenteric artery
metarteriole
MI
mitral valve
monocyte
myocardial infarction
norepinephrine
open circulatory system
pacemaker
papillary muscle
pericardial cavity
peripheral resistance
plasma
platelet
precapillary sphincter

prothrombin
pulmonary artery
pulmonary circulation
pulmonary trunk
pulmonary valve
pulmonary vein
pulse
Purkinje fiber
red blood cell
renal artery
renal vein
renin
right lymphatic duct
SA node
semilunar valve

septum (-ta)
serum (-ra)
sinoatrial node
sinus (-ses)
sinus venosus
spleen
Starling's law of the heart
stroke volume
subclavian vein
superior vena cava
systemic circulation
systole
thoracic duct
thrombin
thrombocyte

thrombus (-bi)
thymus gland
tissue fluid
tonsil
tunica intima
valve
vasomotor center
vein
ventricle
ventricular fibrillation
venule
WBC count
white blood cell

SELF TEST

Multiple Choice: Some questions may have more than one correct answer.

1. Prothrombin
   a. is produced in liver.
   b. is a precursor to thrombin.
   c. catalyzes conversion of fibrinogen to fibrin.
   d. requires vitamin K for its production.
   e. is a toxic by-product of metabolism.

2. The thrombocytes of mammals are
   a. WBCs.
   b. RBCs.
   c. called platelets.
   d. anucleate.
   e. derived from megakaryocytes.

3. Erythrocytes are
   a. also called RBCs.
   b. spherical.
   c. produced in bone marrow.
   d. one kind of leucocyte.
   e. carriers of oxygen.

4. The white blood cells that contain an anticlotting chemical
   a. contain histamine.
   b. carry oxygen.
   c. stain blue with basic dyes.
   d. transport carbon dioxide.
   e. contain hemoglobin.

5. Monocytes
   a. are RBCs.
   b. are WBCs.
   c. are produced in the spleen.
   d. are large cells.
   e. can become macrophages.

6. An antibody is
   a. a plasma lipid.
   b. in serum.
   c. a protein.
   d. a gamma globulin.
   e. involved in clotting.

7. Fibrinogen is
   a. a plasma lipid.
   b. in serum.
   c. a protein.
   d. a gamma globulin.
   e. involved in clotting.

8. Very small vessels that deliver oxygenated blood to capillaries are called
   a. arterioles.
   b. venules.
   c. arteries.
   d. veins.
   e. setum vessels.

9. The blood-filled cavity of an open circulatory system is called the
   a. lymphocoel.
   b. hemocoel.
   c. atracoel.
   d. sinus.
   e. ostium.

10. Blood enters the right atrium from the
    a. right ventricle.
    b. inferior vena cava.
    c. superior vena cava.
    d. pulmonary artery.
    e. jugular vein.

11. Pulmonary arteries carry blood that is
    a. low in oxygen.
    b. low in CO_2.
    c. high in oxygen.
    d. high in CO_2.
    e. on its way to the lungs.

12. Phenomena that tend to be associated with hypertension include
    a. obesity.
    b. increased vascular resistance.
    c. increased workload on the heart.
    d. decrease in ventricular size.
    e. deteriorating heart function.

13. Ventricles receive stimuli for contraction directly from the
    a. SA node.
    b. AV node.
    c. Perkinje fibers.
    d. atria.
    e. intercalated disks.

14. A heart murmur may result when
    a. the heart is punctured.
    b. systole is too high.
    c. semilunar valves don't close tightly.
    d. diastole is too low.
    e. blood backflows into ventricles.

15. If a person's heart rate is 65 and one ventricle pumps 75 ml with each contraction, what is that person's cardiac output in liters?
    a. 3.575
    b. 4.000
    c. 4.525
    d. 4.875
    e. 5.350

16. The semilunar valve(s) is/are found between a ventricle and
    a. an atrium.
    b. another ventricle.
    c. the aorta.
    d. the pulmonary vein.
    e. the pulmonary artery.

17. The pericardium surrounds
    a. atria only.
    b. ventricles only.
    c. blood vessels.
    d. the heart.
    e. clusters of some blood cells.

18. One would expect an ostrich to have a heart most like a
    a. lizard.
    b. crocodile.
    c. earthworm.
    d. human.
    e. guppy.

19. Three chambered hearts generally consist of the following number of atria/ventricles:
    a. 2/1.   d. 0/3.
    b. 1/2.   e. 3/0.
    c. 1/1 and an accessory chamber.

20. In general, blood circulates through vessels in the following order:
    a. veins, venules, capillaries, arteries, arterioles.   d. arteries, arterioles, capillaries, venules, veins.
    b. venules, veins, capillaries, arterioles, arteries.   e. arteries, arterioles, capillaries, veins, venules.
    c. arteries, arterioles, venules, capillaries, veins.

VISUAL FOUNDATIONS

Color the parts of the illustration below as indicated.

RED   ❑   artery and arrows indicating flow of blood away from the heart

GREEN   ❑   capillary bed

YELLOW ❑   lymph capillaries

BLUE   ❑   vein and arrows indicating flow of blood to the heart

ORANGE ❑   lymphatic and lymph duct

BROWN ❑   lymph node

TAN   ❑   loose connective tissue

PINK   ❑   arteriole

VIOLET ❑   venule

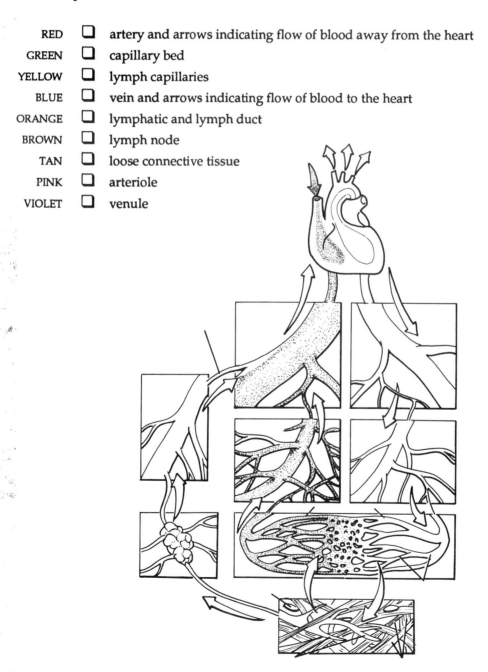

Color the parts of the illustration below as indicated.

RED ❑ arteriole
GREEN ❑ tunica intima
YELLOW ❑ tunica media
BLUE ❑ venule
ORANGE ❑ precapillary sphincter
BROWN ❑ smooth muscle cell
TAN ❑ tunica adventitia
PINK ❑ capillary

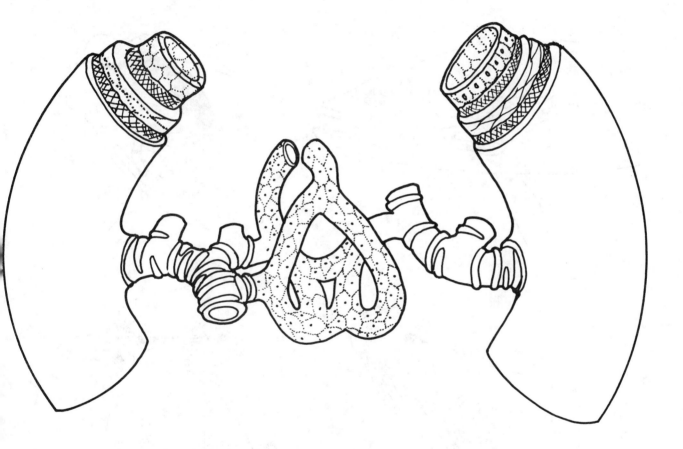

Color the parts of the illustration below as indicated.

RED ☐ aorta and pulmonary artery
GREEN ☐ ventricle
YELLOW ☐ atrium
BLUE ☐ vein from body
ORANGE ☐ valve
BROWN ☐ partition
TAN ☐ conus
PINK ☐ sinus venosus

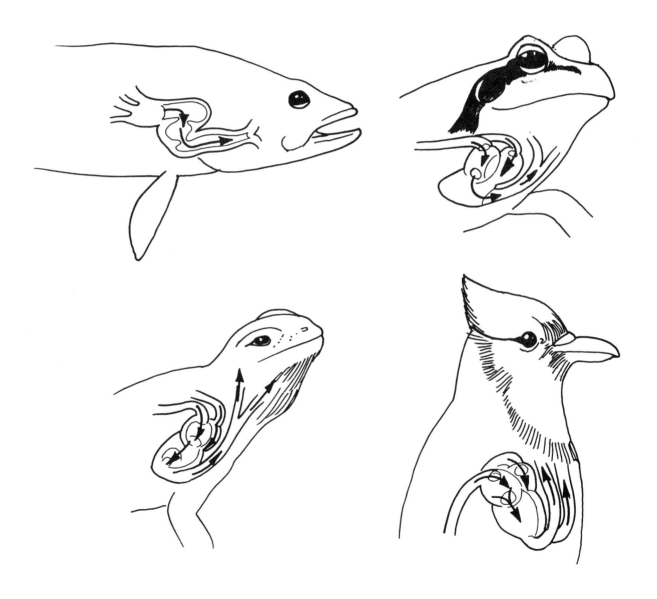

Color the parts of the illustration below as indicated.

RED ❑ chordae tendineae
GREEN ❑ tricuspid valve
YELLOW ❑ pulmonary valve
BLUE ❑ mitral valve
ORANGE ❑ aortic semilumar valve
BROWN ❑ papillary muscle
TAN ❑ atrium
PINK ❑ ventricle

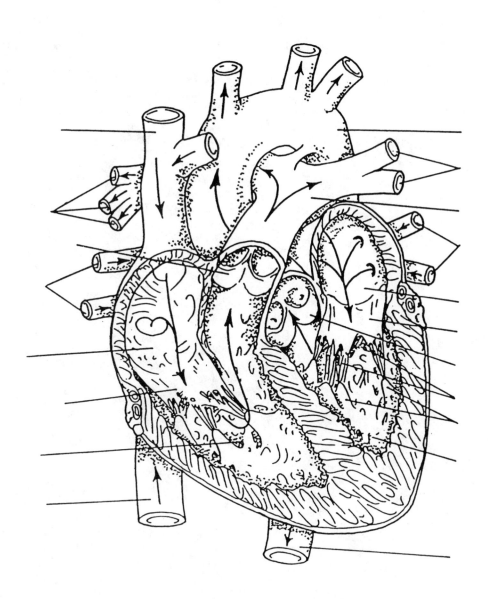

❑

CHAPTER 43

Internal Defense

OUTLINE

Internal defense mechanisms in invertebrates
Internal defense mechanisms in vertebrates
Nonspecific defense mechanisms
Immune responses

All animals have the ability to prevent disease-causing microorganisms (pathogens) from entering their body. Because these external defense mechanisms sometimes fail, most animals have developed internal defense mechanisms as well. Before an animal can attack an invader, it must be able to recognize its presence, to distinguish between self (its own cells) and nonself (foreign matter). Whereas most animals are capable of nonspecific responses, such as phagocytosis and the inflammatory response, only vertebrates are capable of specific responses, such as the production of antibodies that target specific pathogens. This type of internal defense is characterized by a more rapid and intense response the second time the organism is exposed to the pathogen. In this manner, the organism develops resistance to the pathogen. Resistance to many pathogens can be induced artificially by either the injection of the pathogen in a weakened, killed, or otherwise altered state, or the injection of antibodies produced by another person or animal. The immune system is constantly surveying the body for abnormal cells and destroys them whenever they arise; failure results in cancer. This same surveillance system is responsible for graft rejection; the system simply reads foreign cells as nonself cells and destroys them. Sometimes the surveillance system fails and the immune system falsely reads self cells as nonself and destroys them; the resulting conditions are known as autoimmune diseases. Allergies are another example of abnormal immune responses.

CHAPTER OUTLINE AND CONCEPT REVIEW (Fill in the blanks)

I. INTRODUCTION

■ 1 A substance that stimulates an immune response is an _____.

■ 2 The two broad categories of defense mechanisms are, first,
(a)_____ _____, which are those that prevent pathogens from entering the body and attack those that do enter and, second, the "tailor-made," precisely directed
(b)_____.

■ 3 _____ is the study of specific defense mechanisms.

■ 4 _____ are highly specific proteins that help destroy antigens.

II. INVERTEBRATES HAVE INTERNAL DEFENSE MECHANISMS THAT ARE MAINLY NONSPECIFIC

■ 5 Most invertebrates are capable only of nonspecific responses as in "eating" invading cells, or (a)_____, and carrying out aspects of the (b)_____, which is characterized by localized heat, swelling, and discoloration.

III. VERTEBRATES CAN LAUNCH BOTH NONSPECIFIC AND SPECIFIC IMMUNE RESPONSES

IV. NONSPECIFIC DEFENSE MECHANISMS INCLUDE MECHANICAL AND CHEMICAL BARRIERS AGAINST PATHOGENS

■ 6 The first line of defense in animals is the (a)_____. Other nonspecific defense mechanisms that prevent entrance of pathogens include the skin, (b)_____ in the stomach, and the (c)_____ of the respiratory passageways.

A. Interferons help defend the body against viral infections

■ 7 Interferons are proteins that inhibit (a)_____ and activate the lymphocytes known as (b)_____.

B. Inflammation is a protective mechanism

■ 8 When pathogens invade tissues, they trigger an inflammatory response, which brings needed phagocytic cells and antibodies to the infected area. Injured cells release _____ that dilates blood vessels in the area.

■ 9 The four clinical manifestations of inflammation are _____ _____.

■ 10 A fever may result when leukocytes release (a)_____ that act to reset the body's thermostat in the (b)_____ gland. The most potent of these regulatory proteins is (c)_____ released by macrophages.

C. Phagocytes destroy pathogens

■ 11 (a)_____ are two leukocytes that phagocytize and destroy bacteria. Once "eaten," the invader is sequestered in a cytoplasmic vesicle called a (b)_____, which then fuses with (c)_____ that contain powerful hydrolytic enzymes.

V. SPECIFIC DEFENSE MECHANISMS INCLUDE ANTIBODY-MEDIATED IMMUNITY AND CELL-MEDIATED IMMUNITY

A. Cells of the immune system include lymphocytes and phagocytes

■ 12 T cells are a type of agranular lymphocyte that originate from the (a)_____ cells that are found in (b)_____.

■ 13 The three main types of T cells are: the (a)_____ T cells, or killer T cells that kill cells with antigens on their surfaces, the (b)_____ T cells that secrete substances that initiate or enhance the immune response, and the (c)_____ T cells that inhibit the immune response.

■ 14 Macrophages are sometimes referred to as APCs, which stands for _____ _____.

B. The thymus "instructs" T cells and produces hormones

■ 15 The thymus gland somehow "instructs" or otherwise effects immunological competence upon T cells. It also secretes _____, a hormone that is thought to stimulate mature T cells.

C. The major histocompatibility complex permits recognition of self

D. Antibody-mediated immunity is a form of chemical warfare

■ 16 In antibody-mediated immunity, also called humoral immunity, the _____ _____ are activated when specific antigens are presented by a macrophage.

■ 17 _____ is the process whereby one or a few selected B cells respond to a specific antigen by repeatedly dividing, thereby forming a population of virtually identical cells, all from the one or few cells.

■ 18 The sequence of events that take place between invasion of the body by a pathogen and the production of antibodies can be summarized as

pathogen enters body —–> (a)_____ phagocytizes
pathogen —–> (b)_____ complex is displayed —–>
(c)_____ cell binds with complex —–> (d)_____
cell binds with complex —–> (e)_____ cell secretes
interleukens —–> (f)_____ cells are activated, forming a
clone —–> B cells differentiate to form (g)_____ —–>
antibodies are produced

■ 19 Antibodies are highly specific proteins called (a)_____, abbreviated as (b)_____. These molecules "recognize" specific amino acid sequences on antigens called (c)_____. The antibody molecule is Y-shaped, the two arms functioning as (d)_____ _____.

■ 20 Antibodies are grouped into five principle isotypes, which are abbreviated as (a)_____. Of these, about 75% are (b)_____ in humans. (c)_____ predominate in secretions, (d)_____ is an antigen receptor on the membranes of B lymphocytes, and (e)_____ is the mediator of allergic responses.

■ 21 The process of coating pathogens to make them less "slippery," so macrophages and neutrophils can phagocytize them, is known as _____.

E. Cell-mediated immunity provides cellular warriors

■ 22 After a cytotoxic T cell combines with an antigen on the surface of a target cell, it secretes _____ that contain a variety of lytic proteins.

■ 23 _____ released by cytotoxic T cells are soluble proteins that are toxic to cancer cells.

■ **24** Cell-mediated immune response can be summarized as

pathogen enters body ——> (a)_____ phagocytizes
pathogen ——> (b)_____ complex is displayed ——>
(c)_____ cell activated by complex ——> clone of
(d)_____ secretes interleukens ——>
(e)_____ develop and migrate to the infected area ——>
proteins are released ——> target cells destroyed

F. A secondary immune response is more rapid than a primary response

■ **25** The principal antibody synthesized in the primary response is (a)_____. Second exposure to an antigen evokes a secondary immune response, which is more rapid and more intense than the primary response. The principal antibody synthesized in the secondary response is (b)_____.

G. Active immunity follows exposure to antigens

■ **26** Active immunity develops as a result of exposure to antigens; it may occur naturally or may be artificially induced by _____.

■ **27** Immunity conferred on an individual by antibodies produced in another individual is called _____.

H. The body normally defends itself against cancer

■ **28** The theory that the body's immune system destroys abnormal cells as they arise is known as the _____.

■ **29** Antibodies that combine with cancer cell antigens, thereby preventing T cells from adhering to and destroying them, are called _____.

I. Graft rejection is an immune response against transplanted tissue

■ **30** Transplanted human tissues possess protein markers known as (a)_____ that stimulate graft rejection, an immune response launched mainly by (b)_____ that destroy the transplant.

■ **31** Tissue transplanted from one location to another in the same individual is called an (a)_____, while transplantation between different individuals is an (b)_____.

J. Certain body sites are immunologically privileged

K. Immunological tolerance can be induced

L. In an autoimmune disease the body attacks its own tissues

M. Allergic reactions are inappropriate immune responses

■ **32** In an allergic response, an (a)_____ stimulates production of (b)_____ type antibody, which combines with the receptors on mast cells; the mast cells then release (c)_____ and other substances, causing inflammation and other symptoms of allergy.

N. AIDS is an immune disease caused by a retrovirus

■ 33 AIDS is an acronym for (a)_____
which is caused by infection with the retrovirus (b)_____
_____. The virus compromises the immune system of its
victims by destroying (c)_____ cells.

KEY TERMS

Frequently Used Prefixes and Suffixes: Use combinations of these prefixes and suffixes to generate
terms for the definitions below.

Prefixes	The Meaning	Suffixes	The Meaning
anti-	against, opposite of	-cyte	cell
auto-	self, same	-some	body
leuko-	white (without color)		
lyso-	loosening, decomposition		
mono-	alone, single, one		
phago-	eat, devour		

Prefix	Suffix	Definitions
_____	-body	1. A specific protein that acts against pathogens and helps destroy them.
_____	-histamine	2. A drug that acts against (blocks) the effects of histamine.
coelomo-	_____	3. A wandering phagocytic cell of the coelom.
_____	_____	4. A vesicle (or "body") within a phagocyte that contains and devours engulfed pathogens.
lympho-	_____	5. A white blood cell strategically positioned in the lymphoid tissue; the main "warrior" in specific immune responses.
_____	_____	6. A white or colorless blood cell that functions in resistance to infection.
_____	-clonal	7. An adjective pertaining to a single clone of cells.
_____	-graft	8. A tissue or organ that is grafted into a new position on the body of the same individual from which it was removed.
_____	-immune	9. An adjective pertaining to the situation wherein the body reacts immunologically against its own tissues (against "self").
_____	-zyme	10. An enzyme that lyses bacteria by degrading the cell wall.

Other Terms You Should Know

active immunity	allergy (-gies)	antigen-antibody complex
affinity (-ties)	allograft	(-xes)
AIDS	antibody-mediated immunity	antigen-presenting cell
allergen	antigen	antigenic determinant
allergic asthma		APC

autoimmune disease
B cell
B lymphocyte
blocking antibody (-dies)
bursa of Fabricius
C region
cell-mediated immunity
clonal selection
competent B cell
competent T cell
constant segment
cytotoxic T cell
decline phase
desensitization therapy
 (-pies)
edema
epitope
fever
graft rejection
hapten
helper T cell
histamine
HIV
hives

HLA
human immunodeficiency
 virus
humoral immunity
Ig
IgA
IgD
IgE
IgG
IgM
immune response
immunization
immunocompetent
immunoglobulin
immunology
interferon
interleukin
J region
junctional segment
latent period
logarithmic phase
lymphotoxin
lysozyme
macrophage

MHC
monoclonal antibody (-dies)
NK cell
nonspecific defense
 mechanism
passive immunity
pathogen
phagosome
plasma cell
primary response
pyrogen
secondary response
specific defense mechanism
suppressor T cell
systemic anaphylaxis
T cell
T lymphocyte
T-cell receptor
theory of immune
 surveillance
thymosin
thymus gland
V region
variable segment

SELF TEST

Multiple Choice: Some questions may have more than one correct answer.

1. T cells
   a. are lymphocytes.
   b. are called LGLs.
   c. are also called T lymphocytes.
   d. are agranular and mononucleated.
   e. are involved in specific cellular immunity.

2. An allergic reaction involves
   a. killer T cells.
   b. mast cells.
   c. interaction between an allergen and mast cells.
   d. production of IgE.
   e. histocompatibility antigens.

3. The histocompatibility complex is
   a. found in cell nuclei.
   b. called HLA in humans.
   c. different in each individual.
   d. the same on individuals comprising a species.
   e. a group of proteins.

4. Which of the following is/are true of Ig?
   a. They are antibodies.
   b. They contain a C region.
   c. They are produced in response to specific antigens.
   d. They contain an antigenic determinant.
   e. They are also called immunoglobulins.

5. The millions of different types of antibodies produced by the immune system are most likely due to
   a. a one gene-one antibody ratio.
   b. gene mutations.
   c. a combination of many germ line V genes and mutations.
   d. spacers between V and C regions of DNA.
   e. many C regions.

6. The functions of antibodies are determined by their
   a. C region.
   b. V region.
   c. J region.
   d. constant segment.
   e. variable segment.

7. Which of the following immunoglobulins would you expect to be most involved in protecting you from air borne pathogens?
   a. IgG
   b. IgM
   c. IgA
   d. IgD
   e. IgE

8. Treatment with thymosin might be appropriate when the patient has
   a. an underdeveloped thymus gland.
   b. too few B cells.
   c. too many immunologically active T cells.
   d. dysfunction of the MHC system.
   e. certain types of cancer.

9. Complement
   a. is a system of several proteins.
   b. is highly antigen-specific.
   c. is stimulated into action by an antibody-antigen complex.
   d. helps destroy pathogens.
   e. is an antibody.

10. Active immunity can be artificially induced by
    a. transfusions.
    b. injecting attenuated virus.
    c. passing maternal antibodies to a fetus.
    d. injecting gamma globulin.
    e. stimulating macrophage growth.

11. Nonspecific defense mechanisms in vertebrates include
    a. skin.
    b. acid secretions.
    c. inflammation.
    d. phagocytes.
    e. antibody-mediated immunity.

12. The secondary response is due to
    a. killer T cells.
    b. memory cells.
    c. plasma cells.
    d. macrophages.
    e. helper T cells.

13. B cells
    a. are granular.
    b. are lymphocytes.
    c. clone after contacting its targeted antigen.
    d. are derived from plasma cells.
    e. include many antigen-binding forms.

14. Judging from the distribution of antibody classes among vertebrate taxa, one would logically conclude that _____ evolved first.
    a. IgG
    b. IgM
    c. IgA
    d. IgD
    e. IgE

15. T-cell receptors
    a. bind antigens.
    b. have no known function.
    c. are identical on all T cells.
    d. are found on killer cells.
    e. stimulate antibody production.

16. The body's thermostat is reset during fever by the action of
    a. a peptide.
    b. phagosomes.
    c. prostaglandins.
    d. a substance released by macrophages.
    e. interleukin 1.

17. Which of the following is true of AIDS?
   a. HIV infects helper T cells.
   b. AIDS is not spread by casual contact.
   c. Both heterosexuals and homosexuals are at risk.
   d. AIDS related complex (ARC) may lead to AIDS.
   e. There is currently no cure for AIDS.

VISUAL FOUNDATIONS

Color the parts of the illustration below as indicated. Label cell-mediated immunity, antibody-mediated immunity, T cell, and B cell.

RED ☐ bone marrow
GREEN ☐ antigen stimulation
YELLOW ☐ T cell
BLUE ☐ migration to lymph node
ORANGE ☐ thymus
BROWN ☐ bursa
TAN ☐ B cell
PINK ☐ plasma cell

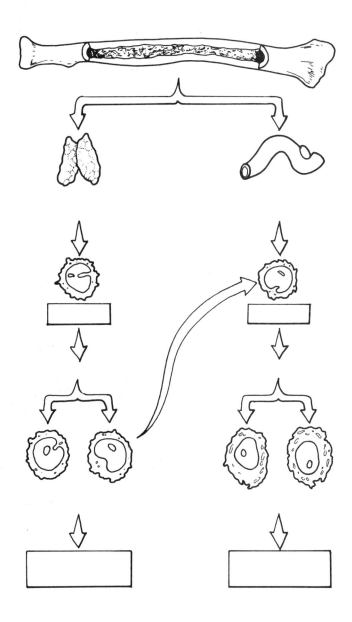

Color the parts of the illustration below as indicated. Label variable portion and constant portion.

RED ☐ antigenic determinants
GREEN ☐ antigen
YELLOW ☐ antibody-heavy chain
BLUE ☐ disulfide bond
ORANGE ☐ antibody-light chain
BROWN ☐ binding site
TAN ☐ antigen-antibody complex

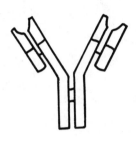

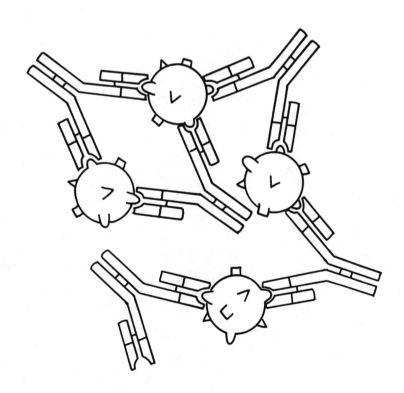

CHAPTER 44

□

Gas Exchange

O U T L I N E

Adaptations for gas exchange
Air versus water
Respiratory pigments
The human respiratory system
Effects of inhaling pollutants

In small, aquatic organisms, gases diffuse directly between the environment and all body cells. In larger, more complex organisms, specialized respiratory structures, such as tracheal tubes, gills, and lungs, are required. Gas exchange with air is more efficient than gas exchange with water. This is because air contains more oxygen than water and oxygen diffuses more rapidly in air than in water. Respiratory pigments, such as hemoglobin and hemocyanin, greatly increase respiratory efficiency by increasing the capacity of blood to transport oxygen. The human respiratory system consists of two lungs and a system of tubes through which air reaches them. The largest tube, the trachea, branches into two bronchi, one extending into each lung. Each bronchus branches repeatedly, giving rise ultimately to tiny bronchioles. The bronchioles, in turn, branch repeatedly, giving rise to clusters of alveoli. Oxygen diffuses from the alveoli into pulmonary capillaries, and carbon dioxide diffuses from the blood into the alveoli, each gas moving from a region of its greater concentration to a region of its lower concentration. Although breathing is an involuntary process controlled by respiratory centers in the brain, it can be consciously influenced. Respiration can be adversely affected by hyperventilation, flying, diving, smoking, and air pollution.

CHAPTER OUTLINE AND CONCEPT REVIEW (Fill in the blanks)

I. INTRODUCTION

■ 1 The exchange of gases between an organism and the medium in which it lives is called (a)_____, while actively moving air or water over the respiratory surface is (b)_____.

II. ANIMALS HAVE EVOLVED SEVERAL DIFFERENT ADAPTATIONS FOR GAS EXCHANGE

■ 2 Specialized respiratory structures must have (a)_____ to facilitate the transfer of adequate volumes of gases, (b)_____ surfaces that can dissolve oxygen and carbon dioxide, and a rich supply of (c)_____ to transport respiratory gases.

■ 3 _____ are the four principal types of respiratory structures.

A. The body surface may be adapted for gas exchange

■ 4 Animals that rely entirely on gas exchange across the body surface are small and therefore have a large (a)_____ ratio and a low (b)_____.

B. Tracheal tube systems of arthropods deliver air directly to the cells

■ 5 Arthropods have respiratory system consisting of a network of (a)_____ _____ that air enters through small openings called (b)_____. The respiratory network terminates in small fluid-filled (c)_____ where gas exchange takes place.

C. Gills of aquatic animals are specialized respiratory surfaces

■ 6 The gills of bony fish have many thin (a)_____ that extend out into the water, within which are numerous capillaries. Blood flows through these structures in a direction opposite to water movement, an arrangement called the (b)_____, a system that maximizes the difference in oxygen concentrations between the animal's blood and the water.

D. Terrestrial vertebrates exchange gases through lungs

■ 7 Lungs are respiratory structures that develop as ingrowths of the (a)_____ or from the wall of a (b)_____ such as the pharynx. For example, the (c)_____ of spiders are within an inpocketing of the abdominal wall.

III. RESPIRATORY STRUCTURES ARE ADAPTED FOR GAS EXCHANGE IN AIR OR WATER

■ 8 Primarily because of water's greater (a)_____, aquatic animals expend 10 to 20 times more energy than air breathers to move their medium over their respiratory surfaces. Air contains (b)_____(more of less?) oxygen than water, and oxygen diffuses (c)_____(faster or slower?) in air than in water.

IV. RESPIRATORY PIGMENTS INCREASE CAPACITY FOR OXYGEN TRANSPORT

■ 9 The respiratory pigment of most vertebrates is hemoglobin, a compound containing a (a)_____ group bound to the protein (b)_____. Three other respiratory pigments are (c)_____ _____. Of these, the (d)_____ are blue, copper-containing proteins found in some of the mollusks and arthropods.

V. THE HUMAN RESPIRATORY SYSTEM IS TYPICAL OF AIR-BREATHING VERTEBRATES

■ 10 The pathway of air as it passes through the human's "breathing apparatus" can be summarized as

nose and mouth ——> (a)_____ (throat region) ——>
(b)_____ (voice box) ——> (c)_____ (wind pipe) ——>
(d)_____ (tubes leading to lungs) ——> (e)_____
(small branching tubes) ——> (f)_____ (air sacs)

A. The airway conducts air into the lungs

■ 11 The cough reflex occurs when the (a)_____ fails to close off the (b)_____ from the esophagus.

B. Gas exchange occurs in the alveoli of the lungs

■ 12 The lungs are located in the thoracic cavity. The right lung has (a)_____(#?), and the left lung has (b)_____(#?) lobes.

C. Ventilation is accomplished by breathing

■ 13 Breathing involves inhalation, or (a)_____, and exhalation, or (b)_____.

D. The quantity of air respired can be measured

■ 14 (a)_____ is the volume of air that is inhaled and exhaled during "normal" breathing. It averages about (b)_____(volume?). (c)_____ _____ is the maximum volume of air that can be expelled.

E. Gas exchange takes place in the air sacs

■ 15 The factor that determines the direction and rate of diffusion of a gas across a respiratory surface is the (a)_____ of that gas. (b)_____ _____ states that the total pressure in a mixture of gases is the sum of the pressures of the individual gases

F. Oxygen is transported in combination with hemoglobin

■ 16 Oxygen combines with the element (a)_____ in the (b)_____ group of hemoglobin. This "association" can be illustrated as $Hb + O_2 \longrightarrow$ (c)_____. The resultant is a compound that carries oxygen and releases it where it is in lower concentrations.

■ 17 HbO_2 is prone to dissociation in a (a)_____(lower or higher?) pH. A change in the normal HbO_2 dissociation curve caused by a change in pH is known as the (b)_____.

G. Carbon dioxide is transported mainly as bicarbonate ions

■ 18 Most of the carbon dioxide transported in blood is dissolved in plasma as (a)_____ ions, the formation of which is catalyzed by an enzyme in RBCs called (b)_____.

H. Breathing is regulated by respiratory centers in the brain

■ 19 Respiratory centers are groups of neurons in the (a)_____ that regulate the rhythm of ventilation. (b)_____ are specialized nerve endings that are sensitive to changes in hydrogen ion concentration.

I. Hyperventilation reduces carbon dioxide concentration

J. High flying or deep diving can disrupt homeostasis

■ 20 _____ is a deficiency of oxygen that may cause drowsiness, mental fatigue, and headaches.

■ 21 A sudden decrease in environmental pressure may cause "bubbling" in the tissues and/or blood vessels; that is, the release of dissolved nitrogen in the blood in the form of bubbles that block capillaries, causing a very painful syndrome which, in the vernacular, is called "the bends," and in the scientific community as _____.

VI. THE EFFECT OF BREATHING DIRTY AIR IS RESPIRATORY INSULT

A. A variety of defense mechanisms protect the lungs

B. Continued respiratory insult leads to respiratory disease

■ 22 Chronic bronchitis and emphysema are examples of COPDs, which stands for
_____. COPDs are
linked to smoking and air pollution.

KEY TERMS

Frequently Used Prefixes and Suffixes: Use combinations of these prefixes and suffixes to generate terms for the definitions below.

Prefixes	The Meaning	Suffixes	The Meaning
hyper-	over	-ox(ia)	containing oxygen
hyp(o)-	under		

Prefix	Suffix	Definitions
_____	-ventilation	1. Excessive rapid and deep breathing.
_____	_____	2. Oxygen deficiency.

Other Terms You Should Know

air sac
alveolus (-li)
barometric pressure
bicarbonate ion
Bohr effect
breathing
bronchial constriction
bronchiole
bronchus (-hi)
carbonic anhydrase
cellular respiration
chemoreceptor
chloride shift
chronic bronchitis
chronic obstructive
  pulmonary disease
cough reflex (-xes)
countercurrent exchange
  system
decompression sickness

dermal gill
diaphragm
emphysema
epiglottis (-ses)
expiration
external naris (-res)
filament
gill
heme ring
hemocyanin
hyperventilate
inspiration
larynx (-nges or -xes)
lung
lung cancer
nasal cavity (-ties)
nose
operculum (-la)
organismic respiration

oxygen-hemoglobin
  dissociation curve
oxyhemoglobin
parabronchus (-hi)
percent saturation
pharynx (-nges or -xes)
pleura
pleural cavity (-ties)
pleural membrane
respiration
respiratory acidosis
respiratory center
respiratory pigment
spiracle
tidal volume
trachea
tracheal tube
ventilation
vital capacity (-ties)
vocal cord

SELF TEST

Multiple Choice: Some questions may have more than one correct answer.

1. The diaphragm and rib muscles alternately contract and relax, these actions causing respectively
   a. inspiration and expiration.
   b. inhalation and exhalation.
   c. expiration and inspiration.
   d. exhalation and inhalation.
   e. oxygen intake and carbon dioxide output.

2. If the PO_2 in the tissue of our pet dog is 10, atmospheric PO_2 is 150, and arterial PO_2 is 110, one would expect the dog to (units are mm Hg)
   a. function normally.
   b. accumulate carbon dioxide.
   c. have a serious, but not lethal, oxygen deficit.
   d. die.
   e. become dizzy from too much oxygen.

3. If you shared the same parameters as those given for your dog (question 2) and your tissues were consuming 2.5 liters of oxygen per minute, it's likely that
   a. you are asleep.
   b. you are resting.
   c. you and your dog are out for a leisurely walk.
   d. you are exercising vigorously.
   e. you and your dog will both die.

4. If the parameters given in question 3 applied, the color(s) of blood in your arteries and veins respectively would be
   a. pink and light blue.
   b. both blue.
   c. both red.
   d. distinctly red and blue.
   e. distinctly blue and red.

5. We humans do not have to gulp air like a frog because we have
   a. alveoli.
   b. lungs.
   c. a diaphragm.
   d. lower oxygen requirements.
   e. a larger surface to volume ratio.

6. The Heimlich maneuver may be necessary if
   a. the epiglottis does not close.
   b. food enters the esophagus.
   c. a foreign object sticks in the larynx.
   d. air contacts surfactants.
   e. we have severe pleurisy.

7. Most of the mucus produced by epithelial cells in the nasal cavities is disposed of by means of
   a. a large handkerchief.
   b. nose picking.
   c. evaporation.
   d. absorption in surrounding lymph ducts.
   e. swallowing.

8. Characteristics that all respiratory surfaces share in common include
   a. moist surfaces.
   b. thin walls.
   c. specialized structures such as tracheal tubes, gills, or lungs.
   d. large surface to volume ratio.
   e. alveoli or spiracles.

9. We detect odors once they have entered our respiratory systems through the
   a. nostrils.
   b. nasal cavities.
   c. buccal cavity.
   d. external nares.
   e. olfactory os.

10. Organismic respiration in a swordfish relies on
    a. dermal gills.
    b. filaments.
    c. internal gills along edges of slits in the pharynx.
    d. spiracles and tracheal tubes.
    e. countercurrents.

11. A cockroach obtains oxygen for tissues located deep in its body by means of
    a. book gills.
    b. pseudolungs.
    c. circulation of hemolymph.
    d. a countercurrent exchange system.
    e. branching tubes.

12. A respiratory pigment that contains protein but no porphyrin group is called
    a. hemoglobin.
    b. hemolymph.
    c. hemoglobin.
    d. hemocyanin.
    e. erythrocyanin.

13. Air passes through structures of a mammal's respiratory system in the following order:
    a. trachea, pharynx, bronchus, bronchioles.
    b. larynx, trachea, bronchioles, bronchus.
    c. pharynx, larynx, bronchus, bronchioles.
    d. larynx, trachea, bronchus, bronchioles.
    e. trachea, larynx, bronchus, bronchioles.

14. With respect to organismic respiration, some of the advantages of life on land as compared to an aquatic life pertain to
    a. energy conservation.
    b. facilitated diffusion.
    c. maintaining ion homeostasis.
    d. keeping the respiratory surface moist.
    e. maintenance of a higher body temperature.

15. Reasonable rates of expiration and inspiration fall in the range of
    a. 25/min.
    b. 10/min.
    c. 72/hr.
    d. 0.2/sec.
    e. 650/hr.

16. The ability of oxygen to be released from oxyhemoglobin is affected by
    a. temperature.
    b. pH.
    c. concentration of carbon dioxide.
    d. oxygen concentration in tissues.
    e. ambient oxygen concentration.

17. If the action of carbonic anhydrase in RBCs increases, one would expect an increase in
    a. concentration of reduced hemoglobin.
    b. bicarbonate ions in plasma.
    c. diffusion of chloride ions out of cells.
    d. the chloride shift.
    e. bonding of carbon dioxide and hemoglobin.

18. If a patient is found to have inelastic air sacs, trouble expiring air, an enlarged right ventricle, and large alveoli, he/she probably
    a. has lung cancer.
    b. has emphysema.
    c. has chronic obstructive pulmonary disease.
    d. lives at a high altitude.
    e. experiences acute bronchitis periodically.

19. If you are SCUBA diving for a lengthy time at 200 feet, you might get the bends if
    a. you surface quickly.
    b. you are using a helium mixture.
    c. nitrogen in your blood is rapidly absorbed by your tissues.
    d. you come to the surface very slowly.
    e. you stay at 200 feet even longer.

20. Proper CPR procedures include
    a. placing the victim on his/her stomach.
    b. external cardiac compression.
    c. exhalation into the victim's mouth.
    d. pausing 15 seconds between breaths.
    e. extending the victim's neck.

VISUAL FOUNDATIONS

Color the parts of the illustration below as indicated.

RED	☐	brachial plume
GREEN	☐	external gills
YELLOW	☐	trachea
BLUE	☐	internal gills
ORANGE	☐	book lung
TAN	☐	lungs

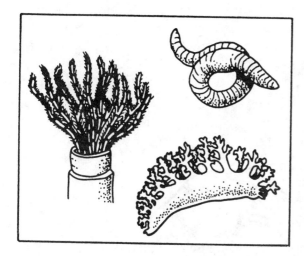
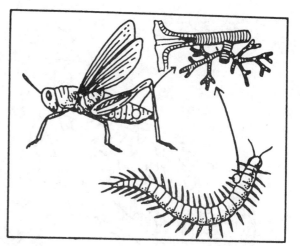
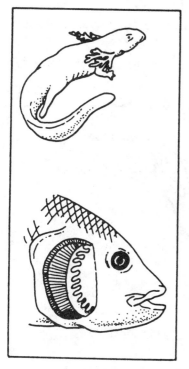
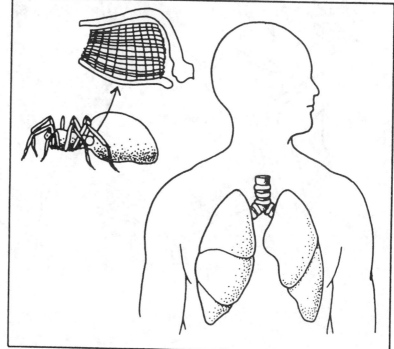

Color the parts of the illustration below as indicated.

RED   ❑   artery

GREEN   ❑   macrophage

YELLOW   ❑   alveolus

BLUE   ❑   vein

ORANGE   ❑   epithelial cell of alveolus

BROWN   ❑   red blood cell

TAN   ❑   bronchiole

PINK   ❑   capillary

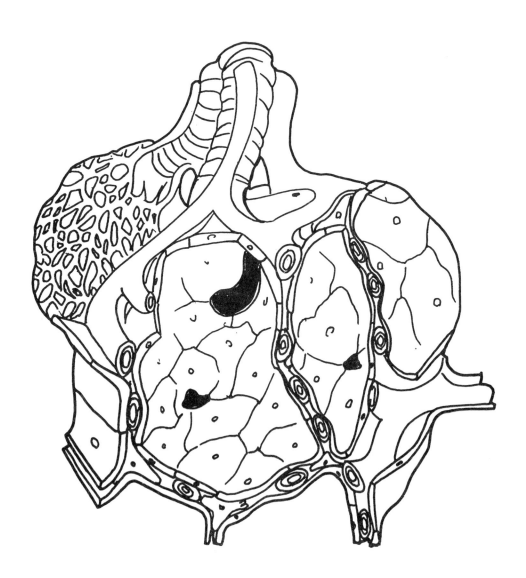

CHAPTER 45

□

Processing Food

O U T L I N E

Overview of food processing
Modes of nutrition
Incomplete digestive systems
Complete digestive systems
The human digestive system
Nutrition and metabolism

The processing of food involves several steps — taking food into the body, breaking it down into its constituent nutrients, absorbing the nutrients, and eliminating the material that is not broken down and absorbed. Animals eat either plants, animals, or both. Some animals have no digestive systems, with digestion occurring intracellularly within food vacuoles. Others have incomplete digestive systems with only a single opening for both food to enter and wastes to exit. Still other animals, most in fact, have complete digestive systems, in which the digestive tract is a complete tube with two openings, a mouth where food enters and an anus where waste is expelled. The human digestive system has highly specialized structures for processing food. All animals require the same basic nutrients — carbohydrates, lipids, proteins, vitamins, and minerals. Carbohydrates are used by the body as fuel. Lipids are also used as fuel, and additionally, as components of cell membranes, and as substrates for the synthesis of steroid hormones and other lipid substances. Proteins serve as enzymes and as structural components of cells. Vitamins and minerals are needed for many biochemical processes. Serious nutritional problems result from eating too much food, eating too little food, or not eating a balanced diet.

CHAPTER OUTLINE AND CONCEPT REVIEW (Fill in the blanks)

I. INTRODUCTION

■ 1 Organisms that obtain their main source of energy from organic molecules synthesized by other organisms are known as _____.

II. FOOD PROCESSING INVOLVES INGESTION, DIGESTION, ABSORPTION, AND ELIMINATION

■ 2 Undigested material is removed from the digestive tracts of simpler animals by a process known as (a)_____, and from more complex animals by (b)_____.

III. ANIMALS ARE ADAPTED TO THEIR MODE OF NUTRITION

■ **3** Primary consumers, or _____, have special digestive processes to break down the plant material they consume.

■ **4** (a)_____ have well-developed canine teeth, digestive enzymes that break down proteins, and an overall (b)_____(shorter or longer?) gastrointestinal tract.

■ **5** Organisms that utilize both animal and plant material are called _____.

IV. SOME INVERTEBRATES HAVE INCOMPLETE DIGESTIVE SYSTEMS

V. MOST INVERTEBRATES AND ALL VERTEBRATES HAVE COMPLETE DIGESTIVE SYSTEMS

■ **6** The vertebrate digestive system is a complete tube extending from the (a)_____ to the (b)_____. The specialized portions of this tube, in order, are the

mouth ——> (c)_____ ——> (d)_____ ——> stomach ——> (e)_____ ——> large intestine ——> anus

VI. THE HUMAN DISGESTIVE SYSTEM HAS HIGHLY SPECIALIZED STRUCTURES FOR PROCESSING FOOD

A. The wall of the digestive tract is composed of four layers

■ **7** The wall of the vertebrate digestive tract consists of four layers: the (a)_____ is densely packed with blood vessels and nerves, the (b)_____ contains goblet cells that secrete mucous and a large folded surface area, the (c)_____ is the outer layer of the system, and the (d)_____ contains muscles that carry out peristalsis.

B. Food processing begins in the mouth

■ **8** Ingestion and the beginning of mechanical and enzymatic breakdown of food take place in the mouth. The teeth of mammals perform varied functions: (a)_____ are designed for biting, (b)_____ for tearing and stabbing (ask Dracula!), and (c)_____ crush and grind food. Each tooth is covered by a coating of very hard (d)_____, under which is the main body of the tooth, the (e)_____, which resembles bone. The (f)_____ contains blood vessels and nerves.

■ **9** The salivary glands of terrestrial vertebrates moisten food and release _____, an enzyme that initiates carbohydrate (starch) digestion.

C. The pharynx and esophagus conduct food to the stomach

■ **10** Waves of muscular contractions called (a)_____ move a lump of food, at this point called a (b)_____, through the esophagus into the stomach.

D. Food is mechanically and enzymatically digested in the stomach

■ **11** In the stomach, food is mechanically broken down, pepsin in gastric juice initiates protein digestion, and food is reduced to chyme. The stomach is lined with (a)_____ cells, which contain many gastric glands and specialized cells. For example, (b)_____ cells secrete hydrochloric acid and (c)_____, a substance needed for absorption of vitamin B, and (d)_____ cells secrete pepsinogen, an enzyme precursor that converts to (e)_____.

E. Most enzymatic digestion takes place inside the small intestine

■ 12 The three regions of the small intestine are (a)_____. Most enzymatic digestion of food in vertebrates takes place in the (b)_____ portion.

■ 13 The surface area of the small intestine is increased by small fingerlike projections called (a)_____ along its surface, and by (b)_____, which are outfoldings of the plasma membranes of cells that line the intestinal lumen.

F. The liver secretes bile, which mechanically digests fats

■ 14 Among the many important functions performed by the liver are the secretion of bile, maintenance of homeostasis, the conversion of glucose to the carbohydrate storage molecule (a)_____, the conversion of excess amino acids to (b)_____ and the detoxifying of drugs and other poisons.

■ 15 Bile produced by the liver is stored in the (a)_____. Bile breaks apart fat droplets in a process called (b)_____.

G. The pancreas secretes digestive enzymes

H. Enzymatic digestion occurs as food moves through the digestive tract

■ 16 Pancreatic juice contains several categories of enzymes, including the (a)_____ that hydrolyze proteins, the (b)_____ that break down fats, and the (c)_____ that act on polysaccharides.

■ 17 Enzymes reduce macromolecular polymers to the small subunits that comprise them. For example, carbohydrates are digested to (a)_____, proteins are reduced to (b)_____, and fats are depolymerized to (c)_____.

■ 18 There are three types of proteolytic enzymes that break down (a)_____. (b)_____ break the bond connecting an end amino acid to a peptide chain, (c)_____ break small peptides into amino acids, and (d)_____ break peptide bonds within peptide chains.

I. Nerves and hormones regulate digestion

J. Absorption takes place mainly through the villi of the small intestine

■ 19 Most digested nutrients are absorbed through the (a)_____ of the small intestine. Monosaccharides and amino acids enter the (b)_____; glycerol, fatty acids, and monoacylglycerols enter the (c)_____.

K. The large intestine eliminates wastes

■ 20 The large intestine absorbs sodium and water, cultures bacteria, and eliminates wastes. It is made up of seven regions, which are, in order: the (a)_____, a blind pouch near the junction of the small and large intestines; the (b)_____ _____; the (c)_____; the

(d)_____; the (e)_____; the
(f)_____, the last portion of the tube; and the (g)_____, the opening at the
end of the tube.

■ 21 (a)_____ involves mainly the filtering out of wastes by the kidneys and
lungs, while (b)_____ involves disposal of wastes that never
participated in metabolism.

VII. ADEQUATE AMOUNTS OF REQUIRED NUTRIENTS ARE NECESSARY TO SUPPORT METABOLIC PROCESSES

A. Carbohydrates are a major energy source in the human diet

■ 22 Glucose, the end-product of carbohydrate digestion, is used by cells mainly as fuel
for _____.

■ 23 The amount of energy in food is given as (a)_____ per gram, which is
equivalent to the unit (b)_____.

■ 24 _____ is a mixture of cellulose and other indigestible carbohydrates derived
from plants.

B. Lipids are used to supply energy and to make needed biological molecules

■ 25 Fat is stored in (a)_____ tissue. When (b)_____ in the blood falls
below the steady state, fat is mobilized and used as an energy source.

■ 26 Macromolecular complexes of cholesterol or triacylglycerols bound to proteins are
called (a)_____. Two types of these important complexes are the
(b)_____, which apparently decrease the risk of
heart disease by transporting excess cholesterol to the liver; and (c)_____
_____, which have been associated with coronary
artery disease.

C. Proteins serve as enzymes and are essential structural components of cells

D. Essential amino acids must be ingested in the diet

■ 27 Excess amino acids are removed from the system by the (a)_____ through a
process known as (b)_____.

E. Vitamins are organic compounds essential for normal metabolism

■ 28 Vitamins are divided into two broad groups: the (a)_____ vitamins
such as A, D, E, and K, and the (b)_____ vitamins that include
(c)_____.

F. Minerals are inorganic nutrients required by cells

■ 29 The essential minerals required by the body are _____
_____.

G. Energy metabolism is balanced when energy input equals energy output

■ 30 When energy input equals energy output, body weight remains constant. When
energy input exceeds energy output, body weight (a)_____(increases or
decreases?); and when energy input is less than output, the body draws on fuel
reserves (fat) and body weight (b)_____(increases or decreases?).

■ 31 (a)_____ reflects the amount of energy an organism must expend to survive. (b)_____ is the sum of basic energy required to survive *and* the energy needed to carry out daily activities.

KEY TERMS

Frequently Used Prefixes and Suffixes: Use combinations of these prefixes and suffixes to generate terms for the definitions below.

Prefixes	The Meaning		Suffixes	The Meaning
ana-	up		-itis	inflammation
cata-	down		-micro(n)	small, "tiny"
epi-	upon, over, on		-vore	eating
micro-	small			
omni-	all			
sub-	under, below			

Prefix	Suffix	Definitions
herbi-	_____	1. An animal that eats plants.
carni-	_____	2. An organism that eats flesh.
_____	_____	3. An organism that eats both plants and animals.
_____	-bolism	4. In living organisms, the "building up" (synthesis) of more complex substances from simpler ones.
_____	-bolism	5. In living organisms, the "breaking down" of more complex substances into simpler ones.
_____	-mucosa	6. A layer of connective tissue below the mucosa that binds it to the muscle layer beneath.
periton-	_____	7. Inflammation of the peritoneum.
_____	-glottis	8. A flap of tissue over the glottis (larynx opening) that prevents food and drink from entering the larynx when swallowing.
_____	-villi	9. Small projections of the cell membrane that increase the surface area of the cell.
chylo-	_____	10. Tiny droplets of lipid that are absorbed from the intestine into the lymph circulation.

Other Terms You Should Know

absorption	bile salts	chief cell
adventitia	BMR	cholecystokinin
amylase	bolus (-luses)	chyme
appendix (-ixes, -ices)	brush border	chymotrypsin
basal metabolic rate	canine	colon
beta-oxidation	carboxypeptidase	commensal
bile	cecum	complete digestive system

complete protein
constipation
defecate
defecation
dentin
deoxyribonuclease
diarrhea
digested
digestion
dipeptidase
duodenum
edema
egestion
elimination
emulsification
enamel
endopeptidase
enterokinase
epiglottis (-ses)
esophagus
essential amino acid
essential fatty acid
excretion
exopeptidase
fat-soluble vitamin
fiber
gallbladder
gastric gland
gastric juice
gastrin
goblet cell
HDL

hepatic portal vein
heterotroph
ileocecal valve
ileum
incisor
incomplete digestive system
ingested
ingestion
intestinal gland
intrinsic factor
jejunum
kwashiorkor
lacteal
large intestine
LDL
lipoprotein
liver
lumen
maltase
metabolic rate
metabolism
micelle
molar
mucosa
muscularis
nutrient
nutrition
obesity
pancreas
pancreatic amylase
pancreatic lipase
parietal cell

parietal peritoneum
pepsin
pepsinogen
peristalsis (-ses)
peristaltic contraction
peritoneal cavity
pharynx (-nges, -nxes)
premolar
primary consumer
protein deficiency (-cies)
pulp
pulp cavity
pylorus (-ri)
rectum (-ta)
reticulum
ribonuclease
ruga (-are)
salivary amylase
salivary gland
small intestine
stomach
taste bud
tooth (teeth)
total metabolic rate
trace element
trypsin
vermiform appendix (-ixes, -ices)
villus (-li)
visceral peritoneum
vitamin
water-soluble vitamin

SELF TEST

Multiple Choice: Some questions may have more than one correct answer.

1. The principal sources of energy in the human diet are
   a. proteins.
   b. lipids.
   c. carbohydrates.
   d. starches and sugars.
   e. meat.

2. Which of the following is true concerning human nutrition?
   a. Water is essential.
   b. Excess nutrients are converted to fat.
   c. We cannot synthesize some required fatty acids.
   d. Proteins cannot be used for energy.
   e. The liver synthesizes amino acids.

3. The total metabolic rate
   a. encompasses the basal metabolic rate.
   b. is less than basal metabolic rate.
   c. refers to metabolic rate after exercise.
   d. is the rate when at rest.
   e. is a nonexistent phrase.

4. A herbivore would likely have
   a. well-developed claws.   d. a short gut.
   b. flattened molars.   e. symbiotic microorganisms.
   c. large, sharp canines.

5. Glands near the ears that initiate starch digestion are the
   a. submaxillaries.   d. parotids.
   b. parietals.   e. sublinguals.
   c. pancreatic glands.

6. Glucose can result from
   a. glycogenesis.   d. glycolysis.
   b. glycogenolysis.   e. B-oxidation.
   c. gluconeogenesis.

7. Nerves and blood vessels in teeth are located in the
   a. pulp.   d. cementum.
   b. enamel.   e. canines.
   c. dentin.

8. All animals are
   a. omnivores.   d. consumers.
   b. carnivores.   e. heterotrophs.
   c. primary consumers.

9. Bile is principally involved in the digestion of
   a. carbohydrates.   d. nucleic acids.
   b. lipids.   e. proteins.
   c. starch.

10. Which of the following indicate the correct order of layers in the mammalian digestive tract?
    a. muscularis, mucosa, submucosa, adventitia.   d. mucosa, submucosa, adventitia, muscularis.
    b. mucosa, submucosa, muscularis, adventitia.   e. adventitia, muscularis, submucosa, mucosa.
    c. muscularis, submucosa, mucosa, adventitia.

11. A herbivore is a/an
    a. plant consumer.   d. omnivore.
    b. primary consumer.   e. mutualistic symbiote.
    c. carnivore.

12. The principal function(s) of villi is/are to
    a. absorb nutrients.   d. stimulate digestion.
    b. secrete enzymes.   e. secrete hormones.
    c. increase surface area.

13. The single most crucial intermediate molecule in the metabolism of most nutrients is
    a. keto acid.   d. NAPH.
    b. ATP.   e. glucose.
    c. acetyl CoA.

14. Which of the following is true regarding cellulose in the human diet?
    a. It's harmful.   d. It's an important carbohydrate nutrient.
    b. It's a source of bulk.   e. It's an important source of fiber.
    c. We can't digest it.

15. A lacteal is a
    a. lymph vessel.   d. milk protein enzyme.
    b. capillary bed.   e. salivary enzyme.
    c. digestive gland.

16. Which of the following has/have a complete digestive system?
    a. birds.   d. jelly fish.
    b. earthworms.   e. fish.
    c. sponges.

17. Maltase is principally involved in the digestion of
    a. carbohydrates.   d. nucleic acids.
    b. lipids.   e. proteins.
    c. starch.

18. Goblet cells and chief cells are found in the
    a. mucosa.   d. stomach.
    b. submucosa.   e. adventitia.
    c. epithelium.

19. Peristalsis results from activity of tissues in the
    a. submucosa.   d. muscularis.
    b. mucosa.   e. peritoneum.
    c. adventitia.

20. Pepsin is principally involved in the digestion of
    a. carbohydrates.   d. nucleic acids.
    b. lipids.   e. proteins.
    c. starch.

21. Some important functions of minerals include their role in/as
    a. cofactors.   d. maintaining fluid balance.
    b. nerve impulses.   e. digestive hormones.
    c. neurotransmitters.

22. The smallest molecular units would be obtained by the action of
    a. exopeptidases.   d. bile salts.
    b. endopeptidases.   e. amylase.
    c. dipeptidases.

23. Most absorption of macromolecular subunits occurs in the
    a. stomach.   d. small intestine.
    b. rectum.   e. colon.
    c. large intestine.

24. Which of the following is true of vitamins in human nutrition?
    a. Some are needed in large quantities.   d. Vitamin A can be stored in fat tissue.
    b. Megadoses can be harmful.   e. Vitamin D must be ingested.
    c. Antibiotics may cause a vitamin deficiency.

25. Which of the following is/are essential nutrient(s) for humans?
    a. lysine.   d. glucose.
    b. starch.   e. linoleic acid.
    c. cellulose.

26. The relative amounts of nutrient types consumed by impoverished societies, as compared to affluent societies, would likely be
    a. more protein, less carbohydrate.
    b. more lipid, less carbohydrate.
    c. more carbohydrate, less protein.
    d. more carbohydrate, less lipid.
    e. more vegetable matter, less animal matter.

27. Lipids are used to
    a. maintain body temperature.
    b. build membranes.
    c. provide energy.
    d. produce hormones.
    e. transport vitamins.

28. Which of the following would likely be used for energy as a last resort?
    a. carbohydrate
    b. lipid
    c. protein
    d. glycogen
    e. fatty acid

29. Products of complete fat digestion include
    a. fatty acids.
    b. amino acids.
    c. chylomicrons.
    d. glycerol.
    e. triacylglycerol.

Visual Foundations on next page ☞

VISUAL FOUNDATIONS

Color the parts of the illustration below as indicated.

RED	❑	liver
GREEN	❑	gall bladder
YELLOW	❑	esophagus
BLUE	❑	pancreas
ORANGE	❑	stomach
BROWN	❑	large intestine
TAN	❑	small intestine
PINK	❑	rectum and anus
VIOLET	❑	vermiform appendix

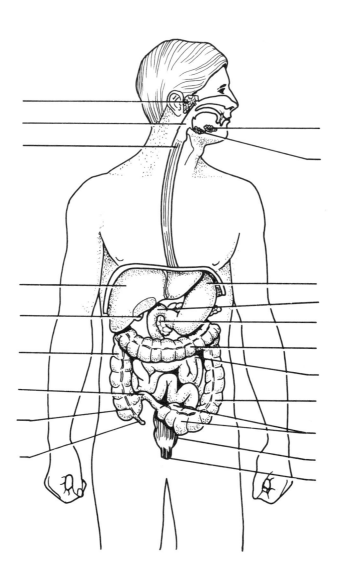

Color the parts of the illustration below as indicated.

RED	☐	artery
GREEN	☐	lymph
YELLOW	☐	nerve
BLUE	☐	vein
ORANGE	☐	goblet cell
BROWN	☐	intestinal gland
TAN	☐	villus
PINK	☐	epithelial cell of villus

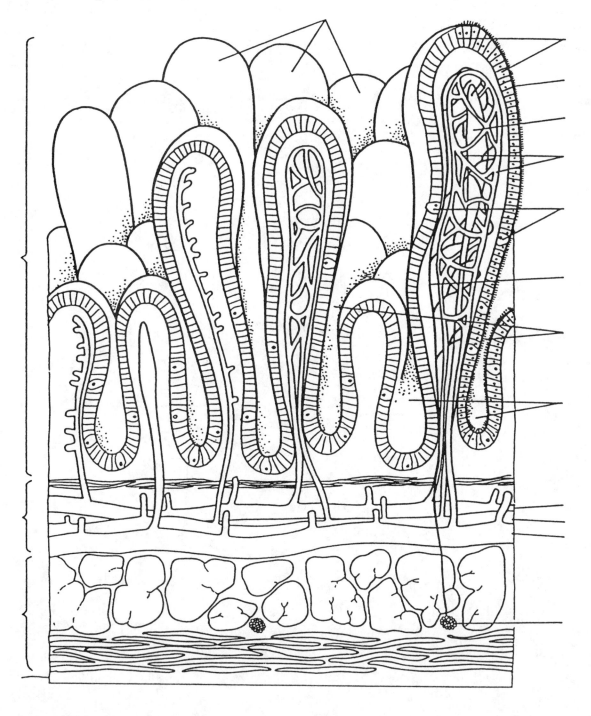

CHAPTER 46

Osmoregulation and Disposal of Metabolic Wastes

O U T L I N E

Functions of the excretory system
Metabolic waste products
Osmoregulation and waste disposal in invertebrates
Osmoregulation and waste disposal in vertebrates
Fluid balance and waste disposal in humans

The water content of the animal body, as well as the concentration and distribution of ions in body fluids is carefully regulated. Most animals have excretory systems that function to rid the body of excess water, ions, and metabolic wastes. The excretory system collects fluid from the blood and interstitial fluid, adjusts its composition by reabsorbing from it the substances the body needs, and expels the adjusted excretory product, e.g., urine in humans. The principal metabolic wastes are water, carbon dioxide, and nitrogenous wastes. Excretory systems among invertebrates are diverse and adapted to the body plan and lifestyle of each species. The kidney, with the nephron as its functional unit, is the primary excretory organ in vertebrates. There are over one million nephrons in each human kidney.

CHAPTER OUTLINE AND CONCEPT REVIEW (Fill in the blanks)

I. INTRODUCTION

II. EXCRETORY SYSTEMS HELP MAINTAIN HOMEOSTASIS

■ 1 Two important functions of the excretory system, aside from the excretion of metabolic wastes, are _____.

III. THE PRINCIPAL METABOLIC WASTE PRODUCTS ARE WATER, CARBON DIOXIDE, AND NITROGENOUS WASTES

■ 2 Principal metabolic wastes in most animals are water, carbon dioxide, and nitrogenous wastes in the form of _____.

■ 3 The urea cycle is a series of steps that produces urea from the toxic nitrogenous compound (a)_____ and (b)_____.

■ 4 Many desert animals conserve water by excreting nitrogenous wastes as the semi-solid compound _____.

IV. INVERTEBRATES HAVE SOLVED PROBLEMS OF OSMOREGULATION AND METABOLIC WASTE DISPOSAL WITH A VARIETY OF MECHANISMS

■ 5 Animals whose body fluids are at equilibrium with their environments are called (a)_____, in contrast to (b)_____ that have adaptations enabling them to live in hypertonic or hypotonic media.

A. Nephridial organs are tubules specialized for osmoregulation and/or excretion in some invertebrates

■ 6 The nephridial organs of many invertebrates consist of tubes that open to the outside of the body through (a)_____. In flatworms, these excretory organs are called (b)_____, and in annelids and mollusks they are known as (c)_____.

B. Antennal glands are important in osmoregulation in crustaceans

■ 7 Antennal glands, or (a)_____, as they are sometimes called, are the principal excretory organs among the (b)_____. They consist of a (c)_____ sac, a (d)_____ chamber, an excretory tubule, and an exit duct.

C. Malpighian tubules are an important adaptation for conserving water in insects

■ 8 Malpighian tubules have blind ends that lie in the (a)_____. Wastes are transferred from blood to the tubule by (b)_____.

V. THE KIDNEY IS THE KEY VERTEBRATE ORGAN OF OSMOREGULATION AND EXCRETION

■ 9 The vertebrate kidney functions in excretion and osmoregulation and is vital in maintaining homeostasis. It typically functions by using three processes, namely,

_____.

■ 10 In most vertebrates, the _____ all aid the kidneys in ridding the body of wastes and maintaining fluid balance.

A. Freshwater animals must rid themselves of excess water

B. Marine animals must replace lost fluid

C. The mammalian kidney is vital in maintaining homeostasis

VI. THE KIDNEYS, URINARY BLADDER, AND THEIR DUCTS MAKE UP THE HUMAN URINARY SYSTEM

■ 11 The urinary system is the principal excretory system in human beings and other vertebrates. Urine produced in the kidneys is transported to the urinary bladder through two tubes called (a)_____, then it passes from the bladder to the outside through the (b)_____.

A. The nephron is the functional unit of the kidney

■ 12 The functional units of the kidneys are the nephrons. Each nephron consists of a cup-shaped (a)_____ and a (b)_____. The filtrate passes through structures in the nephron in this sequence:

glomerulus ——> (c)_____ ——> proximal convoluted tubule ——> (d)_____ ——> distal convoluted tubule ——> (e)_____ ——> ureters

B. Urine is produced by filtration, reabsorption, and secretion

■ 13 Urine formation is accomplished by filtration of plasma, reabsorption of needed materials, and secretion of a few substances into the renal tubule. (a)_____ is not a selective process, whereas (b)_____ is highly selective.

■ 14 Plasma is filtered out of the glomerular capillaries and into _____ _____. Needed materials as well as wastes and excess substances become part of the filtrate.

■ 15 Most of the filtrate is reabsorbed from the _____ back into the blood in order to return needed materials to the blood and adjust the composition of the filtrate.

■ 16 _____ is the adjusted filtrate consisting of water, nitrogenous wastes, salts, and other substances.

C. Urine concentration depends on a countercurrent mechanism

■ 17 There are two types of nephrons in kidneys. The "ordinary variety" in the cortex are called (a)_____ nephrons and those with an exceptionally long loop of Henle that extends deep into the medulla are called (b)_____ nephrons.

■ 18 Filtrate is concentrated as it moves downward through the (a)_____ loop and diluted as it moves upward through the (b)_____ loop. This movement of filtrate in opposite directions, called the (c)_____ _____, helps maintain a hypertonic interstitial fluid that draws water out of the collecting ducts.

■ 19 The _____ are efferent arterioles that collect water from interstitial fluid.

D. Urine is composed of water, nitrogenous wastes, and salts

E. Urine volume is regulated by the hormone ADH

■ 20 Urine volume is regulated by the hormone (a)_____. It is released by the (b)_____.

■ 21 The "thirst center" that responds to dehydration is located in the _____.

F. Sodium reabsorption is regulated by the hormone aldosterone

■ 22 Aldosterone secreted by the (a)_____ stimulates distal tubules and collecting ducts to reabsorb more (b)_____.

KEY TERMS

Frequently Used Prefixes and Suffixes: Use combinations of these prefixes and suffixes to generate terms for the definitions below.

Prefixes	The Meaning	Suffixes	The Meaning
juxta-	beside, near	-cyte	cell
podo-	foot		
proto-	first, earliest form of		

Prefix	Suffix	Definitions
_____	-nephridium	1. The flame cell excretory organs of lower invertebrates and of some larval higher animals; the earliest form of specialized excretory organ.
_____	_____	2. A specialized epithelial cell possessing elongated foot processes, which cover the surfaces of most of the glomerular capillaries.
_____	-medullary	3. Pertains to nephrons situated nearest the medulla of the kidney.

Other Terms You Should Know

ADH
afferent arteriole
aldosterone
ammonia
antennal gland
antidiuretic hormone
Bowman's capsule
collecting duct
countercurrent mechanism
deamination
diabetes insipidus
distal convoluted tubule
efferent arteriole
elimination
excretion
excretory system
filtration
filtration barrier
flame cell

glomerular filtrate
glomerulus (-li)
green gland
juxtamedullary nephron
kidney
loop of Henle
Malpighian tubule
metanephridium (-ia)
nephridial organ
nephron
osmoregulation
osmotic conformer
osmotic regulator
peritubular capillary (-ries)
proximal convoluted tubule
reabsorption
renal corpuscle
renal cortex
renal medulla

renal pelvis
renal threshold
renal tubule
renin
secretion
slit pore
Tm
tubular transport maximum
urea
urea cycle
ureter
urethra
uric acid
urinalysis
urinary bladder
urinary system
urination
urine
vasa recta

SELF TEST

Multiple Choice: Some questions may have more than one correct answer.

1. The amount of needed substances that can be reabsorbed from renal tubules is a function of
   a. Tm.
   b. blood pH.
   c. tubular transport maximum.
   d. concentration of urea in forming urine.
   e. renal threshold.

2. The principal functional unit(s) in the vertebrate kidneys is/are
   a. nephridia.
   b. nephrons.
   c. Bowman's capsules.
   d. antennal glands.
   e. Malpighian tubules.

3. Water is reabsorbed by interstitial fluids when osmotic concentration in interstitial fluid is increased by
   a. salts from filtrate.
   b. urea from filtrate.
   c. free ions.
   d. concentration of filtrate.
   e. deamination in kidney cells.

4. Urea is a principal nitrogenous waste product produced by
   a. amphibians.
   b. the kidneys.
   c. nephrons.
   d. mammals.
   e. the liver.

5. Most reabsorption of filtrate in the kidney takes place at the
   a. ureter.
   b. loop of Henle.
   c. proximal convoluted tubule.
   d. urethra.
   e. collecting duct.

6. Urea is synthesized
   a. from uric acid.
   b. in kidneys.
   c. from ammonia and carbon dioxide.
   d. in the urea cycle.
   e. in aquatic invertebrates.

7. Relative to sea water, fluids in the bodies of marine organisms
   a. are isotonic.
   b. are hypotonic.
   c. are hypertonic.
   d. lose water.
   e. gain water.

8. The duct in humans that leads from the urinary bladder to the outside is the
   a. ureter.
   b. loop of Henle.
   c. proximal convoluted tubule.
   d. urethra.
   e. collecting duct.

9. Which of the following statements most accurately describes changes in the concentration of filtrate in the two portions of the loop of Henle?
   a. decreases in both
   b. increases in both
   c. increases in descending/decreases in ascending
   d. decreases in descending/increases in ascending
   e. remains essentially the same in both

10. A potato bug would excrete wastes by means of
    a. nephridia.
    b. nephrons.
    c. a pair of kidney-like structures.
    d. green glands.
    e. Malpighian tubules.

11. Relative to fresh water, fluids in the bodies of aquatic organisms
    a. are isotonic.
    b. are hypotonic.
    c. are hypertonic.
    d. lose water.
    e. gain water.

12. Excretion is specifically defined as
    a. maintaining water balance.
    b. homeostasis.
    c. removal of metabolic wastes from the body.
    d. the elimination of undigested wastes.
    e. the concentration of nitrogenous products.

13. The main way(s) that *any* excretory system maintains homeostasis in the body is/are to
    a. excrete metabolic wastes.
    b. eliminate undigested food.
    c. regulate body fluid constituents.
    d. regulate salt and water.
    e. concentrate urea in urine.

14. Crustacean excretory organs include
    a. flame cells.
    b. antennal glands.
    c. nephridia.
    d. Bowman's capsules.
    e. filtration of fluid from blood into a coelomic sac.

15. The group(s) of animals that have no specialized excretory systems include
    a. insects.
    b. cnidarians.
    c. annelids.
    d. sponges.
    e. sharks and rays.

16. The principal forms of nitrogenous waste products in various animal groups include(s)
    a. uric acid.
    b. carbon dioxide.
    c. amino acids.
    d. ammonia.
    e. urea.

17. The principal nitrogenous waste product(s) in human urine is/are
    a. uric acid.
    b. carbon dioxide.
    c. amino acids.
    d. ammonia.
    e. urea.

18. The first step in the catabolism of amino acids
    a. is conversion of ammonia to uric acid.
    b. is deamination.
    c. produces ammonia.
    d. is removal of the amino group.
    e. is accomplished with peptidases.

19. The type(s) of excretory organ(s) found in animals that collect wastes in flame cells is/are
    a. nephridia.
    b. nephrons.
    c. branching tubes that open to the outside through pores.
    d. green glands.
    e. Malpighian tubules.

20. Conservation of body fluids in a monarch butterfly is accomplished by
    a. ingestion of large volumes of fluids.
    b. concentrating urea in wastes.
    c. reabsorption of water from the digestive tract.
    d. reabsorption of water into the hemocoel.
    e. reabsorption of water from excretory tubules.

VISUAL FOUNDATIONS

Color the parts of the illustration below as indicated.

RED   ❑ abdominal aorta
GREEN   ❑ adrenal gland
YELLOW   ❑ ureter and renal pelvis
BLUE   ❑ inferior vena cava
ORANGE   ❑ urethra
BROWN   ❑ kidney
TAN   ❑ urinary bladder
PINK   ❑ renal artery
VIOLET   ❑ renal vein

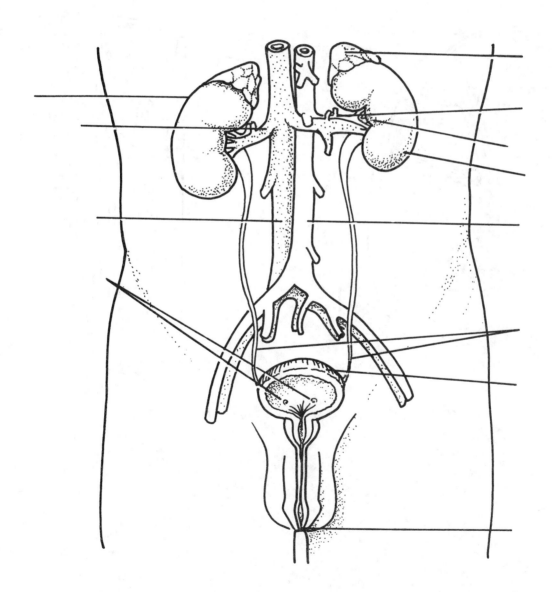

Color the parts of the illustration below as indicated.

RED	❑	artery
GREEN	❑	renal pelvis
YELLOW	❑	collecting tubule
BLUE	❑	vein
ORANGE	❑	loop of Henle
BROWN	❑	medulla
TAN	❑	cortex
PINK	❑	glomerulus
VIOLET	❑	proximal convoluted tubule and distal convoluted tubule

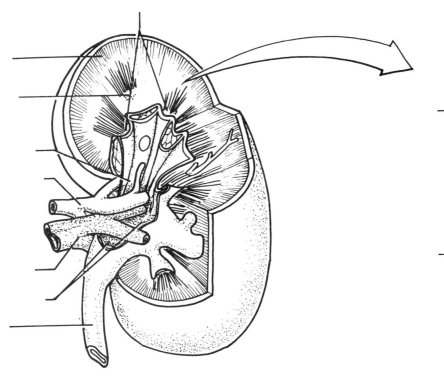

a

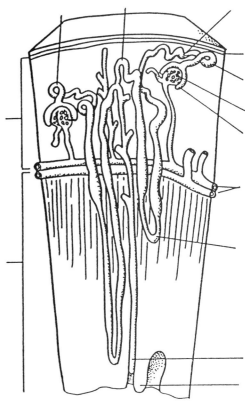

b

CHAPTER 47

□

Animal Hormones: Endocrine Regulation

OUTLINE

The chemical nature of hormones
Regulation of hormone secretion
Mechanisms of hormone action
Invertebrate hormones
Vertebrate hormones

The endocrine system is a diverse collection of glands and tissues that secrete hormones — chemical messengers that are transported by the blood to target tissue where they stimulate a physiological change. The endocrine system works closely with the nervous system to maintain the steady state of the body. Hormones may be steroids, peptides, proteins, or derivatives of amino acids or fatty acids. They diffuse from the blood into the interstitial fluid and then combine with receptor molecules on or in the cells of the target tissue. Some hormones activate genes that lead to the synthesis of specific proteins. Others activate a second messenger that relays the hormonal message to the appropriate site within the cell. Most invertebrate hormones are secreted by neurons rather than endocrine glands. They regulate growth, metabolism, reproduction, molting, and pigmentation. In vertebrates, hormonal activity is controlled by the hypothalamus, which links the nervous and endocrine systems. Vertebrate hormones regulate growth, reproduction, salt and fluid balance, and many aspects of metabolism. The hypothalamus secretes hormones that target the pituitary gland. The pituitary, in turn, secretes hormones that regulate a variety of body functions. Hormones produced by the thyroid gland regulate metabolic rate. The parathyroid glands regulate the calcium level in the blood. The islets of the pancreas secrete insulin, which acts to lower the glucose level in the blood, and glucagon, which acts to elevate it. The adrenal glands secrete hormones that help the body cope with stress. Malfunction of any of the endocrine glands can lead to specific disorders.

CHAPTER OUTLINE AND CONCEPT REVIEW (Fill in the blanks)

I. INTRODUCTION

■ 1 The endocrine system is a collection of ductless (a)_____ that works closely with the (b)_____ system.

■ 2 Hormones are transported throughout the body by blood, however, they affect only specific areas known as _____.

■ 3 Hormones secreted by neurons are (a)_____ and the cells that secrete them are called (b)_____ cells.

363

II. HORMONES CAN BE ASSIGNED TO FOUR CHEMICAL GROUPS

■ **4** The four chemical groups of hormones _____
_____.

■ **5** In vertebrates, testes and ovaries secrete steroids synthesized from
(a)_____. Steroids are classified as (b)_____ molecules.

III. HORMONE SECRETION IS REGULATED BY NEGATIVE FEEDBACK MECHANISMS

IV. HORMONES COMBINE WITH SPECIFIC RECEPTOR PROTEINS OF TARGET CELLS

A. Some hormones activate genes

■ **6** Steroid hormones pass through the plasma membrane of a target cell and combine
with specific receptors in the cell. These hormone-receptor complexes combine with
(a)_____ in the nucleus, which activates certain genes to synthesize
(b)_____ that in turn codes for specific proteins.

B. Some hormones work through second messengers

■ **7** Many protein hormones combine with receptors in the cell membrane of the target
cell and act by way of a second messenger. Two common second messengers are
(a)_____, which is derived from ATP, and (b)_____,
which, for example, takes a role in disassembling microtubules.

■ **8** Second messengers act by altering the functions of _____.

C. Prostaglandins are local chemical mediators

■ **9** Prostaglandins may help regulate hormone action by regulating the formation of
(a)_____. They are sometimes called (b)_____
because they usually act only on nearby cells.

■ **10** Prostaglandins are chemically classified as PGA through PGI, according to their
structure. They are also assigned a numerical subscript. For example, PGF_2 means
that PGF has two _____.

V. INVERTEBRATE HORMONES REGULATE GROWTH, METABOLISM, REPRODUCTION, MOLTING, AND PIGMENTATION

■ **11** Most invertebrate hormones are secreted by _____ rather than by endocrine
glands. They help to regulate regeneration, molting, metamorphosis, reproduction,
and metabolism.

A. Color change in crustaceans is regulated by hormones

B. Insect development is regulated by hormones

■ **12** Hormones control development in insects. Neurosecretory cells in the brain of
insects produce the "brain hormone" called (a)_____,
which stimulates (b)_____ glands to produce the molting hormone
(c)_____.

VI. VERTEBRATE HORMONES REGULATE GROWTH, METABOLISM, AND REPRODUCTION

A. Endocrine disorders may result from hyposecretion or hypersecretion

B. Nervous system and endocrine regulation are integrated by the hypothalamus

■13 The hypothalamus regulates the anterior lobe of the pituitary gland by producing several _____ hormones that regulate secretion of pituitary hormones.

C. The posterior lobe of the pituitary gland releases two hormones

■14 The hypothalamus produces (a)_____, which causes collecting ducts in the kidney to reabsorb more water. It also produces (b)_____, which stimulates the breasts of nursing mothers. Although produced in the hypothalamus, these hormones are released by the (c)_____.

D. The anterior lobe of the pituitary gland regulates growth and several other endocrine glands

■15 _____ hormones stimulate other endocrine glands.

■16 _____ stimulates mammary glands to produce milk.

E. Growth hormone stimulates protein synthesis

■17 Among hormones that influence growth are growth hormone (GH), also called (a)_____, secreted by the (b)_____, and thyroid hormones, secreted by the thyroid gland. The fundamental function of both GH and the thyroid hormones at the molecular level is to promote (c)_____.

■18 Release of growth hormone is regulated by a negative feedback mechanism that signals the (a)_____. When GH blood titers are high, this structure secretes (b)_____, which slows the release of GH from the pituitary. When GH titers are low, (c)_____ is secreted, so the pituitary releases more GH.

■19 Hypersecretion of GH during childhood may cause a disorder known as (a)_____. In adulthood, hypersecretion can cause "large extremities," or (b)_____.

F. Thyroid hormones stimulate metabolic rate

■20 One thyroid hormone, (a)_____, is also known as T_4 because each molecule contains four atoms of (b)_____.

■21 Thyroid secretion is regulated by a negative feedback system between the thyroid gland and the (a)_____ gland, which releases more (b)_____ when thyroid hormone blood titers are too low.

■22 Hyposecretion of thyroid hormones during infancy and childhood may result in retarded development, a condition known as _____.

G. The parathyroid glands regulate calcium concentration

■ 23 Parathyroid hormone regulates calcium levels in body fluids by stimulating release of calcium from (a)_____ and calcium reabsorption by the (b)_____ _____. A negative feedback system induces the (c)_____ gland to secret (d)_____, which inhibits parathyroid hormone activity.

H. The islets of the pancreas regulate glucose concentration

■ 24 Glucose concentration in the blood is regulated mainly by "islet hormones." Numerous clusters of cells in the pancreas, named (a)_____, secrete hormones that regulate glucose concentration. When blood glucose levels are high, (b)_____ cells release the hormone (c)_____; when it is low, (d)_____ cells release (e)_____.

■ 25 Insulin lowers the concentration of glucose in the blood by stimulating uptake of glucose by (a)_____; and by promoting (b)_____, the formation of glycogen.

■ 26 Glucagon raises the blood sugar level by stimulating the liver to convert glycogen to glucose, a process called (a)_____, and by (b)_____, the synthesis of glucose from other metabolites.

■ 27 In the carbohydrate metabolism disorder known as _____, cells are unable to utilize glucose properly and turn to fat and protein for fuel.

I. The adrenal glands help the body adapt to stress

■ 28 The adrenal medulla and the adrenal cortex secrete hormones that help the body cope with stress. The adrenal medulla secretes the two hormones _____ _____ (also known as adrenaline and noradrenaline) which increase heart rate, metabolic rate, and strength of muscle contraction, and reroute the blood to organs that require more blood in time of stress.

■ 29 The adrenal cortex produces three types of hormones in appreciable amounts. (a)_____. Of these, the (b)_____ known as (c)_____ is converted to the masculinizing hormone testosterone.

■ 30 Kidneys reabsorb more sodium and excrete more potassium in response to the mineralocorticoid hormone _____.

■ 31 The main function of _____ is to enhance gluconeogenesis in the liver.

■ 32 Stress stimulates secretion of CRF, which is the abbreviation for (a)_____ _____, from the hypothalamus. CRF stimulates the anterior pituitary to release (b)_____, which regulates the secretion of glucocorticoids and aldosterones.

J. Many other hormones are known

■ 33 The thymus gland produces the hormone _____.

■ 34 _____ from the kidneys helps regulate blood pressure.

■ 35 The heart secretion ANF, which stands for _____,
lowers blood pressure.

■ 36 The _____ gland in the brain produces the hormone _____ which
influences the onset of sexual maturation.

KEY TERMS

Frequently Used Prefixes and Suffixes: Use combinations of these prefixes and suffixes to generate
terms for the definitions below.

Prefixes	The Meaning		Suffixes	The Meaning
hyper-	over		-megaly	enlargement
hypo-	under			
neuro-	nerve			

Prefix	Suffix	Definitions
_____	-hormone	1. A hormone secreted by a nerve cell.
_____	-secretory	2. Refers to a nerve cell that secretes a neurohormone.
_____	-secretion	3. An excessive secretion; over secretion.
_____	-secretion	4. A diminished secretion; under secretion.
_____	-thyroidism	5. A condition resulting from an overactive thyroid gland.
_____	-glycemia	6. An abnormally low level of glucose in the blood.
_____	-glycemia	7. An abnormally high level of glucose in the blood.
acro-	_____	8. An abnormal condition characterized by enlargement of the head, and sometimes other structures.

Other Terms You Should Know

acromegaly
ACTH
Addison's disease
adenylate cyclase
ADH
adrenal cortex (-tices)
adrenal gland
adrenal medulla (-as)(-ae)
adrenocorticotropic hormone
alarm reaction
aldosterone
alpha cell
amine
anterior pituitary
beta cell
BH
brain hormone
calcitonin

calmodulin
cAMP
catecholamine
corpus (-pora)
  allatum (-ta)
corpus (-pora)
  cardiacum (-ca)
corticotropin releasing factor
cortisol
cretinism
CRF
Cushing's disease
cyclic AMP
diabetes mellitus
ecdysone
endocrine gland
endocrine system
endocrinology

epinephrine
exocrine gland
G protein
GDP
GH
GHIH
GHRH
gigantism
glucagon
glucocorticoid
goiter
growth hormone
growth hormone-inhibiting
  hormone
GTP
guanosine diphosphate
guanosine triphosphate
hormone

insulin	oxytocin	release-inhibiting hormone
insulin shock	parathyroid gland	releasing hormone
islets of Langerhans	parathyroid hormone	second messenger
juvenile hormone	peptide	somatomedin
ketosis	pheromone	target tissue
local hormone	pineal gland	thymus gland
melatonin	pituitary dwarf	thyroid gland
mineralocorticoid	pituitary gland	thyroid hormone
molting hormone	posterior lobe of pituitary	thyroid-stimulating hormone
myxedema	gland	thyroxine
negative feedback	prolactin	triiodothyronine
neurosecretory cell	prostaglandin	tropic hormone
norepinephrine	protein kinase	TSH
optic gland	prothoracic gland	

SELF TEST

Multiple Choice: Some questions may have more than one correct answer.

1. A person with a fasting level of 750 mg glucose per 100 ml of blood
   a. is hyperglycemic.
   b. is hypoglycemic.
   c. is about normal.
   d. probably has cells that are not using enough glucose.
   e. has too much insulin.

2. Diabetes in an adult with ample numbers of adequately functioning beta cells
   a. is type I diabetes.
   b. is type II diabetes.
   c. indicates that target cells are not using insulin.
   d. indicates that not enough insulin is being produced.
   e. is unusual since this condition usually occurs in children.

3. Control of hormone secretions by a negative feedback mechanism generally includes
   a. hyperactivity.
   b. hypoactivity.
   c. change in a direction that tends to maintain homeostasis.
   d. increased hormone secretion with increase in hormone effect.
   e. decreased hormone secretion with increase in hormone effect.

4. A three-year old cretin and an adult suffering from myxedema most likely have
   a. a low metabolic rate.
   b. Cushing's disease.
   c. Addison's disease.
   d. a low titer of thyroid hormones.
   e. a high titer of thyroid hormones.

5. The molecule PGF_2
   a. has two double bonds.
   b. is a neurohormone.
   c. probably affects cyclic AMP formation.
   d. is a calcium-binding protein.
   e. is a prostaglandin.

6. Small, hydrophobic hormones that form a hormone-receptor complex with intranuclear receptors include
   a. steroids.
   b. cyclic AMP.
   c. chemicals that activate genes.
   d. thyroid hormones.
   e. prostaglandins.

7. A chemical produced by one cell that has a specific regulatory effect on another cell is the definition of a
   a. neurohormone.
   b. hormone.
   c. exocrine gland secretion.
   d. pheromone.
   e. prohormone.

8. The hormone and target tissue that are involved in raising glucose concentration in blood by glycogenolysis and gluconeogenesis are
   a. insulin and pancreas.
   b. glucagon and liver.
   c. thyroid-stimulating hormone and thyroid gland.
   d. adrenocorticotropic hormone and adrenal cortex.
   e. thyroxine and various metabolically active cells.

9. Increased skeletal growth results directly and/or indirectly from activity of
   a. aldosterone.
   b. growth hormone.
   c. hormones from the hypothalamus.
   d. somatomedins.
   e. epinephrine and/or norepinephrine.

10. Activity of the anterior lobe of the pituitary gland is controlled by
    a. the hypothalamus.
    b. ADCH.
    c. epinephrine and norepinephrine.
    d. releasing hormones.
    e. release-inhibiting hormones.

VISUAL FOUNDATIONS

Color the parts of the illustration below as indicated.

RED ☐ receptor molecule
GREEN ☐ ribosome
YELLOW ☐ hormone
BLUE ☐ DNA
ORANGE ☐ mRNA
BROWN ☐ endocrine gland cell
TAN ☐ nucleus
PINK ☐ blood vessel
VIOLET ☐ protein molecule

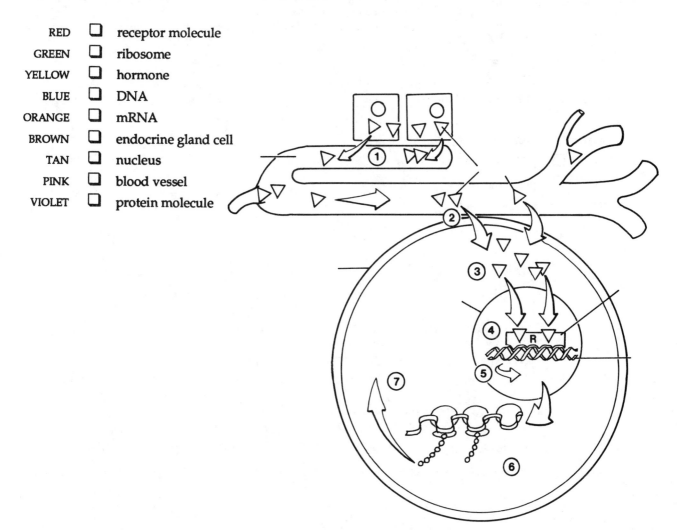

Color the parts of the illustration below as indicated.

RED	☐	receptor
GREEN	☐	cytoplasm
YELLOW	☐	hormone
BLUE	☐	extracellular fluid
ORANGE	☐	second messenger
BROWN	☐	plasma membrane

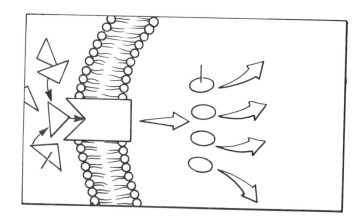

Color the parts of the illustration below as indicated.

RED	☐	pituitary gland
GREEN	☐	pineal gland
YELLOW	☐	thyroid gland
BLUE	☐	parathyroid gland
ORANGE	☐	hypothalamus
BROWN	☐	thymus gland
TAN	☐	pancreas
PINK	☐	testis and ovary
VIOLET	☐	adrenal gland

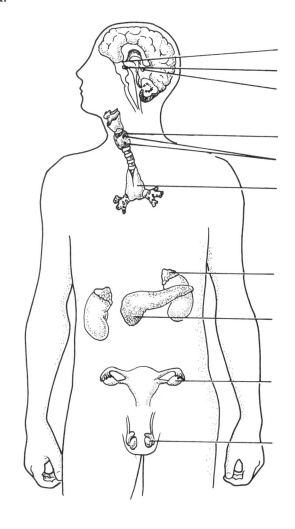

CHAPTER 48

Reproduction

OUTLINE

Asexual reproduction
Sexual reproduction
Reproductive variations: Metagenesis, parthenogenesis,
and hermaphroditism
Human reproduction: The male
Human reproduction: The female
Human sexual response
Fertilization
Infertility
Birth control
Abortion
Sexually transmitted diseases

For a species to survive, it is essential that it replace individuals that die. Animal species do this either asexually or, more commonly, sexually. In asexual reproduction, a single parent splits, buds, or fragments, giving rise to two or more offspring that are genetically identical to the parent. This method of reproduction is advantageous for sessile animals that cannot move about to search for mates. In sexual reproduction, sperm contributed by a male parent, and an egg contributed by a female parent, unite, forming a fertilized egg, or zygote, that develops into a new organism. This method of reproduction has the advantage of promoting genetic variety among the offspring, and therefore, giving rise to individuals that may be better able to survive than either parent. Depending on the species, fertilization occurs either outside the body (most aquatic animals) or inside the body (most terrestrial animals). Some animals have both asexual and sexual stages. In some, the unfertilized egg may develop into an adult animal. In still others, both male and female reproductive organs may occur in the same individual. More typically, animals have only a sexual stage, fertilization is required for development to occur, and male and female reproductive organs occur only in separate individuals. The complex structural, functional, and behavioral processes involved in reproduction in vertebrates are regulated by hormones secreted by the brain and gonads. Sexual stimulation in humans results in increased blood flow to reproductive structures and the skin, and increased muscle tension. Sexual response includes sexual desire, excitement, orgasm, and resolution. There are a number of causes of infertility in humans. There are a variety of effective methods of birth control. Next to the common cold, sexually transmitted diseases are the most prevalent communicable diseases in the world.

CHAPTER OUTLINE AND CONCEPT REVIEW (Fill in the blanks)

I. INTRODUCTION

II. ASEXUAL REPRODUCTION IS COMMON AMONG SOME ANIMAL GROUPS

■ 1 In asexual reproduction, a single parent endows its offspring with a set of genes identical with its own. Sponges and cnidarians can reproduce by _____, involving the production of a new individual from a group of expendable cells that separate from the "parent's" body.

■ 2 _____ is a form of asexual reproduction whereby a "parent" breaks into several pieces, each piece giving rise to a new individual.

III. SEXUAL REPRODUCTION IS THE MOST COMMON TYPE OF ANIMAL REPRODUCTION

■ 3 In sexual reproduction, each of two parents contributes a sex cell or _____ containing half of the offspring's genetic endowment.

IV. ANIMALS HAVE EVOLVED INTERESTING REPRODUCTIVE VARIATIONS

A. Metagenesis is characteristic of some animal groups

■ 4 Metagenesis means that an organism has both _____ reproductive stages.

B. Fertilization does not occur in parthenogenesis

■ 5 Parthenogenesis refers to the production of a complete organism from a(n) _____.

C. In hermaphroditism, one individual produces sperm and eggs

■ 6 Most hermaphroditic organisms copulate and fertilize each other, however, a few, such as the _____, for example, are capable of self fertilization.

V. HUMAN REPRODUCTION: THE MALE PROVIDES SPERM

A. The testes produce sperm

■ 7 The testes, housed in an external sac called the (a)_____, contain the (b)_____, where sperm are produced, and the (c)_____, which secrete the male hormone (d)_____.

■ 8 The initial development of a sperm begins with an undifferentiated stem cell called a (a)_____, which enlarges to form the (b)_____ _____ that goes through meiosis.

■ 9 The _____ at the front of a sperm contains enzymes that assist the sperm in penetrating an egg.

■ 10 The _____ are the passage ways that testes take as they descend through the abdominal wall into the scrotum.

B. A series of ducts transports sperm

■ 11 Sperm complete their maturation and are stored in the (a)_____, from which they are moved up into the body through the (b)_____.

■ **12** During ejaculation, sperm pass through the (a)_____
into the (b)_____, which passes through the penis.

C. The accessory glands produce the fluid portion of semen

■ **13** Human semen contains about (a)_____(#?) sperm suspended in
secretions that are mostly produced by two glands — the (b)_____
_____.

D. The penis transfers sperm to the female

■ **14** The penis consists of an elongated (a)_____ that terminates in an expanded
portion called the (b)_____, which is partially covered by a fold of skin called
the (c)_____. An operation that removes this "extra" skin is
known as (c)_____.

■ **15** The penis contains three columns of erectile tissue referred to as (a)_____
_____. When they become engorged with (b)_____, the
penis erects.

E. Reproductive hormones promote sperm production and maintain masculinity

■ **16** The two gonadotropic hormones (a)_____, stimulate sperm production and
secretion of the hormone (b)_____, which is responsible for
establishing and maintaining primary and secondary sex characteristics in the male.

VI. HUMAN REPRODUCTION: THE FEMALE PRODUCES OVA AND INCUBATES THE EMBRYO

■ **17** The principal organs of the female reproductive system are the _____
_____.

A. The ovaries produce ova and sex hormones

■ **18** Formation of ova is called (a)_____. In this process, stem cells in the
fetus, the (b)_____, enlarge, forming the (c)_____,
which begin meiosis before birth.

■ **19** A developing ovum and the cells immediately around it comprise the
(a)_____. At puberty, the hormone (b)_____ stimulates
development of these cells, causing some of them to complete meiosis. One of the
resulting cells, the (c)_____, continues development to
form the ovum.

■ **20** After ovulation, the portion of the follicle that remains in the ovary develops into an
endocrine gland called the _____.

B. The uterine tube transports the ovum

■ **21** After ovulation, the ovum enters the _____,
where it may be fertilized.

C. The uterus incubates the embryo

■ **22** Embryos normally implant in the (a)_____, which is the inner
layer of cells of the uterus. This signals the beginning of pregnancy. If there is no
embryo, or the embryo fails to implant, the bleeding phase known as
(b)_____ commences.

■ 23 The lower portion of the uterus, the _____, forms the back wall of the vaginal tube.

D. The vagina receives sperm

E. The vulva are external genital structures

■ 24 The structures that comprise the vulva are the _____
_____.

F. The breasts function in lactation

■ 25 (a)_____ is the production of milk. It begins as a result of stimulation by the hormone (b)_____ secreted by the (c)_____
_____.

G. The menstrual cycle is regulated by hormones

■ 26 Events of the menstrual cycle are coordinated by the _____
_____ hormones.

■ 27 The first day of menstrual bleeding marks the first day of the menstrual cycle. _____ occurs at about day 14 in an average, typical 28-day menstrual cycle.

■ 28 FSH stimulates follicle development. The developing follicles release (a)_____, which stimulates development of the endometrium. The two hormones (b)_____ stimulate ovulation, and (d)_____ also promotes development of the corpus luteum.

■ 29 The corpus luteum secretes _____, which stimulates final preparation of the uterus for possible pregnancy.

H. In menopause the ovaries secrete less estrogens and progesterone

VII. SEXUAL RESPONSE INVOLVES PHYSIOLOGICAL CHANGES

■ 30 _____ are two basic physiological responses to sexual stimulation.

■ 31 The cycle of sexual response includes four phases: the _____
_____.

VIII. FERTILIZATION IS THE FUSION OF SPERM AND EGG TO PRODUCE A ZYGOTE

IX. INFERTILITY IS THE INABILITY TO ACHIEVE CONCEPTION

■ 32 Male infertility is usually the result of _____, which means lack of or insufficient quantities of viable sperm.

X. BIRTH CONTROL METHODS ALLOW INDIVIDUALS TO CHOOSE

A. Hormone contraceptives prevent ovulation

■ 33 Most oral contraceptives are combinations of the synthesized hormones _____
_____.

B. Use of the intrauterine device (IUD) has declined

C. Other common contraceptive methods include the diaphragm and the condom

■ 34 The condom is the only contraceptive device that affords some protection against

_____.

D. Sterilization renders an individual incapable of producing offspring

■ 35 Male sterilization is by (a)_____ and female sterilization is by (b)_____.

XI. THERE ARE THREE TYPES OF ABORTION

XII. SEXUALLY TRANSMITTED DISEASES ARE SPREAD BY SEXUAL CONTACT

KEY TERMS

Frequently Used Prefixes and Suffixes: Use combinations of these prefixes and suffixes to generate terms for the definitions below.

Prefixes	The Meaning	Suffixes	The Meaning
acro-	extremity, height	-ectomy	excision
circum-	around, about	-gen(esis)	production of
contra-	against, opposite, opposing	-some	body
endo-	within		
intra-	within		
oo-	egg		
partheno-	virgin		
post-	behind, after		
pre-	before, prior to, in advance of, early		
spermato-	seed, "sperm"		
vaso-	vessel		

Prefix	Suffix	Definitions
_____	_____	1. The production of an adult organism from an egg in the absence of fertilization; virgin development.
_____	_____	2. The production of sperm.
_____	_____	3. The organelle (body) at the extreme tip of the sperm head that helps the sperm penetrate the egg.
_____	-cision	4. The removal of all or part of the foreskin by cutting all the way around the penis.
_____	_____	5. The production of eggs or ova.
_____	-metrium	6. The lining within the uterus.
_____	-ovulatory	7. Pertains to a period of time after ovulation.
_____	-menstrual	8. Pertains to a period of time prior to menstruation.
vas-	_____	9. Male sterilization in which the vas deferentia are partially excised.
_____	-congestion	10. The engorgement of vessels or sinusoids with blood.

_____ -uterine   11. Located or occurring within the uterus.

_____ -ception   12. Birth control; techniques or devices opposing conception.

Other Terms You Should Know

abortion
alveolus (-li)
asexual reproduction
breast
budding
bulbourethral gland
castration
cavernous body (-dies)
cervix (-xes)(-ices)
clitoris
coitus
colustrum
conception
condom
contraceptive diaphragm
contraceptive sponge
corona radiata
corpus (-pora) cavernosum (-sa)
corpus (-pora) luteum (-ea)
ejaculation
ejaculatory duct
epididymis (-ides)
erectile dysfunction
erectile tissue
estrogen
eunuch
excitement phase
external fertilization
female pronucleus (-li)
fertilization
first polar body
follicle
follicle-stimulating hormone
foreskin
fragmentation
FSH
gamete
glans
GnRH

gonadotropin-releasing hormone
hCG
hermaphroditic
human chorionic gonadotropin
hymen
infertility
inguinal canal
inguinal hernia
internal fertilization
interstitial cell
intrauterine device
IUD
labium (-ia) majus (-jora)
labium (-ia) minus (-nora)
lactation
LH
ligament of Cooper
luteinizing hormone
male pronucleus (-li)
mammary gland
menopause
menstrual cycle
menstrual phase
menstruation
metagenesis
mons pubis
oogonium (-ia)
oral contraceptive
orgasm phase
ovary (-ries)
oviduct
ovulation
ovum (-va)
penis (-nes)(-ses)
PMS
postovulatory phase
premenstrual syndrome
preovulatory phase

prepuce
primary oocyte
primary spermatocyte
progesterone
prolactin
prostate gland
resolution phase
scrotum (-ta)(-ms)
secondary oocyte
secondary sexual characteristic
secondary spermatocyte
semen
seminal vesicle
seminiferous tubule
sertoli cell
sexual reproduction
shaft
sperm duct
spermatic cord
spermatid
spermatogonium (-ia)
spontaneous abortion
sterile
sterility
sterilization
stroma
testis (-tes)
testosterone
therapeutic abortion
tubal ligation
urethra (-ae)(-as)
uterine tube
uterus (-ri)
vagina (-as)(-ae)
vas deferens
vulva (-ae)(-as)
zona pellucida
zygote

SELF TEST

Multiple Choice: Some questions may have more than one correct answer.

1. The tube(s) that transport(s) gametes away from gonads is/are the
   a. vas deferens.
   b. seminal vesicle.
   c. sperm cord.
   d. vagina.
   e. oviduct.

2. The normal, periodic flow of blood from the vagina is associated with
   a. ovulation.
   b. orgasm.
   c. sloughing of the endometrium.
   d. removal of the inner lining of the vagina.
   e. miscarriages.

3. Human female breast size is primarily a function of the amount of
   a. alveoli.
   b. milk glands.
   c. fat.
   d. milk the woman can produce after giving birth.
   e. colostrum.

4. Which of the following is/are true of the human penis?
   a. It has three erectile tissue bodies.
   b. Circumcision involves removal of the prepuce.
   c. The glans is an expanded portion of one of the cavernous bodies.
   d. It has a small bone that articulates with the pelvis.
   e. Erection results from increased blood pressure.

5. A human female at birth already has
   a. primary oocytes in the first prophase.
   b. preovulatory Graafian follicles.
   c. all the formative oogonia she will ever have.
   d. some secondary oocytes in meiosis II.
   e. zona pellucida.

6. The *basic* physiological responses that result from effective sexual stimulation are
   a. vasocongestion.
   b. erection.
   c. vaginal lubrication.
   d. muscle tension.
   e. orgasm.

7. Human chorionic gonadotropin (hCG)
   a. is produced by embryonic membranes.
   b. is secreted by the endometrium.
   c. affects functioning of the corpus luteum.
   d. is originally produced in the pituitary.
   e. causes menstrual flow to begin.

8. The hormone(s) primarily responsible for secondary sexual characteristics is/are
   a. luteinizing hormone.
   b. estrogen.
   c. follicle-stimulating hormone.
   d. testosterone.
   e. human chorionic gonadotropin.

9. The corpus luteum is directly or indirectly involved in
   a. producing estradiol.
   b. producing progesterone.
   c. stimulating glands to produce nutrients for an implanted embryo.
   d. responding to signals from embryonic membrane secretions.
   e. maintaining a newly implanted embryo.

10. Follicle-stimulating hormone (FSH) stimulates development of
    a. seminiferous tubules.
    b. interstitial cells.
    c. sperm cells.
    d. the corpus luteum.
    e. ovarian follicles.

11. The refractory period typically occurs
    a. in men, not women.   d. in the orgasmic phase.
    b. in both men and women.   e. in the resolution phase.
    c. in women, not men.

12. When used properly, the most effective form of birth control in a fertile person is
    a. condoms.   d. the rhythm method.
    b. douching.   e. spermicidal sponges.
    c. oral contraception.

13. The first measurable response to effective sexual stimulation
    a. is orgasm.   d. is penile erection.
    b. is ejaculation.   e. is vaginal lubrication.
    c. occurs in the excitement phase.

14. The form(s) of birth control that can definitely deter contraction of sexually transmitted diseases is/are
    a. condoms.   d. diaphragms.
    b. douching.   e. spermicides.
    c. oral contraception.

15. A count of 400 million sperm in a single ejaculation would be considered
    a. clinical sterility.   d. about average.
    b. above average.   e. impossible.
    c. below average.

16. A Papanicolaou test (Pap smear)
    a. can detect early cancer.   d. looks at cells from the menstrual flow.
    b. requires an abdominal incision.   e. examines cells scraped from the cervix.
    c. involves examination of cells taken from the lower portion of the uterus.

17. Primitive, undifferentiated stem cells that line the seminiferous tubules are called
    a. sperm.   d. secondary spermatocytes.
    b. spermatids.   e. spermatogonia.
    c. primary spermatocytes.

18. The external genitalia of the human female are collectively known as the
    a. vagina.   d. major and minor lips.
    b. clitoris.   e. vulva.
    c. mons pubis.

19. Which of the following is/are likely true of PMS?
    a. It causes fatigue.   d. It probably has a psychological basis.
    b. It causes anxiety.   e. Its symptoms usually occur just after the woman's period.
    c. The cause may be due to estradiol-progesterone imbalance.

20. In gametogenesis in the human male, the cell that develops after the first meiotic division is the
    a. sperm.   d. secondary spermatocyte.
    b. spermatid.   e. spermatogonium.
    c. primary spermatocyte.

21. Alternation of sexual and asexual generations is called
    a. parthenogenesis.   d. fertilization.
    b. metagenesis.   e. gametogenesis.
    c. hermaphroditism.

22. The portion of the ovarian follicle that remains behind after ovulation
    a. develops into the corpus luteum.
    b. becomes a gland.
    c. is called the Graafian follicle.
    d. is surrounded by the zona pellucida.
    e. transforms into the endometrium.

23. The term gonad refers to the
    a. penis.
    b. testes.
    c. vagina.
    d. ovaries.
    e. gametes.

24. Human sperm cell production occurs in the
    a. epididymis.
    b. sperm tube.
    c. testes.
    d. accessory glands.
    e. seminiferous tubules.

VISUAL FOUNDATIONS

Color the parts of the illustrations below as indicated. Label mitosis, first meiotic division, and second meiotic division.

RED   ☐   sperm
PINK   ☐   large chromosome
YELLOW   ☐   spermatogonia
BLUE   ☐   small chromosome
ORANGE   ☐   primary spermatocyte
BROWN   ☐   secondary spermatocyte
TAN   ☐   spermatids

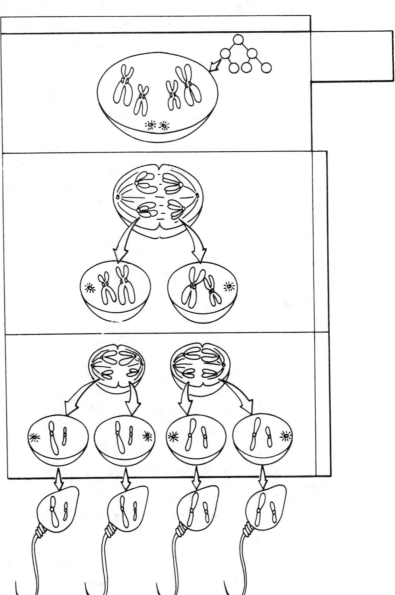

Color the parts of the illustrations below as indicated. Label mitosis, first meiotic division, and second meiotic division.

RED ☐ ovum

GREEN ☐ large chromosome

YELLOW ☐ oogonium

BLUE ☐ small chromosome

ORANGE ☐ primary oocyte

BROWN ☐ secondary oocyte

TAN ☐ polar body

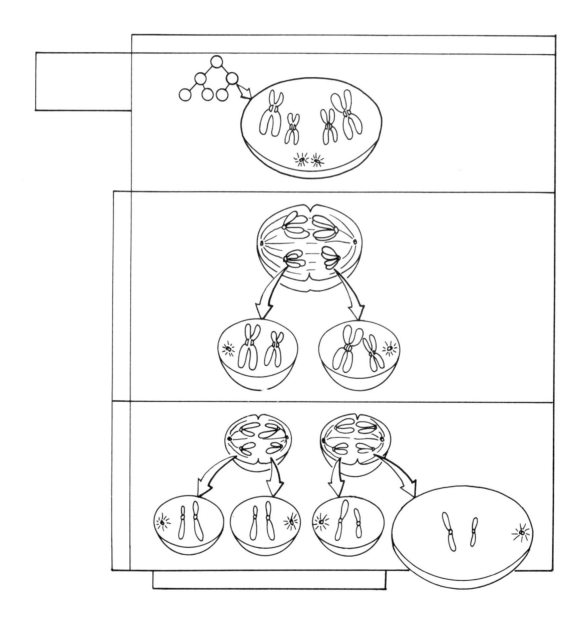

Development

O U T L I N E

Developmental processes
Fertilization
Cleavage: From one cell to many
Gastrulation: Formation of germ layers
Development of organs
Extraembryonic membranes and placenta
Human development
The human life cycle
The aging process

Development encompasses all of the changes that take place during the life of an organism from fertilization to death. Guided by instructions encoded in the DNA of the genes, the organism develops from one cell to billions, from a formless mass of cells to an intricate, highly specialized and organized organism. Development is a balanced combination of several interrelated processes: (1) cell division, (2) increase in the number and size of cells, (3) cellular movements that arrange cells into specific structures and appropriate body forms, and (4) biochemical and structural specialization of cells to perform specific tasks. The stages of early development, which are basically similar for all animals, include fertilization, cleavage, gastrulation, and organogenesis. All terrestrial vertebrates have four extraembryonic membranes that function to protect and nourish the embryo. Environmental factors, such as nutrition, vitamin and drug intake, cigarette smoke, and disease-causing organisms, can adversely influence developmental processes. Aging is a developmental process that results in decreased functional capacities of the mature organism.

CHAPTER OUTLINE AND CONCEPT REVIEW (Fill in the blanks)

I. INTRODUCTION

■ 1 (a)_____, a theory of development popular in the 17th century, postulated that each human egg contained a fully formed miniature person. The theory of (b)_____ opposes this concept, holding that development proceeds from one cell to billions.

II. DEVELOPMENT IS A BALANCED COMBINATION OF SEVERAL PROCESSES

■ 2 During development, the number of cells (a)_____(increases or decreases?), resulting in growth. Cells undergo (b)_____, a process during which they organize into specific structures. The process by which cells become specialized is called (c)_____.

III. FERTILIZATION RESTORES THE DIPLOID NUMBER OF CHROMOSOMES

■ 3 Fertilization restores the (a)_____ number of chromosomes, during which time the (b)_____ of the individual is determined.

A. The first step in fertilization involves contact and recognition

■ 4 The jelly coat, or (a)_____, in mammals is contacted by the sperm, causing the (b)_____ to release enzymes that hydrolyze the jelly coat down to the (c)_____ membrane.

■ 5 _____, a species-specific protein of the acrosome, adheres to a receptor on the vitelline membrane.

B. Sperm entry is regulated

■ 6 Microvilli on the egg elongate, forming the (a)_____, which contracts, thereby drawing the (b)_____ into the egg.

C. Sperm and egg pronuclei fuse

■ 7 _____ draw the sperm nucleus toward the egg nucleus.

D. Fertilization activates the egg

■ 8 The increase in (a)_____ ions within egg cytoplasm is associated with the (b)_____ reaction. It blocks polyspermy and triggers metabolic activity in the egg.

IV. DURING CLEAVAGE THE ZYGOTE DIVIDES, GIVING RISE TO MANY CELLS

■ 9 The zygote undergoes _____, forming a morula and then a blastula.

A. Cleavage provides building blocks for development

■ 10 During cleavage, cells increase in (a)_____, but they do not increase in (b)_____; consequently, the morula is a/the (c)_____(same, larger, smaller) size compared to the zygote.

B. The amount of yolk determines the pattern of cleavage

■ 11 The isolecithal eggs of (a)_____ undergo (b)_____ cleavage, that is, the entire egg divides into cells of about the same size. (c)_____ cleavage of isolecithal eggs is common in deuterostomes, while some protostomes undergo (d)_____ cleavage.

■ 12 The (a)_____ cleavage that occurs in the telolecithal eggs of reptiles and birds is restricted to the (b)_____.

V. THE GERM LAYERS DEVELOP DURING GASTRULATION

■ 13 The blastula becomes a three-layered embryo during (a)_____; these layers, collectively called (b)_____, include the _____ _____.

A. Each germ layer has a specific fate

■ 14 (a)_____ gives rise to the lining of the gut, the lungs, and the liver. (b)_____ forms epidermis and the nervous system. The (c)_____ gives rise to the skeleton, muscle, circulatory system, excretory system, and reproductive system.

B. The pattern of gastrulation is affected by the amount of yolk present

■ 15 In the sea star or *Amphioxus*, cells from the blastula wall invaginate and form a cavity called the _____.

■ 16 In the amphibian, cells from the animal pole move down over the yolk-rich cells and invaginate, forming the dorsal lip of the _____.

■ 17 In the bird, invagination occurs at the _____ and no archenteron forms.

VI. ORGANOGENESIS BEGINS WITH THE DEVELOPMENT OF THE NERVOUS SYSTEM

■ 18 The (a)_____ induces the formation of the neural plate, from which cells migrate downward thus forming the (b)_____, flanked on each side by neural folds. When neural folds fuse, the (c)_____ is formed. This structure gives rise to the (d)_____.

■ 19 (a)_____ meet the (b)_____ thereby forming branchial arches that give rise to elements of the face, jaws, and neck.

VII. EXTRAEMBRYONIC MEMBRANES AND PLACENTA PROTECT AND NOURISH THE EMBRYO

■ 20 Terrestrial vertebrates have four extraembryonic membranes: _____ _____.

A. The chorion and amnion enclose the embryo

■ 21 The _____ is a fluid-filled sac that surrounds the embryo and keeps it moist.

B. The allantois functions in waste disposal

■ 22 The (a)_____, an outgrowth of the digestive tract, fuses with the chorion, forming the highly vascular (b)_____.

C. The yolk sac encloses the yolk

D. The placenta is an organ of exchange

■ 23 In placental mammals, the _____ tissues give rise to the placenta, the organ of exchange between mother and developing child.

■ 24 The (a)_____ connects the embryo with the placenta. It contains the (b)_____.

VIII. HUMAN PRENATAL DEVELOPMENT REQUIRES ABOUT 266 DAYS

A. Cleavage takes place in the uterine tube

■ 25 The (a)_____ dissolves when the embryo enters the uterus. At this time, the embryo is in the (b)_____ stage, but now begins to differentiate into a (c)_____.

B. The embryo implants in the wall of the uterus

■ 26 The trophoblast cells of the blastocyst secrete enzymes that digest an opening into the uterine wall in a process called (a)_____. The process begins at about the (b)_____ day of development.

■27 Implantation is complete by day (a)_____ of development. At this time, a woman will probably (b)_____(know or not know?) that she is pregnant.

C. Organ development begins during the first trimester

■28 Gastrulation occurs during the _____ weeks of human development.

■29 After the first two months of development, the embryo is referred to as a _____.

■30 Sex (gender) can be determined by external observation by _____.

D. The fetus continues to develop during the second and third trimesters

E. The birth process may be divided into three stages

■31 The normal duration of a human pregnancy, usually called the (a)_____ _____, is (b)_____ days. It culminates in (c)_____, the process of birth.

■32 During the first stage of labor the (a)_____ becomes dilated and effaced; during the second stage the (b)_____; and during the third stage the (c)_____.

F. The neonate must adapt to its new environment

G. Environmental factors influence the embryo

■33 By controlling environmental factors such as _____ _____ _____, to name a few, a pregnant woman can help ensure the well-being of her unborn child.

■34 A sonogram is a picture taken by the technique known as (a)_____. This technique and (b)_____, the analysis of intrauterine fluids, are helpful in anticipating potential birth defects.

IX. THE HUMAN LIFE CYCLE EXTENDS FROM CONCEPTION TO DEATH

■35 The human life cycle can be divided into the following stages: _____ _____.

X. HOMEOSTATIC RESPONSE TO STRESS DECREASES DURING AGING

KEY TERMS

Frequently Used Prefixes and Suffixes: Use combinations of these prefixes and suffixes to generate terms for the definitions below.

Prefixes	The Meaning	Suffixes	The Meaning
arch-	primitive	-blast	embryo
blasto-	embryo	-gen	production of
holo-	whole, entire	-mere	part
iso-	equal		
mero-	part, partial		
neo-	new, recent		
telo-	end		

terato-	monster
tri-	three
tropho-	nourishment

Prefix	Suffix	Definitions
_____	-lecithal	1. Having an accumulation of yolk at one end (vegetal pole) of the egg.
_____	-lecithal	2. Having a fairly equal distribution of yolk in the egg.
_____	_____	3. Any cell that is a part of the early embryo (specifically, during cleavage).
_____	-cyst	4. The blastula of the mammalian embryo.
_____	-pore	5. The opening into the archenteron of an early embryo.
_____	-blastic	6. A type of cleavage in which the entire egg divides.
_____	-blastic	7. A type of cleavage in which only the blastodisc divides; partial cleavage.
_____	-enteron	8. The primitive digestive cavity of a gastrula.
_____	_____	9. The extraembryonic part of the blastocyst that chiefly nourishes the embryo or that develops into fetal membranes with nutritive functions.
_____	-mester	10. A term or period of three months.
_____	-nate	11. A newborn child.
_____	_____	12. A substance or agent that may lead to the production of monsters or abnormal growths.

Other Terms You Should Know

acrosome reaction
adolescence
afterbirth
aging
amniocentesis
amnion
animal pole
bindin
blastocoele
blastodisc
blastula (-as)(-ae)
branchial arch (-hes)
branchial groove
cellular differentiation
cesarean section
childhood
chorioallantoic membrane
chorion
cleavage
conjoined twin

cortical reaction
development
ectoderm
endoderm
extraembryonic membrane
fertilization
fertilization cone
fetus
first trimester
fraternal twin
gastrula
gastrulation
germ layer
gestation period
HCG
Hensen's node
human chorionic
 gonadotropin
identical twin
implant

induce
infancy
inner cell mass
isolecithal
jelly coat
labor
lanugo
mesoderm
morphogenesis
morula (-as)(-ae)
navel
neonatal period
neural crest
neural fold
neural groove
neural plate
neural tube
organogenesis
parturition
pharyngeal pouch (-hes)

placenta
preformation theory
primitive streak
progeria
radial cleavage
sonogram

spiral cleavage
theory of epigenesis
umbilical artery (-ries)
umbilical cord
umbilical vein
vegetal pole

vernix
vitelline membrane
yolk sac
zona (-ae) pellucida (-ae)
zygote

SELF TEST

Multiple Choice: Some questions may have more than one correct answer.

1. The liver, pancreas, trachea, and pharynx
   a. arise from Hensen's node.
   b. are formed prior to the primitive streak.
   c. are derived from endoderm.
   d. are derived from ectoderm.
   e. are derived from mesoderm.

2. A hollow ball of several hundred cells
   a. is a morula.
   b. is a blastula.
   c. surrounds a fluid-filled cavity.
   d. is the gastrula.
   e. contains a blastocoel.

3. The first organ system that begins to form during organogenesis is the
   a. circulatory system.
   b. reproductive system.
   c. excretory system.
   d. nervous system.
   e. muscular system.

4. The release of calcium ions from cortical granules in the egg
   a. follows entry into the egg by a sperm.
   b. is part of the cortical reaction.
   c. is followed by a burst of protein synthesis in the egg.
   d. triggers metabolic changes within the egg.
   e. is part of the mechanism that prevents polyspermy.

5. The duration of pregnancy is referred to as the
   a. parturition.
   b. induction cycle.
   c. gestation period.
   d. neonatal period.
   e. prenatal period.

6. The fertilization membrane
   a. promotes binding of sperm to egg.
   b. facilitates egg-sperm recognition.
   c. prevents polyspermy.
   d. expedites entry of sperm into egg.
   e. blocks entrance of sperm.

7. The cavity within the embryo that is formed during gastrulation is the
   a. archenteron.
   b. gastrulacoel.
   c. blastocoel.
   d. blastopore.
   e. neural tube.

8. An embryo is called a fetus
   a. when the brain forms.
   b. when the limbs forms.
   c. when its heart begins to beat.
   d. after two months of development.
   e. once it is firmly implanted in the endometrium.

9. Embryonic cells are arranged in three distinct germ layers (endoderm, mesoderm, ectoderm) during
   a. cleavage.
   b. gastrulation.
   c. blastulation.
   d. organogenesis.
   e. blastocoel formation.

10. Amniotic fluid
    a. replaces the yolk sac in mammals.
    b. is secreted by the allantois.
    c. is secreted by an embryonic membrane.
    d. fills the space between the amnion and chorion.
    e. fills the space between the amnion and embryo.

11. Recognition of species compatibility between a sperm and an egg is due to the recognition protein
    a. acrosin.
    b. actin.
    c. vitelline.
    d. bindin.
    e. pellucidin.

12. The organ of exchange between a placental mammalian embryo and mother develops from
    a. chorion and amnion.
    b. chorion and allantois.
    c. chorion and uterine tissue.
    d. amnion and uterine tissue.
    e. amnion and allantois.

13. Eggs that have a large amount of yolk concentrated at one pole
    a. are isolecithal.
    b. are telolecithal.
    c. have a metabolically active animal pole.
    d. have a vegetal pole that is metabolically active.
    e. do not have food for the embryo in the egg.

14. _____ gives rise to skeletal tissues, muscle, and the circulatory system.
    a. ectoderm
    b. mesoderm
    c. endoderm
    d. one of the germ layers
    e. the blastocoel

Visual Foundations on next page ☞

VISUAL FOUNDATIONS

Color the parts of the illustration below as indicated.

RED ❑ umbilical artery
GREEN ❑ inner cell mass
YELLOW ❑ yolk sac
BLUE ❑ amniotic cavity and amnion
ORANGE ❑ chorionic cavity and chorion
BROWN ❑ trophoblast
TAN ❑ uterine epithelium
PINK ❑ embryo
VIOLET ❑ placenta

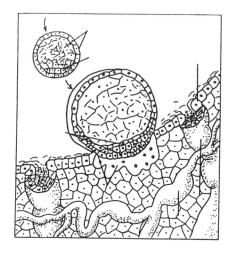
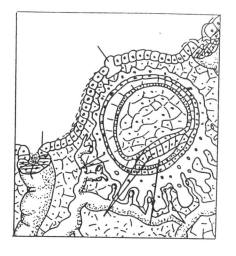
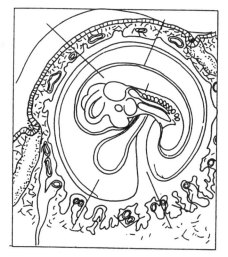
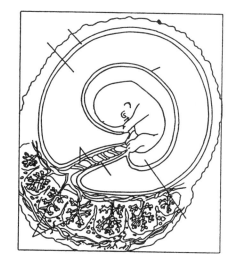

Animal Behavior

O U T L I N E

Behavior is adaptive
Biological rhythms
Inheritance and environmental
modification of behavior
Behavioral ecology
Social behavior

Animal behavior consists of those movements or responses an animal makes in response to signals from its environment; it tends to be adaptive and homeostatic. The behavior of an organism is as unique and characteristic as its structure and biochemistry. An organism adapts to its environment by synchronizing its behavior with cyclic change in its environment. Although behaviors are inherited, they can still be modified by experience. Behavioral ecology focuses on the interactions between animals and their environments and on the survival value of their behavior; migratory and foraging behaviors are studied by behavioral ecologists. Social behaviors, that is, interactions between two or more animals of the same species, have advantages and disadvantages. Advantages include confusing predators, repelling predators, and finding food. Disadvantages include competition for food and habitats, and increased ease of disease transmission. Many species that engage in social behaviors form societies. Some societies are loosely organized, whereas others have a complex structure. A system of communication reinforces the organization of the society. Animals communicate in a wide variety of ways; some use sound, some use scent, and some use pheromones. Examples of various types of social behaviors include the formation of dominance hierarchies, territoriality, courtship, pair bonding, parenting, playing, and altruistic behaviors. Some invertebrate societies exhibit elaborate and complex patterns of social interactions, such as the bees, ants, wasps, and termites. Vertebrate societies are far less rigid.

CHAPTER OUTLINE AND CONCEPT REVIEW (Fill in the blanks)

I. INTRODUCTION

■ 1 Behavior refers to the responses an organism makes to _____ from its environment.

II. BEHAVIOR IS ADAPTIVE

■ 2 _____ is the study of the characteristics, adaptive value, and evolutionary history of innate, species-typical behaviors. Behavioral genetics is concerned with how behavioral traits are inherited.

III. BIOLOGICAL RHYTHMS ANTICIPATE ENVIRONMENTAL CHANGES

A. A variety of behavioral cycles occur among organisms

■ **3** The behavioral cycle of activity for animals depends on when they are most active; they are categorized as (a)_____ when they are active during the day, (b)_____ when active at night, and (c)_____ when active in the twilight hours.

B. Biological rhythms are controlled by an internal clock

■ **4** Most organisms have a number of internal timing mechanisms that regulate their _____.

IV. BEHAVIOR CAPACITY IS INHERITED AND IS MODIFIED BY ENVIRONMENTAL INFLUENCES

A. Behavior develops

■ **5** Behavior is influenced primarily by two systems of the body, namely, the _____ systems.

B. Some behavior patterns have strong genetic components

■ **6** _____ refers to inherited (innate) behaviors.

■ **7** Fixed action pattern (FAP) behaviors are elicited by a _____.

C. Behavior is modified by learning

■ **8** Learning is defined as a change in behavior due to _____.

D. In classical conditioning, a reflex becomes associated with a new stimulus

■ **9** In classical conditioning, an animal makes an association between an irrelevant stimulus and a normal body response. It occurs when a (a)_____ becomes a substitute for (b)_____ stimulus.

E. In operant conditioning, spontaneous behavior is reinforced

■ **10** In operant conditioning, the behavior of the animal is rewarded or punished after it performs a behavior discovered by chance. Repetitive rewards bring about (a)_____, and consistent punishment induces (b)_____.

F. Imprinting is a form of learning that occurs during a critical period

■ **11** In imprinting, an animal forms a _____ to an object during a critical period of its life.

G. Habituation enables an animal to ignore irrelevant stimuli

H. Insight learning uses recalled events to solve new problems

I. Learning abilities may be biased

■ **12** The _____ things appear to be most easily learned.

V. *BEHAVIORAL ECOLOGY EXAMINES INTERACTIONS OF ANIMALS WITH THEIR ENVIRONMENTS*

A. Migration is triggered by environmental changes

■ 13 The components of migration include the (a)_____ and the mechanisms that allow organisms to (b)_____.

B. Efficient foraging behavior contributes to survival

■ 14 _____ refers to an animal's ability to obtain food in the most efficient manner.

■ 15 The method that an animal uses to obtain food is called its _____.

■ 16 The most efficient method of obtaining food reduces three important variables to their minimums. These variable are _____ _____.

VI. *SOCIAL INTERACTION HAS BOTH BENEFITS AND COSTS*

■ 17 A group of cooperating individuals in the same species comprises a _____.

A. Communication is necessary for social behavior

■ 18 _____ are chemical signals that convey information between members of a species.

B. Animals often form dominance hierarchies

■ 19 In some animals, the hormone _____ is the apparent cause of their aggressive behaviors.

C. Many animals defend territory

■ 20 An animal's territory is the portion of its _____ that it defends from members of the same species, and occasionally different species.

D. Sexual behavior is generally social

■ 21 _____ behavior ensures that the male is a member of the same species, and it provides the female with a means of evaluating the male.

■ 22 For some animals, mating is the only social contact they experience; it requires three basic elements of social conduct, namely, _____ _____.

E. Pair bonds establish reproductive cooperation

■ 23 A pair bond is a stable relationship between a male and a female that ensures cooperation in two behaviors that are essential to continuation of the species. These are _____.

F. Play is often practice behavior

G. Highly organized societies occur among insects and vertebrates

■ 24 Elaborate societies with well defined division of labor occurs in the two insect groups _____.

H. Kin selection could produce altruistic behavior

■ 25 More complex social groups tend to exhibit _____,
wherein an individual seemingly acts for the benefit of others rather than for itself.

I. Sociobiology attempts to explain altruism by kin selection

KEY TERMS

Frequently Used Prefixes and Suffixes: Use combinations of these prefixes and suffixes to generate terms for the definitions below.

Prefixes	The Meaning
socio-	social, sociological, society

Prefix	Suffix	Definitions
_____	-biology	1. The school of ethology that focuses on the evolution of social behavior through natural selection.

Other Terms You Should Know

aggregation	FAP	orientation
altruistic behavior	feeding strategy (-gies)	pair bond
behavior	fixed action pattern	pheromone
behavioral ecology	foraging strategy (-gies)	positive reinforcement
biological clock	free-running period	reinforcer
circadian	habituation	releaser
classical conditioning	home range	royal jelly
communication	imprinting	sensitization
conditioned response	insight learning	sign stimulus (-li)
conditioned stimulus (-li)	lunar cycle	signal
conspecific	migration	social behavior
crepuscular	navigation	society (-ties)
critical period	negative reinforcement	territoriality
display	nocturnal	territory (-ries)
dominance hierarchy (-hies)	operant (instrumental)	unconditioned response
ethology	conditioning	unconditioned stimulus (-li)
extinction	optimal foraging	zugunruhe

SELF TEST

Multiple Choice: Some questions may have more than one correct answer.

1. A behavior that resembles a simple reflex in some ways and a volitional behavior in others, and is triggered by a sign stimulus is best described as a/an
   a. releaser.
   b. conditioned reflex.
   c. conditioned stimulus.
   d. FAP.
   e. unconditioned reflex.

2. If your dog perks up its ears when you clap your hands, but stops perking its ears after you have repeatedly clapped your hands for a period of time, your dog has displayed a form of
   a. learning.
   b. operant conditioning.
   c. classical conditioning.
   d. sensory adaptation.
   e. habituation.

3. A change in behavior derived from experience in the environment is the result of
   a. a FAP.
   b. learning.
   c. inherited behavioral characteristics.
   d. redirected behavior.
   e. habituation and/or sensitization.

4. Suppose you decide to perform an experiment with a type of lizard that instantly attacks any lizard with green scales on its sides. You place a live lizard with green scales on its sides in a cage next to a styrofoam block with green paint on its sides. Your experimental lizard ignores the spotted lizard, preferentially attacking the styrofoam block. The styrofoam-attacking lizard is responding to a
   a. releaser.
   b. sign stimulus.
   c. conditioned reflex.
   d. displacement syndrome.
   e. redirected behavioral syndrome.

5. Which of the following is/are used by a worker honey bee to indicate the distance of a food source?
   a. dance speed
   b. frequency of abdominal waggles
   c. angle relative to a north–south axis
   d. angle relative to gravity
   e. angle relative to sun

Use the following list to answer questions 6-10, which are concerned with the different approaches to the study of animal behavior.

   a. ethology
   b. behavioral genetics
   c. social behaviorist
   d. behavioral ecology
   e. neurobiology

6. A field of study that would likely create a model about the behaviors of individuals in the population; a model that is designed to explain how a population is adapted to its environment.

7. A person in this field would probably want to know about the behavior of a hybrid produced from two strains with opposing behavioral characteristics.

8. The study of the behavior of animals in their natural habitats.

9. A person in this field would be interested in behavioral changes that result from destruction of portions of the brain.

10. A person in this field would be interested in interactions between members of the same species.

Visual Foundations on next page ☞

VISUAL FOUNDATIONS

Color the parts of the illustration below as indicated. Label the behavioral cycle and the physiological cycle.

RED   ❑   molting, gonadal development, and physiological cycle
GREEN   ❑   spring
YELLOW   ❑   summer
BLUE   ❑   winter
ORANGE   ❑   autumn
VIOLET   ❑   migration, mating, and behavioral cycle

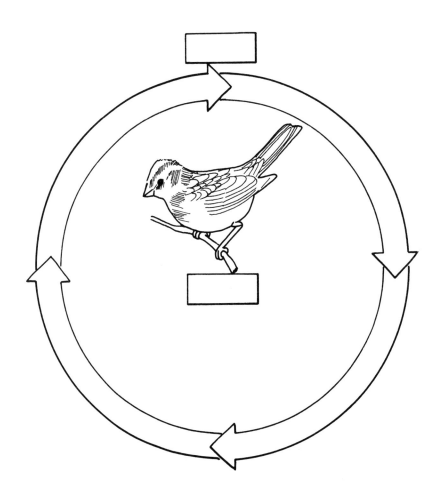

PART VIII

□

Ecology

□

CHAPTER 51

Ecology and the Geography of Life

O U T L I N E

The highest levels of biological organization
Geographic distributions of organisms
Major terrestrial life zones
Aquatic life zones
Life zones interact

Ecology is the study of the relationships among organisms and between organisms and their environment. Ecologists study populations, communities/ecosystems, and the biosphere/ecosphere. Living things are restricted to areas where the environment and potential lifestyles fit their adaptations. The greater the differences among habitats, the greater the differences among the organisms that inhabit them. Major terrestrial life zones (biomes) are large relatively distinct ecosystems characterized by similar climate, soil, plants, and animals, regardless of where they occur on earth. Their boundaries are determined primarily by climate. In terrestrial life zones, temperature and precipitation are the major determinants of plant and animal inhabitants. Examples of biomes include tundra; the taiga, or boreal forests; temperate forests; temperate grasslands; chaparral; deserts; tropical grasslands, or savanna; and tropical rain forests. In aquatic life zones, salinity, dissolved oxygen, light, and mineral nutrients are the major determinants of plant and animal inhabitants. Freshwater ecosystems include flowing water (rivers and streams), standing water (lakes and ponds), and freshwater wetlands. Marine ecosystems include estuaries, the neritic province, and the oceanic province. All terrestrial and aquatic life zones interact with one another.

CHAPTER OUTLINE AND CONCEPT REVIEW (Fill in the blanks)

I. INTRODUCTION

II. ECOLOGISTS STUDY THE HIGHEST LEVELS OF BIOLOGICAL ORGANIZATION

■ 1 A group of individuals in the same species living together in the same area is a (a)_population_____. All of these "groups" interacting in a given area comprise a (b)_community____, and this entity together with its abiotic environmental elements is an (c)_ecosystem___.

III. ORGANISMS HAVE UNIQUE GEOGRAPHIC DISTRIBUTIONS

IV. MAJOR TERRESTRIAL LIFE ZONES, CALLED BIOMES, ARE LARGELY DETERMINED BY CLIMATE

■ **2** Biomes are terrestrial life zones containing characteristic geographic assemblages. _climate_ is the most influential factor in determining the characteristics and boundaries of a biome.

A. Tundra is the northernmost biome

■ **3** Tundra is the northernmost biome. It is characterized by low-growing vegetation, a short growing season, and a permanently frozen layer of subsoil called _permafrost_.

B. Taiga, or boreal forest, is dominated by conifers

C. Temperate forests occur in temperate areas where precipitation is relatively high

■ **4** Coniferous temperate-zones that receive more than 80 inches (over 200 cm) of precipitation annually, like those in northwest North America, are specifically referred to as temperate (a)_rain forest_, whereas the temperate (b)_deciduous forest_ receives 30-49 inches (75-125 cm) of rain and is dominated by broad-leaved trees. _pattern_

D. Temperate grasslands occur in temperate areas of moderate precipitation

■ **5** Moderate precipitation produces temperate grasslands. Most contain mineral-rich soil well suited to grain crops. _Steppes_ are habitats that receive less rain than _process_ the average temperate grassland and more than deserts and are dominated by shorter, drought-resistant grasses. _patter_

E. Chaparral is a thicket of evergreen shrubs and small trees

■ **6** Chaparral is an assemblage of small-leaved shrubs and trees in areas of modest rainfall and mild winters, climatic conditions referred to as (a)_Mediterranean_ climate. Plants in the chaparral often have small, hard, drought-resistant leaves called (b)_sclerophyllous_ leaves.

F. Desert occurs where little precipitation falls

■ **7** Deserts are produced by low rates of precipitation and contain organisms with water-conserving adaptations. Many desert plants secrete toxic chemicals that inhibit growth of nearby competitors, an adaptation known as _allelopathy_.

G. Savanna is a tropical grassland with scattered trees

■ **8** The best known savanna, with its large herds of grazing, hoofed animals and predators, is the (a)_African_ savanna. Smaller savannas also occur on the continents of (b)_South America & northern Australia_

H. Tropical rain forests occur where temperatures are high throughout the year and precipitation falls almost daily

■ **9** Tropical rain forests have about (a)_80-180 in_ inches (cm) of rainfall annually. They have many diverse species and are noted for tall foliage, many epiphytes, and leached soil. Most of the nutrients in tropical rain forests are tied up in the (b)_vegetation_, not in the soil.

V. AQUATIC LIFE ZONES OCCUPY MOST OF THE EARTH'S SURFACE

■ 10 Aquatic life zones differ from terrestrial largely due to the different properties of water and air. The three principal limiting factors in water habitats include _light, minerals, oxygen_.

■ 11 Aquatic life is ecologically divided into the free-floating (a) _plankton_, the strong swimming (b) _nekton_, and the bottom dwelling (c) _benthos_ organisms.

■ 12 The base of the aquatic food chain is composed of the free-floating (a) _phytoplankton_, while the free-floating heterotrophic protozoa and small animals comprise the (b) _zooplankton_.

A. Freshwater ecosystems include rivers, lakes, and freshwater wetlands

■ 13 Rivers and streams are flowing-water ecosystems. The kinds of organisms they contain depends primarily on the _strength of the current_.

■ 14 Lakes and ponds are standing-water ecosystems. Large lakes have three zones: the (a) _littoral_ zone is the shallow water area along the shore; the (b) _limnetic_ zone is the open water area away from shore that is illuminated by sunlight; and the (c) _profundal_ zone, which is the deeper, open water area beneath the illuminated portion. Of these, the (d) _littoral_ zone is the most productive.

■ 15 Vertical thermal layering of standing water is called (a) _thermal stratification_. It is usually marked in the summer by an abrupt temperature transition at some depth called the (b) _thermocline_. When atmospheric temperatures drop in fall, vertical layers mix, a phenomenon known as (c) _fall turnover_. In spring, surface waters sink and bottom waters return to the surface. This is called (d) _spring turnover_.

■ 16 Freshwater wetlands may be (a) _marshes_, where grasslike plants dominate, or (b) _swamps_, where trees and shrubs dominate, or they may be small, shallow ponds called (c) _____, or (d) _____ that are dominated by mosses.

B. Estuaries occur where fresh water and saltwater meet

■ 17 The estuary community is very productive and serves as a nursery for young stages of many aquatic organisms. It often contains _salt marshes_, shallow, swamplike areas dominated by grasses where salinity fluctuates considerably.

C. Marine habitats include the intertidal zone, neritic province, and oceanic province

■ 18 The main marine habitats are similar to those of fresh water but are modified by tides and currents. The marine habitat near the shoreline, where oxygen, light, and food are most abundant is the _interdal zone_. Organisms living there have adaptations enabling them to resist wave action.

■ 19 Sharks, porpoises, and the larger benthic organisms are mostly confined to the shallower waters of the region known as the (a) _neritic province_. The portion of this region that contains sufficient light to support photosynthesis is called the (b) _euphotic_ region.

■ 20 The deeper region of the open ocean, beyond the reach of light, is known as the (a)__oceanic province__. The deepest reaches of the marine environment, far beyond light at the bottom of the sea, are referred to as the (b)__abyss__, in which the animal inhabitants are predators or scavengers subsisting on nutrients that fall into the zone from above.

VI. ALL TERRESTRIAL AND AQUATIC LIFE ZONES INTERACT WITH ONE ANOTHER

KEY TERMS

Frequently Used Prefixes and Suffixes: Use combinations of these prefixes and suffixes to generate terms for the definitions below.

Prefixes	The Meaning
inter-	between, among
phyto-	plant
sub-	under, below
thermo-	heat, hot
zoo-	animal

Prefix	Suffix	Definitions
phyto	-plankton	1. Planktonic plants; primary producers.
zoo	-plankton	2. Planktonic protists and animals.
inter	-tidal	3. Pertains to the zone of a shoreline between high tide and low tide.
thermo	-cline	4. A layer of water marked by an abrupt temperature transition.
sub	-tidal	5. Pertains to the zone of a shoreline below the lowest tide but still shallow enough for vigorous photosynthesis.

Other Terms You Should Know

allelopathy
atmosphere
benthos
biome
biosphere
bloom
boreal forest
chaparral
community (-ties)
desert
ecology
ecosphere
ecosystem
estuary (-ries)
euphotic region

fall turnover
hydrosphere
limnetic zone
lithosphere
littoral zone
Mediterranean climate
nekton
neritic
oceanic province
permafrost
plankton
population
profundal zone
salt marsh (-hes)
savannah

sclerophyllous
spring turnover
steppe
taiga
temperate coniferous forest
temperate deciduous forest
temperate grassland
temperate rain forest
thermal stratification
thermal vent
tropical rain forest
tropics
tundra

SELF TEST

Multiple Choice: Some questions may have more than one correct answer.

1. Water lilies, water striders, and crayfish are found mainly in
   a. the littoral zone.
   b. the limnetic zone.
   c. the profundal zone.
   d. a shallow lake.
   e. the area just below the compensation point.

2. Small or microscopic organisms carried about by currents are
   a. littoral.
   b. all producers.
   c. nektonic.
   d. benthic.
   e. planktonic.

3. The area(s) essentially corresponding to the waters above the continental shelf is/are the
   a. littoral zone.
   b. limnetic zone.
   c. profundal zone.
   d. neritic zone.
   e. benthic zone.

4. Communities dominated by evergreens and containing acid, mineral-poor soil is/are
   a. in South America.
   b. called taiga.
   c. temperate coniferous forests.
   d. typically impregnated with a layer of permafrost.
   e. boreal forest.

5. The two most productive and diverse of communities are the _____ and _____
   a. deciduous forest.
   b. tropical rain forest .
   c. coniferous forest.
   d. coral reef.
   e. grassland.

6. One expects to find organisms with adaptations for retaining moisture and clinging securely to the substrate in the
   a. salt marsh.
   b. intertidal zone.
   c. profundal zone.
   d. littoral zone.
   e. estuary.

7. An aquatic zone at the juncture of a river and the sea is a/an
   a. salt marsh.
   b. intertidal zone.
   c. profundal zone.
   d. littoral zone.
   e. estuary.

8. The factor that determines the nature of the biogeographic realms more than any other single factor is
   a. soil.
   b. climate.
   c. latitude.
   d. types of animals.
   e. altitude.

9. A habitat with short grass and somewhat greater precipitation than moist deserts is a
   a. tropical habitat.
   b. steppes.
   c. boreal forest.
   d. deciduous forest.
   e. muskeg.

10. The relatively moist forests on the western slopes of North America are
    a. boreal forests.
    b. deciduous forests.
    c. temperate communities.
    d. steppes.
    e. typified by Mediterranean climates.

11. The communities containing the most comprehensive coverage of permafrost are called
    a. polar.
    b. steppes.
    c. boreal.
    d. tundra.
    e. muskegs.

12. Most of midwestern North America is
    a. grassland.
    b. deciduous forest.
    c. temperate rain forest.
    d. temperate.
    e. chaparral.

→topsoil, clay rich layers underneath

13. A community with two distinct soil layers abounding in trees, reptiles, and amphibians is
    a. grassland.
    b. deciduous forest.
    c. temperate rain forest.
    d. temperate.
    e. chaparral.

14. An adaptation by which plants discourage competition through secretion of toxic substances is
    a. allelopathy.
    b. typical of vines.
    c. facilitated by secretions of roots or leaves.
    d. typical of tropic plants.
    e. found in deserts.

VISUAL FOUNDATIONS

Color the parts of the illustration below as indicated.

RED ☐ dry air
GREEN ☐ forest
YELLOW ☐ desert
BLUE ☐ moist air

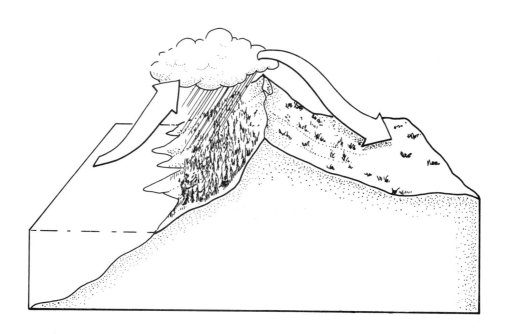

Color the parts of the illustration below as indicated.

RED ☐ pelagic
GREEN ☐ intertidal zone
YELLOW ☐ euphotic zone
BLUE ☐ abyssal
ORANGE ☐ benthic

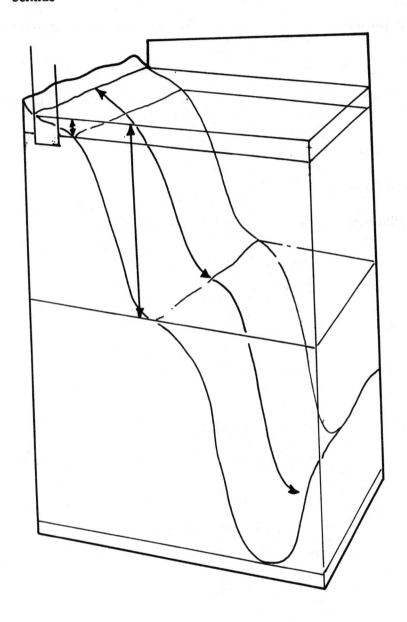

CHAPTER 52

□

Population Ecology

OUTLINE

Density and dispersion
Population growth
Environmental limitations
Reproductive tactics
Human population growth

A population is a group of organisms of a single species living in a particular geographic area at a given time. Populations have properties that do not exist in their component individuals or in the community of which the population is a part. These properties include birth rates, death rates, sex ratios, median ages, and gene pools. Population ecology deals with the numbers of particular organisms that are found in an area and why those numbers change or remain fixed over time. Populations must be understood in order to manage forests, field crops, game, fish, and other populations of economic importance. Important features of a population include the number of individuals living in a given area, its distribution pattern in the community, and its growth. Population growth is determined, in part, by the rate of arrival of organisms through immigration and births, and the rate of departure through emigration and deaths. Population growth curves take a variety of forms. Their shapes are primarily a function of limiting factors and the theoretical maximum rate a population can grow. Population growth is limited by density-dependent factors such as predation, disease, and competition, and density-independent factors, such as hurricanes and fires. Each species has its own survival strategy, that is, its own reproductive characteristics, body size, habitat requirements, migration patterns, and other characteristics that enable it to survive. Most organisms fall into one of two survival strategies: One emphasizes a high rate of natural increase; these organisms often have small body sizes, bear large numbers of young once per lifetime, and inhabit variable environments. The second strategy emphasizes maintenance of a population near the carrying capacity of the environment; these organisms often have large body sizes, repetitive reproductive cycles, long life spans, and inhabit stable environments. The principles of population ecology apply to humans as well as other organisms.

CHAPTER OUTLINE AND CONCEPT REVIEW (Fill in the blanks)

I. INTRODUCTION

■ 1 Populations of organisms have properties that individual organisms do not have.
Four such properties of importance are __boundaries, density & dispersion, age, size & sex structure, population models__.

II. DENSITY AND DISPERSION ARE IMPORTANT FEATURES OF POPULATIONS

■ 2 ___population___ is the number of individuals of a species per unit of habitat area.

■ 3 The area over which a population is distributed is its (a)_habitat_. Individuals within a population may exhibit a characteristic spacing. For example, (b)___random___ exists when individuals are not spaced in a definable, predictable pattern; (c)___uniform___ is even, regular spacing of individuals; and (d)___clumped___ applies when individuals are more-or-less confined to a limited portion of their potential range.

III. MATHEMATICAL MODELS DESCRIBE POPULATION GROWTH

■ 4 Changes in the size of a population result from the difference between the ___birth & death rate___.

■ 5 The rate of growth in a population is expressed as.

$$\frac{\Delta N}{\Delta t} = b - d \quad , \text{where}$$

ΔN is the (a)___change in population___, Δt is the (b)___change in time___, b is the birth rate or (c)___natality___, and d is the death rate or (d)___mortality___.

■ 6 (a)___biotic potential___ is the theoretical maximum rate at which a population can grow. The maximum potential rate of increase cannot be sustained due to the constraints upon growth collectively known as (b)___environment resistance___. The largest population that can be sustained by a particular environment is that environment's (c)___carrying capacity___.

■ 7 Population growth curves take a variety of forms, their shapes primarily a function of biotic potential and limiting factors. The (a)___exponential___ growth curve shows an initial lag phase of population growth, followed by a logarithmic phase, followed by a static or exhaustion phase. The (b)___logistic___ growth curve resembles the logarithmic portion of the s-curve and occurs in the virtual absence of limiting factors.

IV. POPULATION GROWTH IS LIMITED BY THE ENVIRONMENT

A. Density-dependent factors influence populations

■ 8 Density-dependent limiting factors are most effective at ___high___ (high or low?) population densities, and therefore they tend to stabilize populations.

■ 9 ___disease___ is an effective density-dependent limiting factor because transmission is more likely among members of dense populations.

B. Density-independent factors influence populations

■ 10 Density-independent limiting factors are conditions or events that depress population without tending to stabilize it. ___natural disasters___, such as floods or fires are common limiting factors.

V. DIFFERENT SPECIES HAVE EVOLVED DIFFERENT REPRODUCTIVE TACTICS

■ 11 Each species has a life history strategy, The (a)___r-type___ strategy is employed mainly by pioneer organisms. It emphasizes wide and quick dispersal, large numbers of seeds or offspring, and genetic uniformity. A (b)___k-type___ strategy is characteristic of inhabitants of mature communities where replacement is slow.

VI. HUMAN POPULATION GROWTH HAS THE SAME PARAMETERS AS POPULATION GROWTH IN OTHER ORGANISMS

X ■ 12 ZPG stands for ___zero population growth carrying capacity___, which is the point where natality equals mortality.

A. Not all countries have the same rates of population increase

■ 13 The amount of time that it takes for a population to double in size is its (a)___doubling time___. In general, the shorter this time, the (b)___less___(more or less?) developed the country.

■ 14 The number of children that a couple must produce to replace themselves is known as the (a)___replacement level fertility___, a value that is generally relatively (b)___lower___(higher or lower?) in developed countries and (c)___higher___(higher or lower?) in undeveloped and developing countries.

■ 15 The average number of children born to a group of women in their lifetimes is called the (a)___fertility rate___. On a worldwide basis, this value is currently (b)___higher___(higher or lower?) than the replacement value.

B. The age structure of a country can be used to predict its population growth

■ 16 Demographers have identified four stages in the evolution of a country's populations. The (a)___pre industrial___ stage is characterized by high natality and mortality and moderate population growth. The (b)___transmission___ stage is marked by a decline in mortality, relatively constant, high natality, and consequently an accelerated population growth curve. In the (c)___industrial___ stage, birth rates decline and the growth rate slows. Finally, in the (d)___postindustrial___ stage, both natality and mortality are low.

KEY TERMS

Terms You Should Know

age structure	doubling time	lag phase
aggregated distribution	environmental resistance	logistic equation
big bang reproduction	exponential growth	model
biotic potential	industrial stage	mortality rate
carrying capacity (-ties)	infant mortality rate	natality rate
clumped dispersion	instantaneous growth rate	population
demographics	interspecific competition	population crash
density-dependent factor	intraspecific competition	population density
density-independent factor	intrinsic rate of increase	population ecology
developed country (-ries)	K selection	postindustrial stage
dispersion	*k*-strategist	preindustrial stage

r-strategist   replacement-level fertility   transitional stage
random dispersion   S-shaped curve   uniform dispersion
range   total fertility rate   zero population growth

SELF TEST

Multiple Choice: Some questions may have more than one correct answer.

Questions 1-5 pertain to the following equation. Use the list of choices below to answer them.

$$\frac{\Delta N}{\Delta t} = b - d$$

a. growth only ᵃ ×   d. ΔN d   g. value "b" ¹
b. speed of growth   e. Δt   h. value "d" ²
c. change only   f. b – d +   i. ΔN / Δt +

1. Birth rate. g

2. Death rate. h

3. Population change independent of time. d

4. Rate of population change.   f, i

5. The parameter derived from the equation.   b

6. The equation dN/dt = rN is used to derive the
    a. biotic potential (r_m).   d. per capita growth rate (r).
    b. instantaneous growth rate.   e. growth rate at carrying capacity.
    c. decline in growth due to limiting factors.

7. During the exponential growth phase of bacterial growth, the population
    a. increases moderately.   d. decreases dramatically.
    b. increases dramatically.   e. does not change substantially.
    c. decreased moderately.

8. A cultivated field of wheat would most likely display
    a. uniform distribution.   d. no distribution pattern.
    b. random distribution.   e. a form of distribution not found in nature.
    c. clumped distribution.

9. The difference in the biotic potentials of a cow and mouse is largely due to their respective
    a. litter sizes.   d. limiting factors.
    b. habitats.   e. choices of food.
    c. sizes.

10. In the logistic equation, if environmental resistance is sustained, "K" is the
    a. growth rate.   d. carrying capacity.
    b. birth rate.   e. equivalent of the lag phase.
    c. mortality rate.

11. A grove of trees that originated from one seed displays
    a. uniform distribution.   d. no distribution pattern.
    b. random distribution.   e. a form of distribution not found in nature.
    c. clumped distribution.

12. Population growth is most frequently checked by which *one* of the following?
    a. Environmental resistance.   d. Behavioral modifications.
    b. Predation.   e. Resource depletion.
    c. Disease.

13. The inverse of biotic potential is
    a. natality.   d. birth rate minus death rate.
    b. mortality.   e. a population crash.
    c. b – d.

14. r-strategists usually
    a. are mobile animals.   d. survive well in the long term.
    b. have a high "r" value.   e. develop slowly.
    c. inhabit variable environments.

15. Examples of density-dependent limiting factors include
    a. competition.   d. predation.
    b. disease.   e. behavioral mechanisms.
    c. resource depletion.

16. The redwood stands in California are examples of organisms that
    a. use the K-strategy.   d. live in a climax community.
    b. use the r-strategy.   e. pioneer new habitats.
    c. often experience local extinction.

17. The number of individuals of a species per unit of habitat area is the _____ for that species.
    a. distribution curve   d. ΔN
    b. distribution pattern   e. b – d
    c. population density

VISUAL FOUNDATIONS

Color the parts of the illustration below as indicated. Label stable, contracting, and expanding populations.

    RED ☐ post-reproductive
    GREEN ☐ reproductive
    YELLOW ☐ pre-reproductive

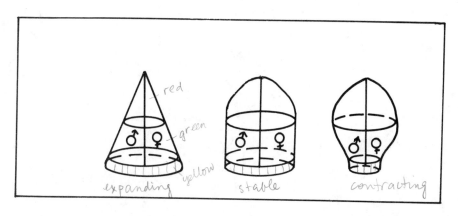

□

CHAPTER 53

Community Ecology

O U T L I N E

A salt marsh community
Producers, consumers, and decomposers
Symbiosis
Ecological niche
Species diversity
Succession
The nature of communities

A community is an association of organisms of different species living and interacting together. Communities interact with and influence one another in numerous ways. Communities exist within communities. For example, a community of microorganisms lives within the gut of a termite that itself is a member of a community of organisms within a rotten log that is part of a forest community, and so on. A community and its physical environment is inseparably linked in what is called an ecosystem. Communities contain producers, consumers, and decomposers. Producers are the photosynthetic organisms that are the basis of most food chains. Consumers feed on other organisms. Decomposers recycle the components of corpses and wastes by feeding on them. Symbiosis is an intimate long-term association between two or more different species. In one form of symbiosis, both partners benefit. In another, one partner benefits and the other neither benefits nor is harmed. In a third, one partner benefits and the other is adversely affected. Every organism has its own ecological niche, that is, its own role within the structure and function of the community. Its niche reflects the totality of its adaptations, its use of resources, and its lifestyle. Communities vary greatly in the number of species they contain. The number is often high where the number of potential ecological niches is great, where a community is not isolated or severely stressed, at the edges of adjacent communities, and in communities with long histories. A community develops gradually over time, through a series of stages, until it reaches a state of maturity; species in one stage are replaced by different species in the next stage. Most ecologists believe that the species composition of a community is due primarily to factors other than biological interactions.

CHAPTER OUTLINE AND CONCEPT REVIEW (Fill in the blanks)

I. INTRODUCTION

■ 1 A biological community consists of a group of organisms that interact and live together. A complete and independent community, interacting with the nonliving environment, makes up an ecosystem .

II. A SALT MARSH COMMUNITY EXHIBITS GREAT DIVERSITY

III. COMMUNITIES CONTAIN PRODUCERS, CONSUMERS, AND DECOMPOSERS

■ 2 The major community roles are those of producer, consumer, and decomposer. Producers, which are also called (a)_autotroph_____, are the photosynthetic organisms that are the basis of most food chains. Consumers, also called (b)_heterotroph_, feed on other organisms. Consumers are further divided into the (c)_primary____ consumers that eat producers, and (d)_secondary_ consumers that eat other consumers. (e)_Omnivores_ eat both plants and animals. Decomposers, which are also known as (f)_detritus_ X, recycle the components of corpses and wastes by feeding on them.

■ 3 ___Detritus x_ _detritivores_ _____ work together with decomposers by consuming the organic matter of plant and animal remains.

IV. LIVING ORGANISMS INTERACT IN A VARIETY OF WAYS

A. In mutualism, benefits are shared

B. Commensalism is taking without harming

C. Parasitism is taking at another's expense

■ 4 Parasites that are usually saprophytic, but parasitic when the opportunity presents itself, are called (a)_____ parasites, whereas the (b)_____ parasite is one that exists only as a parasite.

V. THE NICHE DESCRIBES AN ORGANISM'S ROLE IN THE COMMUNITY

■ 5 The role an organism plays in an ecosystem is its niche. A niche reflects the organism's total adaptational package. The role an organism *could* potentially play in a community is its (a)_fundamental_ niche, while the role it *actually* plays is its (b)_realized_ niche.

A. Competition between two species with identical or similar niches leads to competitive exclusion

■ 6 Competitive exclusion means that one species is excluded from a niche as a result of ___competition interspecific *___, most often for a limited resource.

B. Limiting factors restrict an organism's realized niche

■ 7 The "law of the minimum," discussed earlier in the text, has been revised to the ___law of tolerance___, which adds that organisms may also be limited by a factor that is present in harmful excess.

VI. COMMUNITIES VARY IN SPECIES DIVERSITY

■ 8 Species diversity is inversely related to (a)_____ of a habitat. For example, island communities are generally less diverse than continental communities. The (b)_____ is a phrase used to describe the relatively greater diversity near the borders of a community compared to the interior portions of the same community.

VII. SUCCESSION IS COMMUNITY CHANGE OVER TIME

■ 9 The relatively stable, often long term stage at the end of ecological succession is the _____.

A. Sometimes a community develops in a "lifeless" environment

■ **10** (a)_____ occurs in a habitat that is initially devoid of life. For example, lichens may invade a bare rock, altering its composition enough for other organisms to "take hold." As the first to colonize the "new lifeless" habitat, initial invaders like the lichen, are referred to as (b)_____ communities. In general, the order in which plants are established in a new community is: lichens —> (c)_____ —> grasses —> (d)_____ —> and finally (e)_____.

B. Sometimes a community develops in an environment where a previous community existed

■ **11** _____ begins with an existing community. Abandoned farmlands are examples.

C. Three different hypotheses explain succession

■ **12** The _____ model asserts that succession is due to biological factors, each stage of which changes the environment enough for new organisms to enter it.

■ **13** The _____ model holds that succession does not follow an orderly, predictable pattern, and that the first species to enter a habitat inhibit other species from becoming established there.

■ **14** The _____ model states that species that are better able to compete for limiting resources replace species that entered the environment earlier.

VIII. ECOLOGISTS CONTINUE TO STUDY THE NATURE OF COMMUNITIES

■ **15** The view that the species in a community cooperate like organs in the body of a complex organism is known as the (a)_____ model of a community. The opposing view, namely that biological interactions are less important than abiotic factors in producing communities, and that, in fact, communities may not be biological entities at all, is the (b)_____ model of communities.

KEY TERMS

Frequently Used Prefixes and Suffixes: Use combinations of these prefixes and suffixes to generate terms for the definitions below.

Prefixes	The Meaning	Suffixes	The Meaning
auto-	self, same	-troph	nutrition, growth, "eat"
hetero-	different, other	-vore	eating
omni-	all		

Prefix	Suffix	Definitions
carni-	_____	1. Any animal that eats flesh.
_____	_____	2. Any animal that eats all kinds of foods indiscriminately.
_____	_____	3. An organism that produces its own food.

_____ _____   4. An organism that is dependent upon other organisms for
                                       food, energy, and oxygen.

herbi-   _____   5. Any animal that eats plants.

Other Terms You Should Know

biomass	food chain	pioneer
climax	fundamental niche	predator
climax community	host	primary consumer
commensalism	individualistic model	primary succession
competitive exclusion	inhibition model	producer
consumer	law of the minimum	realized niche
decomposer	law of tolerance	saprophyte
detritivore	limiting factor	secondary consumer
detritus feeder	mutualism	secondary succession
ecological niche	obligate parasite	succession
ecosystem	organismic model	symbiont
edge effect	parasite	symbiosis
facilitation model	parasitism	tolerance model
facultative parasite	pathogen	zooxanthellae

SELF TEST

Multiple Choice: Some questions may have more than one correct answer.

1. Critically important minerals and other essential substances would be permanently lost to organisms if
   it were not for the
   a. producers.   d. consumers.
   b. nitrogen fixers.   e. decomposers.
   c. phosphorus fixers.

2. That growth of an organism is limited by whatever factor essential to it is in shortest supply
   summarizes
   a. the 4–Cs law.   d. the law of the minimum.
   b. normal succession.   e. the exclusion principle.
   c. the concept of a realized niche.

3. The totality of an organism's adaptations, its use of resources, and the lifestyle to which it is fitted are
   all embodied in the term
   a. ecosystem.   d. pyramid.
   b. community.   e. niche.
   c. food web.

4. The nonliving environment together with different interacting species is a/an
   a. ecosystem.   d. pyramid.
   b. community.   e. niche.
   c. food web.

5. An association of different species of organisms interacting and living together is a/an
   a. ecosystem.   d. pyramid.
   b. community.   e. niche.
   c. food web.

6. The major roles of organisms in a community are
   a. producers.
   b. nitrogen fixers.
   c. phosphorus fixers.
   d. consumers.
   e. decomposers.

7. The difference between an organism's potential niche and the niche it comes to occupy may result from
   a. competition.
   b. an inverted pyramid.
   c. factors that exclude it from part of its fundamental niche.
   d. overabundance of resources.
   e. limited space or resources.

8. All food energy in the biosphere is ultimately provided by
   a. producers.
   b. nitrogen fixers.
   c. phosphorus fixers.
   d. consumers.
   e. decomposers.

VISUAL FOUNDATIONS

Color the parts of the illustration below as indicated. Label years after cultivation and the dominant vegetation.

RED	☐	climax forest
GREEN	☐	pine forest
YELLOW	☐	crab grass
BLUE	☐	sky
ORANGE	☐	broomsedge
BROWN	☐	horse weed
TAN	☐	pine seedlings

CHAPTER 54

❑

Ecosystems and the Ecosphere

OUTLINE

Cycles of matter
Energy flow through ecosystems
Environmental factors

Matter, the material of which living things are composed, cycles from the living world to the abiotic physical environment and back again. All materials vital to life are continually recycled through ecosystems and so become available to new generations of organisms. Key cycles include the carbon cycle, the nitrogen cycle, the phosphorus cycle, and the water, or hydrologic, cycle. Although matter is cyclic in ecosystems and the ecosphere, energy flow is not. Energy cannot be recycled and reused. Energy moves through ecosystems in a linear, one-way direction from the sun to producer to consumer to decomposer. Much of this energy is converted to less useful heat as the energy moves from one organism to another. Life is possible on earth because of its unique environment.

CHAPTER OUTLINE AND CONCEPT REVIEW (Fill in the blanks)

I. INTRODUCTION

■ 1 A community *and* its abiotic components together make up an (a)_____, all of these ecological units combined comprise the Earth's (b)_____.

II. MATTER CYCLES THROUGH ECOSYSTEMS

■ 2 The cycling of matter through the biota to the abiota and back again is referred to as (a)_____. Four of these cycles that are of particular importance are the cycles for (b)_____ _____.

A. Carbon dioxide is the pivotal molecule of the carbon cycle

■ 3 Carbon is incorporated into the biota, or "fixed," primarily through the process of (a)_____, and in the short term much of it is returned to the abiotic environment by (b)_____. Some carbon is tied up for long periods of time in wood and fossil fuels, which may eventually be released through combustion.

B. Bacteria are essential to the nitrogen cycle

■ 4 The nitrogen cycle has five steps: (a)_____, which is the conversion of nitrogen gas to ammonia or nitrate; the conversion of ammonia to nitrate, called (b)_____; absorption of nitrates and ammonia by plant roots, called (c)_____; the release of organic nitrogen back into the abiotic environment in the form of ammonia in waste products and by decomposition, a process known as (d)_____; and finally, to complete the cycle, (e)_____ reduces nitrates to gaseous nitrogen.

C. The phosphorus cycle lacks a gaseous component

■ 5 Erosion of (a)_____ releases inorganic phosphorus to the (b)_____ where it is taken up by plant roots.

D. Water circulates in the hydrologic cycle

■ 6 The hydrological cycle results from the evaporation of water from both land and sea and its subsequent precipitation when air is cooled. The movement of water from land to seas is called (a)_____. Water also seeps down through the soil to become (b)_____, which supplies much of the water to streams, soil, and plants.

III. THE FLOW OF ENERGY THROUGH ECOSYSTEMS IS LINEAR

■ 7 Energy flows through ecosystems. It enters in the form of (a)_____ energy, which plants can convert to useful (b)_____ energy that is stored in the bonds of molecules. Since energy conversions are never 100% efficient, when it is converted from one form to another, or moved from the abiota to the biota and back again, some of it is converted to (c)_____ energy and returned permanently to the abiotic environment where organisms cannot use it.

■ 8 Each "link" in a food chain is a (a)_____, the first and most basic of which is comprised of the photosynthesizers, or (b)_____. Once "fixed," materials and energy pass linearly through the chain from photosynthesizers to the (c)_____, and then to the (d)_____.

■ 9 A complex of interconnected food chains is a _____.

A. Ecological pyramids illustrate how ecosystems work

■ 10 The three main types of ecological pyramids are the pyramid of (a)_____, which illustrates how many organisms there are in each trophic level; the pyramid of (b)_____, which is a quantitative estimate of the amount of living material at each level; and the pyramid of (c)_____, a somewhat unique representation since this pyramid, unlike the others, and because of the second law of thermodynamics, can never be inverted.

B. Ecosystems vary in productivity

■ 11 The rate at which carbon compounds are synthesized by photosynthesis is known as the (a)_____ of an ecosystem, and the energy that remains as biomass after energy for respiration is "consumed" is the (b)_____, which basically represents plant growth. This increase in plant biomass is balanced by (c)_____, which is the consumption of plant mass by animals and decomposers.

IV. ENVIRONMENTAL FACTORS INFLUENCE WHERE AND HOW SUCCESSFULLY AN ORGANISM CAN SURVIVE

■ **12** _____ is the average weather conditions, including temperature and rainfall, that prevail in a given location over an extended period of time.

A. The sun warms the Earth

■ **13** Sunlight is almost the sole source of energy available for the biosphere. The quantity of energy that reaches Earth as sunlight is (a)_____. Seasons result from two factors: (b)_____, which varies throughout the year, and most importantly, from (c)_____, the tilt of the Earth's axis relative to the sun.

B. Atmospheric circulation is driven by uneven heating by the sun

■ **14** Local climate is determined largely by the angle at which the solar input falls to earth's surface. Angles are usually greatest and least variable in (a)_____ regions, producing hotter, less variable climates in these regions. Air in this region is heated, causing it to expand and rise, which in turn causes it to cool and sink again. Most of the cool air sinks back into the region from which it came, but some of it flows toward the (b)_____ where it cools and sinks.

■ **15** Winds generally blow from (a)_____(high or low?) pressure areas to (b)_____(high or low?) pressure areas, but they are deflected from these paths by Earth's rotation, a phenomenon known as the (c)_____.

C. Surface ocean currents are driven by winds and by the Coriolis effect

■ **16** Prevailing winds push masses of water along the surfaces of oceans. These movements, or (a)_____, move in large circular patterns called (c)_____.

D. Climate results from patterns of air and water movement

■ **17** The two most important factors in determining temperature and the lengths of days and seasons in a given locale are _____.

■ **18** (a)_____ is greatest where air heavily saturated with water vapor cools to the dew point. (b)_____ develop in the rain shadows of mountain ranges and continental interiors.

■ **19** _____ refers to the relatively long term average weather conditions in a small locale, these conditions usually attributable to the unique topological features in that location.

E. Soil factors affect living organisms

■ **20** A vertical transect through the soil produces a (a)_____, which reveals that soil is organized into horizontal layers called (b)_____. The uppermost layer, the (c)_____, is rich in organic matter. Progressing downward, the second layer is the (d)_____, followed by the underlying subsoil or (e)_____. The (f)_____ borders solid parent rock.

KEY TERMS

Frequently Used Prefixes and Suffixes: Use combinations of these prefixes and suffixes to generate terms for the definitions below.

Prefixes	The Meaning
atmo-	air
bio-	life
hydro-	water
litho-	stone
micro-	small

Prefix	Suffix	Definitions
_____	-sphere	1. The entire zone of air, land, and water at the surface of the earth that is occupied by living things.
_____	-sphere	2. The gaseous envelope surrounding the earth; the air.
_____	-sphere	3. The water on or surrounding the earth, including the oceans and the water in the atmosphere.
_____	-sphere	4. The portion of the earth that is composed of rock.
_____	-climate	5. A localized variation in climate; the climate of a small area.

Other Terms You Should Know

A-horizon
ammonification
assimilation
B-horizon
biogeochemical cycle
biomass
C-horizon
climate
combustion
Coriolis effect
current
denitrification
density (-ties)
ecological pyramid
ecosphere
ecosystem
energy flow
erosion
estuary (-ries)
food chain
food web
fossil fuel
gross primary productivity
ground water
gyre
heterocyst
hydrologic cycle
insolation
net primary productivity
nitrification
nitrogen fixation
nitrogenase
nodule
O-horizon
polar easterly (-lies)
precipitation
prevailing wind
primary consumer
primary producer
pyramid of biomass
pyramid of energy
pyramid of numbers
rain shadow
secondary consumer
soil horizon
soil profile
standing crop
surface runoff
trade wind
trophic level
turnover
water table
westerly (-lies)

SELF TEST

Multiple Choice: Some questions may have more than one correct answer.

1. The rate at which organic matter is incorporated into plant bodies is
   a. significant only in a changing ecosystem.
   b. called gross primary productivity.
   c. called net primary productivity.
   d. equal to the primary producer's metabolic rate.
   e. evident only in a climax community.

2. Which of the following would be positioned at the top of a typical pyramid?
   a. Producers.
   b. Consumers.
   c. The trophic level with the greatest biomass.
   d. Trophic level with the least numbers.
   e. Trophic level with the least energy.

3. Most carbon is fixed _____ and liberated _____
   a. in proteins/as organic compounds.
   b. in CO_2/as complex compounds.
   c. in complex compounds/as CO_2.
   d. in humus/by cyanobacteria.
   e. by plants/by plants.

4. Activities performed by microorganisms involved in the nitrogen cycle include
   a. liberation of nitrogen from nitrate.
   b. oxidization of ammonia to nitrates.
   c. production of ammonia from proteins, urea, or uric acid.
   d. fixation of molecular nitrogen.
   e. continuous cycling of nitrogen.

5. Phosphorus enters aquatic communities by means of
   a. decomposers.
   b. producers.
   c. primary consumers.
   d. secondary consumers.
   e. erosion.

6. The trophic levels found at the two extremes (top and bottom) of a pyramid are
   a. producers.
   b. nitrogen fixers.
   c. phosphorus fixers.
   d. consumers.
   e. decomposers.

7. Compared to the overall regional climate, a microclimate
   a. may be significantly different.
   b. is insignificant to organisms.
   c. is more important to resident organisms.
   d. covers a larger area.
   e. exists for a brief period.

8. A major ocean current that moves northward along the west coast of South America
   a. warms coastal Europe.
   b. warms the U.S.
   c. cools South America.
   d. is called the Gulf Stream.
   e. reaches Alaska.

9. The soil layer richest in organic matter is the
   a. O-horizon.
   b. A-horizon.
   c. B-horizon.
   d. C-horizon.
   e. D-horizon.

10. The totality of the Earth's living inhabitants is the
    a. ecosphere.
    b. lithosphere.
    c. biosphere.
    d. biome.
    e. hydrosphere.

11. The least pronounced seasonal temperature differences are found in the
   a. temperate zones.
   b. equatorial areas.
   c. United States.
   d. polar regions.
   e. subpolar areas.

12. When it is warmest in the United States, it is coolest in
   a. the Southern Hemisphere.
   b. the Northern Hemisphere.
   c. Rio de Janiero.
   d. Europe.
   e. Mexico.

13. A burrow in a hostile desert environment is an example of a/an
   a. subterranean biome.
   b. O-horizon.
   c. macroclimate.
   d. microclimate.
   e. ecosystem.

14. All of Earth's living inhabitants and the abiotic environmental factors comprise the
   a. ecosphere.
   b. lithosphere.
   c. biosphere.
   d. biome.
   e. hydrosphere.

15. The main collector(s) of solar energy used to power life processes is/are the
   a. producers.
   b. atmosphere.
   c. hydrosphere.
   d. green leaves.
   e. photosynthetic organisms.

VISUAL FOUNDATIONS

Color the parts of the illustration below as indicated. Label N_2 in the atmosphere, $NO_3{}^-$ in the soil, and NH_3 in the soil.

RED ☐ nitrogen-fixation
GREEN ☐ plants
YELLOW ☐ ammonification
BLUE ☐ atmosphere
ORANGE ☐ assimilation
BROWN ☐ soil
TAN ☐ animal
PINK ☐ nitrification
VIOLET ☐ denitrification

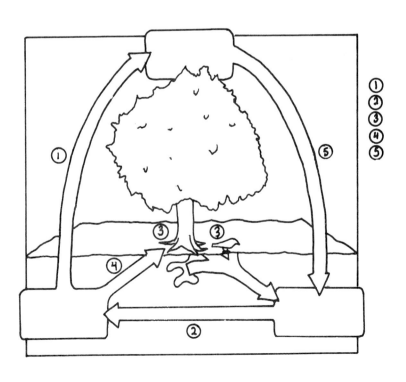

Color the parts of the illustration below as indicated.

RED   ☐   third trophic level
GREEN   ☐   first trophic level
YELLOW   ☐   second trophic level
BLUE   ☐   primary producer
BROWN   ☐   primary consumer
TAN   ☐   secondary consumer

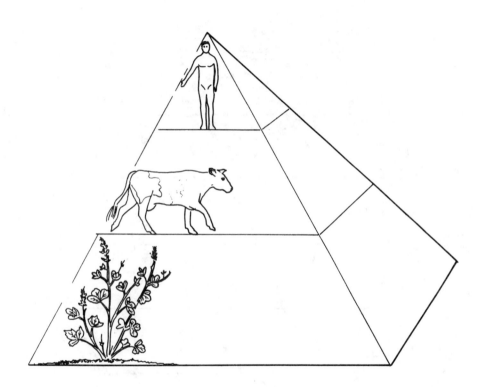

Color the parts of the illustration below as indicated.

RED ☐ equator
GREEN ☐ trade winds
YELLOW ☐ warm air
BLUE ☐ cool air
ORANGE ☐ westerlies
VIOLET ☐ polar easterlies

CHAPTER 55

□

Humans in the Environment

OUTLINE

Declining biological diversity
Deforestation
Global climate change
Ozone depletion in the stratosphere
Interrelationships among environmental problems

Although humans have been on earth a comparatively brief span of time, our biological success has been unparalleled. We have expanded our biological range into almost every habitat on earth. In the process, we have placed great strain on the Earth's resources and resilience, and profoundly affected Other life forms. As a direct result of our actions, many environmental concerns exist today. Species are disappearing from earth at an alarming rate. Conservation biologists estimate that at least one species becomes extinct every day. Although forests provide us with many ecological benefits, including watershed protection, soil erosion prevention, climate moderation, and wildlife habitat, deforestation is occurring at an unprecedented rate. The production of atmospheric pollutants that trap solar heat in the atmosphere threaten to alter the earth's climate. Global warming may cause a rise in sea level, changes in precipitation patterns, death of forests, extinction of animals and plants, and problems for agriculture. It could result in the displacement of thousands or even millions of people. Ozone is disappearing from the stratosphere at an alarming rate; during the 1980s, its rate of depletion was roughly three times the rate in the 1970s. The ozone layer helps to shield Earth from damaging ultraviolet radiation. If the ozone were to disappear, Earth would become uninhabitable for most forms of life.

CHAPTER OUTLINE AND CONCEPT REVIEW (Fill in the blanks)

I. INTRODUCTION

II. SPECIES ARE DISAPPEARING FROM EARTH AT AN ALARMING RATE

■ 1 When the number of individuals in a species is reduced to the point where extinction seems imminent, the species is said to be (a)_____, and when numbers are seriously reduced so that extinction is feared but thought to be less imminent, the species is said to be (b)_____.

A. Species become endangered and extinct for a variety of reasons

■ 2 Although a variety of human activities threaten the continued existence of many organisms, most species facing extinction reach this perilous state as the result of _____.

B. The two types of efforts to save wildlife are *in situ* and *ex situ* conservation

■ 3 *In situ* preservation attempts to preserve species in (a)_____ by establishing reserves (e.g., parks), while *ex situ* conservation tries to preserve species in (b)_____ such as zoos and seed banks.

III. DEFORESTATION IS OCCURRING AT AN UNPRECEDENTED RATE

■ 4 Agriculture replaces diverse natural communities with communities consisting of one or two specially bred species. These artificial communities are unstable, simple, and lack some important mineral cycles, and modern agriculture requires substantial energy and has great environmental impact. For example, most of the surviving tropical forests, which consists mostly of the (a)_____ in South America and the (b)_____ forest in Africa, are being destroyed by individual farmers who clear forest areas by cutting down and burning forest plants. This practice, known as (c)_____ agriculture, ultimately leads to barren, unproductive land.

■ 5 In addition to agriculture, which may be the main culprit, a significant amount of deforestation also results from (a)_____ when it proceeds at a faster rate than is sustainable, as is occurring in the forests of Malaysia, and from the destruction of trees to provide open (b)_____.

IV. THE PRODUCTION OF ATMOSPHERIC POLLUTANTS THAT TRAP SOLAR HEAT IN THE ATMOSPHERE MAY AFFECT EARTH'S CLIMATE

A. Greenhouse gases cause global warming

■ 6 Gases that are accumulating in the atmosphere as a result of human activities include (a)_____. Gas powered vehicles cause three gases to accumulate, namely, (b)_____. Landfills produce (c)_____, and leaking cooling units release (d)_____.

■ 7 Since CO_2 and other gases that are being added to the atmosphere trap heat, global warming is occurring, a phenomenon known as the _____.

B. Global warming could alter food production, destroy forests, submerge coastal areas, and displace millions of people

■ 8 Models developed through computer simulations predict that a doubling of atmospheric CO_2 will raise the average temperature on Earth by (a)_____ degrees Centigrade within (b)_____ years, which will profoundly affect the entire Earth.

C. How should we deal with global warming?

V. OZONE IS DISAPPEARING IN THE STRATOSPHERE

■ 9 Ozone shields Earth from much of the harmful _____ that comes from the sun. Ozone depletion is occurring at an alarming rate.

A. Chemical destruction of ozone in the stratosphere is caused by CFCs and other industrial chemicals

■ 10 CFCs and other human-produced compounds move upward into the stratosphere where ultraviolet radiation breaks them down. For the most part, three elements

result, namely, (a)_____. Of these,
(b)_____ is particularly harmful since a single atom can break down
thousands of ozone molecules.

B. Ozone depletion adversely affects living organisms

VI. ENVIRONMENTAL PROBLEMS ARE INTERRELATED

KEY TERMS

Terms You Should Know

commercial harvest	in situ conservation	sport hunter
commercial hunter	ozone	subsistence agriculture
deforestation	pollution	subsistence hunter
endangered	scrub savannah	threatened
ex situ conservation	seed bank	waste
extinction	slash and burn agriculture	

SELF TEST

We would like to offer you a different kind of *Self Test* in this chapter.

The subject matter covered in this chapter is of critical importance to the continuance of life on earth as we know it. This chapter points out some of the dangers humanity faces if we continue some of our habits. Biologists and other responsible citizens can make a difference. One might even say that we have a responsibility to make a difference. A few genuinely concerned individuals will assume leadership roles in bringing about the social, cultural, and political changes required to protect the environment. All of us can participate in assuring that our children and grandchildren will inherit a clean, healthy, and productive environment.

We encourage you to reflect on what your personal role could be to bring about needed reforms. What might you contribute to the welfare of the planet? What changes are needed to insure that future generations can enjoy a reasonably good quality of life? Think about your lifestyle and aspirations. Are they compatible with a long term investment in the future of the planet?

Visual Foundations on next page ☞

VISUAL FOUNDATIONS

Color the parts of the illustration below as indicated.

RED	❑	heat re-radiated back to Earth
GREEN	❑	Earth
YELLOW	❑	solar energy
BLUE	❑	troposphere
ORANGE	❑	heat escaping to space
PINK	❑	stratosphere
TAN	❑	CO_2 in stratosphere

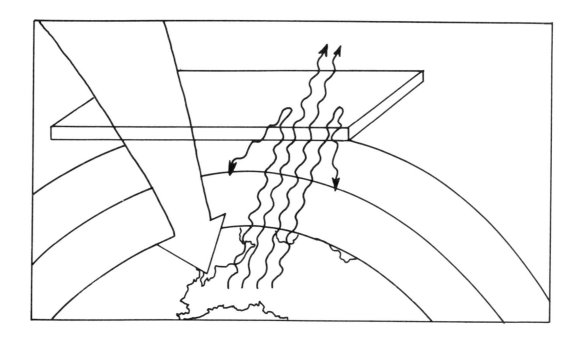

Color the parts of the illustration below as indicated. Label the southern and northern hemispheres.

RED   ☐   antarctic ozone hole

ANSWERS

□

CONCEPT QUESTIONS, DEFINITIONS, SELF TESTS

Chapter 1
Concept Questions: 1. composed of cells, organized, maintain homeostasis, metabolize, move, grow, respond to stimuli, adapt, reproduce 2. cell theory 3. size and/or number 4. development 5. chemical activities 6. homeostasis 7. sexually or asexually 8. evolution 9. genes 10. hormones and neurotransmitters 11. *On the Origin of Species By Means of Natural Selection* 12. (a)vary, (b)survive 13. environmental changes (or pressures) 14. evolution 15. cell 16. population 17. biosphere 18. taxonomy 19. genus and species 20. kingdom 21. (a)prokaryotae, (b)protista, (c)fungi, (d)plantae, (e)animalia 22. cellular respiration 23. producers, consumers, and decomposers 24. scientific method 25. deductive and inductive 26. hypothesis 27. observations and experiments

Definitions: 1. biology 2. ecology 3. asexual 4. homeostasis 5. ecosystem 6. biosphere 7. autotroph 8. heterotroph 9. protozoa

Self Test: 1. a,d,e 2. c 3. a,b,e 4. d 5. d 6. d 7. c,d 8. a 9. b 10. c,e 11. b 12. a 13. e 14. b 15. no 16. yes, bacteria could contaminate the media since there is no effective trap in the T-shaped neck. 17. c,d,e

Chapter 2
Concept Questions: 1. chemical reactions (ordinary chemical means) 2. (a)92, (b)uranium, (c)hydrogen, (d)carbon, hydrogen, oxygen, nitrogen, phosphorus, calcium 3. (a)protons, neutrons, electrons, (b)protons and neutrons, (c)electrons 4. (a)protons, (b)neutrons, (c)electrons 5. atomic number 6. atomic mass 7. (a)atomic mass (or numbers of neutrons), (b)atomic number (or numbers of protons) 8. (a)no (zero), (b)protons and electrons 9. electron shell (or energy level) 10. two or more different elements 11. (a)structural, (b)chemical 12. (a)reactants, (b)products 13. (a)left, (b)right 14. valence 15. molecule 16. potential 17. geometric shape 18. (a)nonpolar, (b)polar 19. (a)cations, (b)anions 20. electronegative 21. polar molecules 22. nonpolar 23. atomic masses of its constituent atoms 24. (a)oxidized, (b)reduced, (c)redox 25. carbon 26. hydrogen 27. hydrogen bonding 28. specific heat 29. hydrogen bonds 30. slight 31. (a)hydroxide ions and a cation, (b)hydrogen ions and an anion 32. (a)0-14, (b)7, (c)base, (d)acid 33. (a)hydrogen atom, (b)mineral 34. (a)buffers, (b)acids or bases

Definitions: 1. neutron 2. radionuclide 3. equilibrium 4. tetrahedron 5. hydration 6. hydrophobic 7. nonelectrolyte

Self Test: 1. b 2. a 3. a,b 4. e 5. c,d 6. c,d,e 7. a,b,d 8. a 9. c,d 10. d 11. b 12. a,e 13. a 14. a,b,d,e 15. b 16. b,d 17. e 18. b 19. a-e 20. b,e

Chapter 3

Concept Questions: 1. carbohydrates, lipids, proteins, nucleic acids   2. single, double, triple   3. tetrahedral   4. structures   5. functional   6. partial charges on atoms   7. polymers   8. carbon, hydrogen, oxygen   9. glucose   10. glycosidic   11. simple sugars (monosaccharides)   12. (a)glycoprotein, (b)glycolipids   13. carbon, hydrogen, oxygen   14. (a)glycerol, (b)fatty acid   15. amphipathic   16. carbon, hydrogen, oxygen, nitrogen, sulfur   17. (a)amino and carboxyl, (b)R group (radical)   18. (a)amino acids, (b)peptide   19. secondary, tertiary, quaternary   20. (a)amino acid, (b)three-dimensional (tertiary)   21. (a)purine or pyrimidine, (b)sugar, (c)phosphate   22. ATP

Definitions: 1. isomer   2. macromolecule   3. monomer   4. hydrolysis   5. monosaccharide   6. hexose   7. disaccharide   8. polysaccharide   9. amyloplast   10. monoglyceride   11. diglyceride   12. triglyceride   13. hydrophilic   14. amphipathic   15. dipeptide   16. polypeptide

Self Test: 1. a,c,e   2. a,d,e   3. b,c,d   4. c,e   5. b,c,e   6. a,b,d   7. a,b,d   8. a,c,d   9. a,d   10. a,c,d   11. c,d,e   12. b,c,d,e   13. b,d   14. d   15. b   16. a,b

Chapter 4

Concept Questions: 1. the activities of life   2. (a)living organisms, (b)preexisting cells   3. plasma membrane   4. (a)surface-to-volume ratio, (b)rate of diffusion (diffusion of molecules)   5. (a)electron, (b)cell fractionation   6. prokaryotic   7. eukaryotic   8. (a)cell walls, (b)plastids, (c)vacuoles   9. genes   10. chromatin   11. endoplasmic reticulum   12. ribosomes   13. transport vesicles   14. Golgi complex   15. lysosomes   16. mitochondria   17. chloroplasts   18. thylakoid   19. (a)stroma, (b)thylakoids   20. shape and movement   21. microfilaments and microtubules   22. centrioles and basal bodies   23. cilia and flagella   24. (a)actin, (b)myosin   25. cellulose   26. glycocalyx

Definitions: 1. chlorophyll   2. chromoplasts   3. cytoplasm   4. cytoskeleton   5. glyoxysome   6. leucoplast   7. lysosome   8. microfilaments   9. microtubules   10. microvilli   11. myosin   12. peroxisome   13. proplastid   14. prokaryotes   13. cytosol

Self Test: 1. a-e   2. a-e   3. a,b,e   4. d,e   5. a,c,e   6. a,c,d   7. a,b,d,e   8. a,b,c,e   9. d,e   10. a,d   11. a,b,c,e   12. c,e   13. b,c   14. prokaryotic   15. nucleoid region   16. plant(eukaryotic)   17. vacuole, chloroplast, mitochondrion, internal membrane system   18. cell wall, prominent vacuole, chloroplast   19. animal (eukaryotic)   20. centriole, lysosome, mitochondrion, internal membrane system   21. lysosomes, centrioles

Chapter 5

Concept Questions: 1. plasma membrane   2. (a)fluid mosaic, (b)lipid bilayer, (c)proteins   3. lipids   4. amphipathic   5. (a)hydrophilic, (b)hydrophobic   6. (a)head groups, (b)fatty acid chains   7. hydrocarbon chains   8. integral membrane proteins   9. peripheral membrane proteins   10. proteins   11. protein   12. selectively permeable   13. size, shape, electrical charge   14. (a)dialysis, (b)osmosis   15. osmotic pressure   16. facilitated diffusion and active transport   17. carrier-mediated active transport   18. desmosomes, tight junctions, gap junctions   19. gap junctions

Definitions: 1. endocytosis   2. exocytosis   3. hypertonic   4. hypotonic   5. isotonic   6. phagocytosis   7. pinocytosis

Self Test: 1. b,e   2. d   3. a   4. c   5. c,e   6. a,c,e   7. d,e   8. a,c,e   9. b,c   10. b,e   11. a,b,d,e   12. b,c,d,e   13. b,d   14. b,c,d,e

Chapter 6

Concept Questions: 1. (a)producers, (b)decomposers 2. work 3. heat, electrical, mechanical, chemical, sound, radiant 4. (a)potential energy, (b)kinetic energy 5. thermodynamics 6. (a)created nor destroyed, (b)transferred 7. randomness or disorder 8. heat 9. chemical and energy transformations 10. total bond energies 11. (a)exothermic, (b)endothermic 12. zero 13. exergonic reactions 14. (a)release, (b)require input of 15. adenine, ribose, three phosphates 16. hydrolysis 17. phosphorylation 18. increase 19. enzyme-substrate complex 20. lock-and-key, induced-fit 21. active sites 22. coenzyme 23. (a)product, (b)substrate 24. shape and amount 25. temperature, pH, salt concentration, chemical inhibition

Definitions: 1. kilocalorie 2. exothermic 3. endothermic 4. exergonic 5. endergonic 6. allosteric

Self Test: 1. a-e 2. b,c,e 3. a,d,e 4. c 5. b,c,e 6. b,c,d 7. a,b,e 8. a,b,c 9. a,b,d 10. a,d 11. b,d,e

Chapter 7

Concept Questions: 1. anabolism 2. catabolism 3. adenosine triphosphate (ATP) 4. aerobic respiration 5. anaerobic respiration 6. fermentation 7. (a)carbon dioxide and water, (b)energy 8. electrons and protons 9. (a)oxidized, (b)reduced 10. NAD^+ and FAD 11. (a)oxidized, (b)reduced 12. dehydrogenations and decarboxylations 13. cytosol (or cytoplasm) 14. glyceraldehyde-3-phosphate (PGAL) 15. pyruvate 16. two 17. (a)NAD^+, (b)CO_2 18. (a)three, (b)two, (c)eight 19. two 20. (a)NAD^+ and FAD, (b)the electron transport system and chemiosmotic phosphorylation 21. (a)electron transport system, (b)molecular oxygen 22. (a)intermembrane space, (b)ATP synthetase 23. fatty acids 24. deamination 25. glycerol and fatty acid 26. beta oxidation 27. organic compound 28. ethyl alcohol 29. lactate 30. (a)two, (b)36-38

Definitions: 1. aerobe 2. anaerobe 3. dehydrogenation 4. decarboxylation 5. deamination 6. glycolysis

Self Test: 1. a,b,e 2. a,d 3. b,e 4. c,d,e 5. a,d,e 6. c,d 7. b,c,d 8. b,c 9. a-e

Chapter 8

Concept Questions: 1. (a)carbon dioxide and water, (b)light (solar), (c)chemical 2. heterotrophs 3. (a)stroma, (b)mesophyll 4. thylakoids 5. (a)chlorophyll, (b)ATP and NADPH 6. (a)water (H_2O), (b)carbon dioxide (CO_2) 7. (a)chlorophyll, (b)electron acceptors 8. (a)ATP, (b)photolysis 9. (a)$NADP^+$, (b)NADPH 10. (a)NADPH and ATP, (b)carbohydrates 11. (a)ground state, (b)excited (energized) 12. (a)fluorescence, (b)electron acceptor 13. blue and red 14. (a)chlorophyll *a*, (b)chlorophyll *b*, (c)carotenoids 15. (a)absorption spectrum, (b)spectrophotometer 16. (a)action spectrum, (b)accessory pigments 17. photophosphorylation 18. (a)NADH, (b)ATP 19. P700 20. (a)thylakoid membrane, (b)electrochemical gradient, (c)chemiosmosis 21. ATP synthetase (CF_0-CF_1) 22. CO_2 fixation 23. (a)glucose, (b)ribulose biphosphate (RuBP), (c)phosphoglycerate (PGA), (d)glyceraldehyde-3-phosphate (PGAL) 24. (a)CO_2, (b)NADPH, (c)ATP 25. oxaloacetate 26. (a)RuBP carboxylate, (b)oxygen

Definitions: 1. mesophyll 2. photosynthesis 3. photolysis 4. photophosphorylation 5. chloroplast 6. chlorophyll

Self Test: 1. b,d,e 2. a 3. c,e 4. a,c,e 5. a,b,d 6. a-e 7. c 8. a,c,d 9. a,c,d 10. a,b,d 11. a,c,d,e 12. c 13. a,c,d 14. c

Chapter 9

Concept Questions: 1. chromatin fibers 2. (a)RNA, (b)polypeptides 3. cell cycle 4. (a)first gap, (b)synthesis, (c)second gap 5. prophase 6. metaphase 7. anaphase 8. telophase 9. asexual 10. (a)sexual, (b)diploid, (c)zygote 11. size and structure 12. (a)homologous chromosome, (b)diploidy 13. (a)gametes or spores, (b)somatic (body) cells 14. (a)prophase, (b)genetic variability 15. anaphase 16. anaphase 17. (a)mitosis, (b)meiosis 18. (a)somatic, (b)gametes (sex cells) 19. (a)sporophyte, (b)gametophyte

Definitions: 1. chromosome 2. interphase 3. haploid 4. diploid 5. centromere 6.gametogenesis 7. spermatogenesis 8. oogenesis 9. gametophyte 10. sporophyte

Self Test: 1. c,d 2. b,c,e 3. a,d 4. c,d,e 5. b 6. b 7. b 8. a,g,h 9. g 10. d 11. j 12. c 13. h 14. a,i 15. e 16. b 17. b,f 18. d 19. g 20. h 21. c 22. b 23. b 24. d 25. d

Chapter 10

Concept Questions: 1. (a)dominant, (b)recessive 2. segregate 3. other pairs of genes 4. locus 5. (a)P or P_1, (b)locus 6. (a)first filial, (b)F_2 or second filial 7. F_1 and homozygous recessive 8. (a)specific events, (b)zero, (c)one 9. sum of their separate probabilities 10. product 11. two loci 12. linked 13. crossing over 14. recombination 15. autosomes 16. (a)XY male, (b)XX female 17. (a)hyperactive, (b)inactivation 18. (a)three or more, (b)locus 19. pleiotropic 20. epistasis 21. polygenic inheritance 22. homozygosity 23. (a)heterozygosity, (b)hybrid vigor

Definitions: 1. monoecious 2. homozygous 3. heterozygous 4. dihybrid 5. monohybrid 6. dioecious 7. hemizygous

Self Test: 1. both j 2. l,m,n,r,s,t 3. b,c,d,e,f 4. RrTT 5. j 6. c/n and d/r 7. b/m and e/s 8. g 9. g 10. d 11. b or d 12. c 13. b 14. e 15. b,c,e 16. b 17. a,b,e 18. a,c,d 19. a,e 20. a,c,d 21. incomplete dominance 22. all pink 23. 1:2:1, red:pink:white

Chapter 11

Concept Questions: 1. Garrod and Beadle and Tatum 2. transformation 3. viruses (bacteriophages) 4. sugar (deoxyribose), phosphate, nitrogenous base 5. X-ray diffraction 6. antiparallel nucleotide strands 7. (a)pyrimidines, (b)cytosine, (c)thymine 8. template 9. (a)5, (b)3, (c)Okazaki 10. DNA polymerases 11. origin(s) of replication 12. prokaryotic 13. (a)nucleosome, (b)histones 14. scaffolding

Definitions: 1. avirulent 2. antiparallel

Self Test: 1. a,c,d 2. a,c,d 3. a-e 4. d 5. a,b,d 6. b,c 7. d 8. b,c,e 9. b 10. a,c 11. d 12. c,e 13. c 14. e 15. a 16. d,e 17. b 18. d

Chapter 12

Concept Questions: 1. (a)transcription, (b)translation 2. codon 3. (a)anticodon, (b)amino acid 4. ribosomal (rRNA), transfer (tRNA), messenger (mRNA) 5. (a)RNA polymerase, (b)ribonucleotide triphosphate 6. promoter or initiator 7. (a)upstream, (b)downstream 8. (a)leader sequence, (b)termination signals 9. aminoacyl-tRNA synthetases 10. (a)anticodons, (b)peptide 11. AUG 12. elongation 13. (a)amino, (b)carboxyl 14. termination or stop codons 15. 7-methyl guanylate 16. (a)intervening sequences, (b)expressed sequences 17. (a)sixty-one, (b)forty 18. nucleotides 19. nucleotide sequence 20. (a)point or base-substitution, (b)missense, (c)nonsense 21. hot spots

Definitions: 1. anticodon 2. ribosome 3. polysome 4. mutagen 5. carcinogen

Self Test: 1. c,d,e 2. d 3. a,c,e 4. a 5. a 6. d,e 7. a,c,d 8. b,c 9. c,e 10. a,c 11. a,c,e 12. a,b 13. b,d 14. b,c,e

Chapter 13
Concept Questions: 1. constitutive genes 2. promoter region 3. operator 4. repressor protein 5. (a)inducer molecule, (b)inducible 6. corepressor 8. regulon 9. feedback inhibition 10. TATA box 11. upstream promoter elements (UPEs) 12. (a)heterochromatin, (b)euchromatin 13. differential nuclear RNA processing 14. chemical modification

Definitions: 1. euchromatin 2. heterochromatin 3. corepressor 4. heterodimer 5. homodimer

Self Test: 1. a,d 2. b,d 3. c 4. e 5. e 6. b 7. d,e 8. b,d 9. a,b,c,e 10. a,d,e 11. a,c,d,e 12. b,c,d 13. a,c,d 14. c,b,d 15. a,e

Chapter 14
Concept Questions: 1. palindromic 2. restriction enzymes 3. plasmids or bacteriophages 4. genetic probes 5. genomic library 6. reverse transcriptase 7. (a)DNA polymerase, (b)heat 8. restriction maps 9. restriction fragment length polymorphisms (RFLPs) 10. promoter and regulator 11. (a)fertilized egg, (b)retroviruses

Definitions: 1. polymorphism 2. transgenic 3. retrovirus

Self Test: 1. a,c,d 2. a 3. b,d,e 4. b,d,e 5. b 6. c,d 7. c,e 8. c,e 9. b,e 10. d 11. d 12. b,c 13. c,d,e 14. a,c,e 15. c,d,e 16. a,c

Chapter 15
Concept Questions: 1. isogenic strains 2. congenital 3. cytogenetics 4. photomicrograph 5. polyploidy 6. (a)aneuploidy, (b)disomic, (c)trisomic, (d)monosomic 7. (a)nondisjunction, (b)Down syndrome 8. (a)nuclear sexing, (b)Barr body 9. Klinefelter syndrome 10. Turner syndrome 11. inborn errors of metabolism 12. sickle cell anemia 13. cystic fibrosis 14. amniocentesis 15. chorionic villus sampling (CVS) 16. severe combined immune deficiency (SCID) 17. (a)I^AI^A and I^Ai^o, (b)I^BI^B and I^Bi^o, (c)I^AI^B, (d)i^oi^o 18. erythroblastosis fetalis 19. consanguineous matings

Definitions: 1. isogenic 2. cytogenetics 3. polyploidy 4. trisomy 5. translocation

Self Test: 1. c,e 2. c 3. b 4. a,b,c 5. a,d 6. b,c,d 7. a,d,e 8. b,c,d 9. a,b,c 10. b,e 11. b 12. e 13. c 14. a 15. d

Chapter 16
Concept Questions: 1. (a)determination, (b)differentiation 2. (a)morphogenesis, (b)pattern formation 3. nuclear equivalence 4. totipotent 5. polytene 6. puffs 7. maternal effect 8. (a)zygotic genes, (b)gap genes (c)homeotic genes 9. homeobox 10. stem or founder cells 11. mosaic 12. chimera 13. regulative 14. genomic rearrangements

Definitions: 1. morphogenesis 2. polytene 3. chronogene

Self Test: 1. a,d 2. a,c,d 3. c,e 4. c,e 5. b 6. c 7. a,c 8. d 9. c,e 10. a 11. c 12. b 13. d 14. c,e 15. b,d

Chapter 17

Concept Questions: 1. descent with modification 2. populations 3. gene pool 4. Jean Baptiste de Lamarck (Lamarck) 5. (a)H.M.S. Beagle, (b)Galapagos Islands 6. artificial selection 7. uniformitarianism 8. Thomas Malthus 9. Alfred Russell Wallace 10. *Origin of Species by Means of Natural Selection* 11. overproduction, variation, competition, and survival to reproduce 12. mutation and recombination 13. (a)geology, (b)paleontology 14. traces or impressions 15. compressions 16. petrifactions 17. index fossils 18. homologous 19. analogous 20. vestigial organs 21. (a)protective coloration, (b)Batesian mimicry, (c)Mullerian mimicry 22. biogeography 23. center of origin 24. antibody-antigen 25. DNA hybridization

Definitions: 1. homologous 2. biogeography

Self Test: 1. c 2. b 3. a 4. a,b 5. a 6. c,d 7. c 8. e 9. b,e 10. d 11. c 12. a 13. b 14. c,e 15. d 16. b

Chapter 18

Concept Questions: 1. microevolution 2. breeding population 3. (a)$p^2 + 2pq + q^2$, (b)Hardy-Weinberg equilibrium 4. large population, isolation, no mutations, no selection, random mating 5. gene or allele frequencies 6. genetic bottleneck 7. founder effect 8. (a)demes, (b)gene flow 9. mutation 10. natural selection 11. coadapted gene complex 12. polymorphism 13. (a)more common, (b)less common

Definitions: 1. microevolution 2. allozymes

Self Test: 1. a 2. c,d 3. b 4. c 5. a 6. c 7. c,e 8. c 9. a,e 10. d 11. b,d,e 12. g 13. k 14. h 15. c 16. j 17. y

Chapter 19

Concept Questions: 1. Linnaeus 2. (a)reproductively, (b)gene pool 3. fertilization 4. (a)temporal isolation, (b)ecological isolation, (c)behavioral isolation, (d)mechanical isolation, (e)gametic isolation 5. hybrid inviability 6. hybrid sterility 7. speciation 8. geographically isolated 9. hybridization and polyploidy 10. (a)polyploidy, (b)allopolyploidy 11. punctuated equilibrium 12. gradualism 13. (a)allometric growth, (b)paedomorphosis 14. (a)adaptive zones, (b)adaptive radiation 15. (a)background, (b)mass

Definitions: 1. polyploidy 2. allopatric 3. allopolyploidy 4. macroevolution 5. allometric

Self Test: 1. d 2. e 3. c 4. b,d 5. b,d 6. a,c,e 7. a 8. c,d 9. e 10. a,c,e 11. a,b,e

Chapter 20

Concept Questions: 1. (a)10-20 billion, (b)4.6 billion 2. absence of free oxygen, energy, chemical building blocks, time 3. (a)Oparin, (b)Haldane 4. (a)Miller and Urey, (b)hydrogen (H_2), methane (CH_4), ammonia (NH_3) 5. protobionts 6. (a)anaerobic, (b)prokaryotic 7. (a)RNA, (b)ribosomes 8. metabolism 9. (a)heterotrophs, (b)autotrophs 10. oxygen 11. endosymbiont 12. (a)archean, (b)3.5 billion, (c)proterozoic, (d)1.5 billion 13. (a)570 million, (b)cambrian explosion, (c)ordovician, (d)ostracoderm 14. (a)248 million, (b)triassic, jurassic, cretaceous, (c)triassic, (d)jurassic 15. 65 million

Definitions: 1. proteinoid 2. protobiont 3. liposome 4. precambrian

Self Test: 1. b,c,d 2. c,d,e 3. a,d 4. e 5. d 6. a,d 7. a,e 8. c,e 9. d 10. a,b,c,d 11. b,d 12. d,e 13. d 14. c

Chapter 21

Concept Questions: 1. (a)prosimians, (b)anthropoids 2. (a)prosimian, (b)oligocene 3. hominoids 4. gibbons (*Hylobates*), orangutans (*Pongo*), gorillas (*Gorilla*), chimpanzees (*Pan*) 5. 4 to 5 million 6. 3.8 million 7. *Australopithecus afarensis* 8. 1.9 million 9. 1.5 to 1.6 million 10. Africa 11. knowledge 12. development of hunter/gatherer societies, development of agriculture and the industrial revolution

Definitions: 1. anthropoid 2. quadrupedal 3. bipedal 4. prosimian 5. hominoid 6. supraorbital

Self Test: 1. b,e 2. c,e 3. d 4. b,c,e 5. e 6. d,e 7. a,d,e 8. b 9. c,d,e 10. a

Chapter 22

Concept Questions: 1. taxonomy 2. (a)binomial system, (b)genus, (c)species 3. species 4. (a)family, (b)class, (c)division, (d)phylum 5. taxon 6. Prokaryotae, Protista, Fungi, Animalia, Plantae 7. evolutionary relationships 8. (a)monophyletic, (b)clade 9. polyphyletic 10. (a)ancient, (b)recent 11. DNA and protein 12. phenetics, cladistics, classical evolutionary taxonomy 13. phenotypic characteristics 14. monophyletic

Definitions: 1. subspecies 2. monophyletic 3. polyphyletic 4. prokaryote 5. eukaryote

Self Test: 1. c,e 2. b 3. d 4. d 5. d 6. c 7. a,d 8. b,d 9. a 10. a,c,d

Chapter 23

Concept Questions: 1. (a)nucleic acid (DNA or RNA), (b)capsid 2. (a)envelope, (b)proteins, lipids, carbohydrates, metals 3. smallpox virus 4. (a)virulent (or lytic), (b)temperate (or lysogenic) 5. attachment, penetration, replication, assembly, release 6. (a)prophage, (b)lysogenic 7. lysogenic conversion 8. transduction 9. adsorptive endocytosis 10. retroviruses 11. oncogenes 12. HTLV 13. Epstein-Barr 14. viroids 15. "bits" of nucleic acids that "escaped" from cells 16. peptidoglycan 17. mesosomes 18. plasmids 19. (a)saprobes, (b)chemosynthetic autotrophs (chemoautotrophs) 20. (a)commensals, (b)parasites, (c)mutualists 21. cyanobacteria, green sulfur bacteria, purple sulfur bacteria, green nonsulfur bacteria, purple nonsulfur bacteria 22. (a)transformation, (b)transduction, (c)conjugation 23. bacteriorhodopsin 24. carbon dioxide and hydrogen 25. (a)cocci, (b)diplococci, (c)streptococci, (d)staphylococci, (e)bacilli, (f)spirilla 26. (a)peptidoglycan, (b)crystal violet 27. axial filaments 28. (a)cyanobacteria, (b)photosynthetic lamellae 29. arthropods 30. streptococci 31. staphylococci 32. clostridia 33. (a)actinomycetes, (b)actinospores

Definitions: 1. bacteriophage 2. viroid 3. oncogene 4. cyanobacteria 5. archaebacteria 6. halophile 7. methanogen 8. phycocyanin 9. phycoerythrin 10. endospore 11. eubacteria 12. pathogen 13. anaerobe

Self Test: 1. c 2. a,b 3. a,c,e 4. e 5. d 6. c 7. a,b 8. b 9. a,b,d 10. a 11. a,b,e 12. b 13. c 14. a,c 15. d 16. d 17. b,d,e 18. b,c,d,e 19. a

Chapter 24

Concept Questions: 1. coenocytic 2. contractile vacuoles 3. plankton 4. pseudopodia 5. *Entamoeba histolytica* 6. axopods 7. *Trypanosoma* 8. choanoflagellates 9. trichocysts 10. (a)micronuclei, (b)macronucleus 11. conjugation 12. *Plasmodium* 13. Zooxanthellae 14. diatomaceous earth 15. (a)chlorophyll *a*, chlorophyll *b*, carotenoids, (b)paramylon 16. lichen 17. (a)isogamous, (b)anisogamous, (c)oogamous 18. (a)holdfast, (b)phycoerythrin, (c)phycocyanin 19. (a)blades, (b)stipes 20. sporangia 21. (a)swarm cell, (b)myxamoeba 22. pseudoplasmodium (slug) 23. (a)mycelium, (b)zoospores, (c)oospores 24. flagellates 25. green algae

Definitions: 1. syngamy 2. protozoa 3. cytopharynx 4. pseudopod 5. trichocyst 6. micronucleus 7. macronucleus 8. pseudoplasmodium 9. isogamy

Self Test: 1. b 2. b,d,e 3. c 4. b 5. c,d 6. a,c,e 7. b 8. c,d,e 9. a,d,e 10. b 11. a

Chapter 25

Concept Questions: 1. (a)saprophytes, (b)carbon dioxide, (c)minerals (plant nutrients) 2. mycorrhizae (fungus-roots) 3. (a)yeasts, (b)molds 4. chitin 5. (a)aerial hyphae, (b)sporocarps 6. (a)dikaryotic, (b)monokaryotic 7. (a)sexual spores and fruiting bodies, (b)Zygomycota, Ascomycota, Basidiomycota, Deuteromycota 8. (a)*Rhizopus stolonifer*, (b)heterothallic 9. (a)asci, (b)conidia, (c)conidiophores 10. ascocarps 11. (a)basidium, (b)basidiospores 12. (a)button, (b)basidiocarp 13. sexual stage 14. (a)crustose, (b)foliose, (c)fruticose 15. (a)green alga or cyanobacterium, (b)ascomycete 16. soredia 17. (a)yeasts, (b)fruit sugars, (c)grains, (d)carbon dioxide 18. (a)*Penicillium*, (b)*Aspergillus tamarii* 19. (a)cyclopeptides, (b)*Amanita* 20. psilocybin 21. (a)*Penicillium notatum*, (b)penicillin 22. (a)ergot, (b)ergotism 23. (a)cutinase, (b)haustoria 24. Deuteromycetes

Definitions: 1. saprophyte 2. coenocytic 3. sporocarp 4. monokaryotic 5. heterothallic 6. conidiophore 7. homothallic

Self Test: 1. d 2. b,d 3. c 4. b,d 5. b 6. a,c,e 7. a-e 8. e 9. a,c,e 10. a,b,d 11. a-e 12. c,d,e 13. a,d 14. a-e

Chapter 26

Concept Questions: 1. (a)green algae; (b)chlorophylls *a* and *b*, carotenes, xanthophylls; (c)starch, (d)cellulose, (e)cell plate 2. (a)mosses and other bryophytes, (b)ferns and their allies, (c)gymnosperms, (d)angiosperms (flowering plants) 3. (a)waxy cuticle, (b)stomata 4. (a)gametangia, (b)antheridium, (c)archegonium 5. lignin 6. (a)zygote (fused gametes), (b)spores 7. mosses, liverworts, hornworts 8. vascular tissues 9. (a)*Sphagnum*, (b)peat mosses 10. gemmae 11. (a)microphyll, (b)megaphyll 12. (a)rhizome, (b)fronds 13. (a)sporangia, (b)sori, (c)prothallus 14. (a)upright stem, (b)dichotomous, (c)fungus 15. *Equisitum* 16. (a)homospory, (b)heterospory 17. (a)microsporangia, (b)microsporocytes, (c)microspores, (d)megaspores, (e)megasporocytes

Definitions: 1. xanthophyll 2. gametangium 3. archegonium 4. gametophyte 5. sporophyte 6. microphyll 7. megaphyll 8. sporangium 9. homospory 10. heterospory 11. megaspore 12. microspore 13. bryophyte

Self Test: 1. a,c,d 2. a,b,c 3. a,e 4. a,c,d 5. b 6. a-e 7. b,d 8. c,d 9. a,e 10. a,b,c 11. b,d 12. b,d 13. b,c,e 14. b,e 15. a,c,e

Chapter 27

Concept Questions: 1. (a)gymnosperms, (b)angiosperms 2. (a)xylem, (b)phloem 3. Coniferophyta (conifers), Ginkgophyta, Cycadophyta, Gnetophyta 4. monoecious 5. pines, spruces, firs, larch, cypress, hemlock, etc. 6. (a)sporophylls, (b)microsporocytes, (c)male gametophytes (pollen grains) 7. (a)megasporangia, (b)haploid megaspores, (c)female gametophyte 8. dioecious 9. vessels 10. Magnoliophyta 11. (a)Liliopsida, (b)Magnoliopsida 12. (a)three, (b)one, (c)endosperm 13. (a)four or five, (b)two, (c)cotyledons 14. (a)palms, grasses, orchids, lilies, etc. (b)oaks, roses, cacti, blueberries, sunflowers, etc. 15. (a)sepals, petals, stamens, carpels; (b)stamens; (c)carpels; (d)perfect 16. (a)androecium, (b)gynoecium, (c)anther, (d)ovary 17. (a)megaspores, (b)embryo sac, (c)polar nuclei 18. (a)microspores, (b)male gametophyte (pollen grain), (c)sperm nuclei 19. zygote, endosperm tissue, seed coat 20. coevolution 21. (a)progymnosperms, (b)seed ferns, (c)gymnosperms

Definitions: 1. angiosperm 2. gymnosperm 3. gynoecium 4. monoecious 5. dioecious 6. androecium

Self Test: 1. a–e 2. b,e 3. b 4. b,c 5. a,b,d,e 6. a,c,d,e 7. a–e 8. a,e 9. b 10. c,d,e

Chapter 28

Concept Questions: 1. (a)sea (salt) water, (b)fresh water, (c)land (terrestrial) 2. (a)parazoa, (b)eumetazoa 3. (a)Radiata, (b)Bilateria 4. (a)anterior, (b)posterior, (c)dorsal, (d)ventral, (e)medial, (f)lateral, (g)cephalic, (h)caudal 5. (a)acoelomates, (b)pseudocoelomates, (c)coelomates 6. (a)mouth, (b)anus, (c)radial cleavage, (d)spiral cleavage 7. choanoflagellates 8. (a)Calcispongia, (b)Hexactinellida, (c)Demospongia, (d)spongin 9. (a)spongocoel, (b)osculum 10. (a)cnidocytes, (b)epidermis, (c)gastrodermis, (d)mesoglea 11. (a)Hydrozoa, (b)Scyphozoa, (c)Anthozoa 12. (a)cnidocytes, (b)nematocysts 13. (a)medusa, (b)polyp 14. (a)biradially, (b)eight, (c)glue cells 15. (a)bilateral, (b)three, (c)protonephridia 16. (a)Turbellaria, (b)Trematoda, (c)Cestoda 17. pharynx 18. (a)suckers, (b)anterior 19. (a)scolex (head), (b)proglottid 20. proboscis 21. (a)three, (b)pseudocoelom, (c)bilateral 22. hookworm 23. (a)trichina worms, (b)cysts 24. pinworms 25. (a)cell constancy, (b)cilia 26. mastax

Definitions: 1. pseudocoelom 2. ectoderm 3. mesoderm 4. protostome 5. deuterostome 6. schizocoely 7. enterocoely 8. spongocoel 9. amoebocyte 10. gastrodermis

Self Test: 1. a,d,e 2. d 3. a,c,e 4. c,e 5. c,e 6. c 7. d,e 8. a,b,c 9. d 10. c 11. a,d 12. e 13. a 14. b

Chapter 29

Concept Questions: 1. (a)mouth, (b)mesoderm   2. independent movement of digestive tract (food) and body, coelomic fluid bathes coelomic cells and organs (transports food, oxygen, wastes), acts as hydrostatic skeleton, space for development and function of organs   3. Arthropoda   4. (a)desiccation (fluid loss), (b)gravity   5. (a)mantle, (b)Arthropoda   6. hemocoel   7. (a)trochophore, (b)veliger   8. eight   9. insects   10. torsion   11. (a)mantle, (b)calcium carbonate   12. adductor muscle   13. (a)tentacles, (b)ten, (c)eight   14. septa   15. setae   16. metanephridia   17. parapodia   18. prostomium   19. (a)Annelida, (b)Oligochaeta, (c)*Lumbricus terrestris*   20. (a)crop, (b)gizzard   21. (a)hemoglobin, (b)metanephridia, (c)skin   22. (a)jointed appendages, (b)exoskeleton, (c)chiton   23. head, thorax, abdomen   24. tracheae   25. (a)Chelicerata, (b)Crustacea, (c)Uriramia   26. (a)trilobites, (b)chelicerates   27. (a)Merostomata, (b)Arachnida   28. (a)cephalothorax, (b)six, (c)four   29. book lungs   30. spinnerets   31. (a)mandibles, (b)biramous, (c)two, (d)five   32. barnacles   33. Decapoda   34. (a)maxillae, (b)maxillipeds, (c)chelipeds, (d)walking legs   35. (a)reproductive, (b)swimmeretes   36. (a)jointed, (b)possess tracheal tubes, (c)have six feet   37. (a)three, (b)one or two, (c)one, (d)malpighian tubules   38. (a)Chilopoda, (b)one, (c)Diplopoda, (d)two

Definitions: 1. endoskeleton   2. exoskeleton   3. bivalve   4. biramous   5. uniramous   6. trilobite   7. cephalothorax   8. hexapod   9. arthropod

Self Test: 1. b,d,e   2. b,c,e   3. b   4. d   5. b,d,e   6. a,b,d   7. b,d   8. a,c,e   9. a,c,d   10. a,d

Chapter 30

Concept Questions: 1. (a)anus, (b)mouth   2. (a)spiny, (b)bilateral, (c)radial (pentaradial), (d)coelom   3. dorsal (upper)   4. (a)central disk, (b)tube feet   5. test   6. tube feet   7. bilateral   8. (a)notochord, (b)nerve cord, (c)pharyngeal gill slits   9. cellulose   10. *Amphioxus*   11. (a)oral hood, (b)pharynx   12. (a)vertebral column, (b)cranium, (c)cephalization   13. (a)Agnatha, (b)Chondrichthyes, (c)Osteichthyes, (d)Amphibia, (e)Reptilia, (f)Aves, (g)Mammalia   14. ostracoderms   15. placoid scales   16. (a)lateral line organ,   (b)electroreceptors,   17. (a)oviparous,   (b)ovoviviparous,   (c)viviparous   18. (a)crossopterygians, (b)coelacanths   19. (a)ray-finned, (b)sarcopterygians, (c)teleosts, (d)swim bladders   20. (a)Urodela, (b)Anura, (c)Apoda   21. neoteny   22. (a)moist skin, (b)three, (c)mucous glands   23. (a)three, (b)uric acid, (c)ectothermic   24. (a)Chelonia, (b)Squamata, (c)Crocodilia   25. (a)Mesozoic, (b)Cretaceous   26. (a)four, (b)endothermic, (c)uric acid, (d)feathers   27. (a)crop, (b)proventriculus, (c)gizzard   28. *Archaeopteryx*   29. hair, mammary glands, differentiated teeth   30. (a)therapsids, (b)Triassic period   31. monotremes   32. (a)marsupials, (b)marsupium   33. oxygen and nutrients

Definitions: 1. agnatha   2. anuran   3. apodal   4. chondrichthyes   5. osteichthyes   6. tetrapod   7. urodela   8. echinoderm

Self Test: 1. c   2. d,e   3. a,c,d   4. d,e   5. a,b   6. c,e   7. c,e   8. b,c   9. d   10. b,e   11. c,d,e   12. c,d,e   13. b,c,d   14. a–e   15. d   16. d   17. a,b   18. a,d   19. c,e   20. b,c,d   21. d   22. a,b,d,e

Chapter 31

Concept Questions: 1. (a)tissues, (b)organs   2. (a)zygote, (b)embryo   3. (a)suspensor, (b)proembryo   4. (a)radicle, (b)epicotyl, (c)hypocotyl   5. oxygen, water, temperature, light   6. abscisic acid   7. (a)root (radicle), (b)root cap   8. (a)indeterminate growth, (b)determinate growth   9. (a)meristems, (b)division, elongation, differentiation (specialization)   10. (a)primary growth, (b)secondary growth   11. (a)area of cell division, (b)area of cell elongation, (c)area of cell maturation   12. (a)leaf primordia, (b)bud primordia

13. (a)vascular cambium, (b)cork cambium   14. (a)perennials, (b)annuals, (c)biennials
15. photosynthesis, storage, secretion   16. support   17. lignin   18. (a)water and dissolved minerals, (b)parenchyma cells, (c)fibers   19. food   20. (a)sieve tube members, (b)sieve plates
21. (a)protection, (b)cuticle, (c)stomata   22. protection   23. (a)taproot, (b)fibrous
24. (a)node, (b)buds, (c)terminal bud, (d)lateral buds   25. (a)blade, (b)petiole   26. hormones
27. callus

Definitions: 1. trichome   2. internode   3. proembryo

Self Test: 1. c   2. c,d,e   3. a,b,c,e   4. a–e   5. a   6. e   7. d   8. e   9. c   10. a,d   11. b
12. a,b   13. a–e   14. a,d   15. a,b,e

Chapter 32

Concept Questions: 1. (a)mesophyll, (b)palisade, (c)spongy   2. (a)xylem, (b)phloem, (c)bundle sheath   3. (a)netted or palmate, (b)parallel, (c)bulliform, (d)subsidiary   4. amount of light, [CO_2], a circadian rhythm, humidity   5. (a)vacuoles, (b)guard, (c)open, (d)close, (e)chlorophyll, flavoprotein   6. (a)guttation, (b)transpiration, (c)temperature, light, wind, relative humidity   7. hormones   8. (a)abscission zone, (b)fibers, (c)suberin, (d)middle lamella   9. (a)spines, (b)tendrils

Definitions: 1. mesophyll   2. photoreceptor   3. epiphyte

Self Test: 1. a,c,d   2. a–e   3. d   4. a,b,e   5. a,b,c   6. a,d,e   7. b,c,e   8. c   9. b,c,e   10. b

Chapter 33

Concept Questions: 1. support, conduct, produce new stem tissue   2. (a)vascular bundles, (b)xylem, phloem, (c)vascular cambium, (d)pith, (e)cortex   3. ground tissue (parenchyma)
4. (a)epidermis, (b)cortical parenchyma, (c)ground parenchyma   5. (a)vascular cambium, (b)cork cambium   6. (a)rays, (b)parenchyma   7. (a)cork cells, (b)cork parenchyma
8. (a)heartwood, (b)sapwood, (c)dicots, (d)gymnosperms   9. (a)springwood, (b)late summerwood   10. (a)water potential, (b)less, (c)more, (d)less   11. soil and root cells
12. (a)tension-cohesion mechanism, (b)transpiration   13. (a)pressure flow, (b)source, (c)sink

Definitions: no definitions in Chapter 33.

Self Test: 1. a,b,d   2. a,b,c,d   3. a,c,e   4. a,b,d   5. a,d   6. b   7. b,d,e   8. c   9. d   10. a–e

Chapter 34

Concept Questions: 1. anchorage, absorption, conduction, storage   2. (a)adventitious, (b)prop, (c)contractile   3. (a)endodermis, (b)casparian strip   4. (a)pericycle, (b)vascular
5. (a)cell walls, (b)cellulose, (c)endodermis, (d)epidermis, (e)endodermal cells, (f)xylem
6. (a)storage, (b)pericycle   7. minerals, organic matter, organisms, soil air, soil water
8. (a)sand and silt, (b)clay, (c)humus, (d)soil air and water   9. hydroponics   10. (a)essential, (b)carbon, oxygen, hydrogen, nitrogen, potassium, phosphorus, sulfur, magnesium, calcium, (c)iron, boron, manganese, copper, zinc, molybdenum, chlorine   11. (a)concept of limiting factors, (b)nitrogen, phosphorus, potassium

Definitions: 1. hydroponics   2. macronutrient   3. micronutrient

Self Test: 1. c   2. a,b,c   3. b,d   4. a,b,d,e   5. a,b,d   6. a,c,e   7. a,d   8. c   9. c   10. b
11. c   12. c   13. b,e

Chapter 35

Concept Questions: 1. (a)flower, (b)seeds and fruits 2. vegetative propagation 3. (a)rhizomes, (b)tubers, (c)bulbs, (d)corms, (e)stolons 4. suckers 5. apomixis 6. simple, aggregate, multiple, accessory 7. (a)simple, (b)berries, (c)drupes, (d)dehiscent, (e)indehiscent, (f)follicle, (g)legumes, (h)capsules 8. (a)aggregate, (b)multiple 9. (a)accessory, (b)receptacle, (c)floral tube 10. wind, animals, water, explosive dehiscence 11. photoperiodism 12. phytochrome 13. vernalization

Definitions: 1. phytochrome

Self Test: 1. d 2. d 3. b,d,e 4. b,d,e 5. b,d,e 6. e 7. e 8. d 9. a,c,d 10. a,e 11. a,e

Chapter 36

Concept Questions: 1. auxins, gibberellins, cytokinins, ethylene, abscisic acid 2. (a)pulvinus, (b)potassium, (c)turgor movements 3. solar tracking 4. circadian rhythms 5. (a)phototropism, (b)gravitropism, (c)thigmotropism 6. (a)curvature, (b)indoleacetic acid (IAA), (c)shoot tip, (d)polar transport 7. (a)acid-growth, (b)apical meristem 8. flowering and germination 9. (a)cell division and differentiation, (b)senescence 10. fruit ripening and leaf abscission 11. water

Definitions: 1. phototropism 2. gravitropism 3. organogenesis

Self Test: 1. a–e 2. e 3. c 4. a,d 5. e 6. a,b,c 7. a 8. b 9. a 10. d

Chapter 37

Concept Questions: 1. number 2. (a)tissue, (b)organs, (c)organ systems 3. (a)sheets, (b)basement membrane 4. protection, absorption, secretion, sensation 5. (a)simple, (b)stratified, (c)squamous, (d)cuboidal, (e)columnar 6. loose, dense, elastic, reticular, adipose, cartilage, bone, blood, lymph, blood cell-producing cells 7. (a)intercellular substance, (b)fibers, (c)matrix 8. intercellular substance 9. (a)collagen, (b)elastin, (c)collagen and glycoprotein 10. (a)fibers and matrix, (b)macrophages 11. (a)mast cells, (b)plasma cells 12. (a)areolar, (b)subcutaneous layer 13. (a)collagen, (b)regular dense, (c)tendons, (d)irregular dense 14. arteries 15. (a)stroma, (b)liver, spleen, lymph nodes 16. energy storage, cushioning internal organs 17. skeleton (endoskeleton) 18. (a)cartilage, (b)bone, (c)Chondrichthyes (sharks, rays, etc.) 19. (a)chondrocytes, (b)matrix, (c)lacunae 20. (a)matrix, (b)osteocytes, (c)vascularized 21. (a)hydroxyapatite crystals, (b)canaliculi 22. (a)osteon, (b)Haversian canal, (c)lamellae 23. plasma 24. (a)erythrocytes, (b)hemoglobin, (c)biconcave discs 25.leukocytes 26. bone marrow 27. fibers 28. (a)actin, (b)myosin, (c)myofibrils 29. (a)skeletal, (b)smooth, (c)cardiac, (d)smooth, cardiac, (e)skeletal 30. (a)neurons, (b)glial 31. synapses 32. nerve 33. (a)cell body, (b)dendrite(s), (c)axon(s) 34. integumentary, skeletal, muscle, nervous, circulatory, digestive, respiratory, urinary, endocrine, reproductive

Definitions: 1. multicellular 2. pseudostratified 3. fibroblast 4. intercellular 5. macrophage 6. chondrocyte 7. osteocyte 8. myofibril

Self Test: 1. a,c,d 2. a,b,c,e 3. d 4. a,c,e 5. b 6. d 7. a,c,d 8. b,e 9. b 10. c 11. a,c,d 12. c,e 13. a,c 14. c 15. a 16. b,c,d 17. d 18. a,b,e 19. a,c,e 20. c

Chapter 38

Concept Questions: 1. skeletal 2. epithelial 3. secretory 4. skin 5. skin 6. (a)epidermis, (b)strata 7. stratum basale 8. (a)keratin, (b)strength, flexibility, waterproofing 9. (a)connective tissue, (b)subcutaneous 10. muscle 11. exoskeleton 12. (a)hydrostatic skeleton, (b)longitudinally, (c)circularly, (d)outer, (e)inner 13. septa 14. (a)shell, (b)protection 15. (a)chitin, (b)molting 16. (a)endoskeletons, (b)chordates 17. echinoderms 18. (a)axial, (b)appendicular 19. (a)skull, (b)vertebral column, (c)rib cage, (d)centrum, (e)neural arch 20. (a)pectoral girdle, (b)pelvic girdle, (c)limbs, (d)opposable digit (thumb) 21. (a)periostium, (b)epiphyses, (c)diaphysis, (d)metaphysis, (e)epiphyseal line, (f)marrow cavity, (g)endosteum, (h)compact bone, (i)cancellous 22. (a)endochrondral, (b)membranous 23. (a)joints, (b)immovable joints, (c)slightly movable joints, (d)freely movable joints 24. (a)actin, (b)myosin 25. (a)smooth, (b)striated 26. fascicles 27. (a)actin and myosin, (b)sarcomere 28. inward 29. (a)acetylcholine, (b)action potential 30. (a)calcium, (b)actin, (c)myosin 31. shortening 32. (a)ATP, (b)creatine phosphate, (c)glycogen 33. (a)tendons, (b)antagonistically, (c)antagonist 34. (a)smooth, (b)cardiac, (c)skeletal, (d)slow-twitch, (e)fast-twitch

Definitions: 1. epidermis 2. epicuticle 3. periosteum 4. endochondral 5. intramembranous 6. osteoblast 7. osteoclast 8. myofibril 9. myofilament

Self Test: 1. c 2. a,b 3. e 4. c,d 5. a 6. d 7. d 8. c,e 9. a 10. a,b,d 11. b,c 12. a,c 13. b,e 14. b 15. b,e

Chapter 39

Concept Questions: 1. (a)endocrine, (b)nervous 2. stimuli 3. (a)reception, (b)afferent, (c)transmit, (d)interneurons, (e)integration, (f)efferent, (g)effectors 4. neuron 5. neuroglia 6. (a)dendrites, (b)cell body, (c)axon 7. (a)axon terminals, (b)synaptic knobs, (c)neurotransmitter 8. (a)Schwann cells, (b)cellular sheath, (c)myelin sheath, (b)myelin 9. (a)nerve, (b)tract (pathway) 10. (a)ganglia, (b)nuclei 11. (a)resting potential, (b)-70 mV 12. inner 13. (a)inside, (b)outside 14. (a)sodium-potassium pump, (b)ion-specific channels, (c)protein anions 15. (a)action potential, (b)sodium, (c)+35 mV 16. (a)voltage-activated ion pumps, (b)threshold level, (c)-55 mV 17. (a)spike, (b)voltage-activated sodium channels 18. wave of depolarization 19. repolarization 20. node of Ranvier 21. all-or-none 22. (a)synapse, (b)neuroglandular, (c)neuromuscular (motor end plates) 23. (a)presynaptic, (b)postsynaptic 24. (a)electrical synapses, (b)chemical synapses, (c)neurotransmitter molecules, (d)synaptic cleft 25. synaptic transmission 26. (a)receptors, (b)chemically activated ion channels 27. (a)excitatory postsynaptic potential (EPSP), (b)inhibitory postsynaptic potential (IPSP) 28. gradual potentials 29. (a)summation, (b)temporal summation, (c)spatial summation 30. (a)acetylcholine, (b)neuromuscular 31. (a)adrenergic neurons, (b)catecholamine 32. further apart 33. dendrites and cell body 34. convergence, divergence, facilitation

Definitions: 1. interneuron 2. neuroglia 3. neurilemma 4. neurotransmitter 5. multipolar 6. postsynaptic 7. presynaptic

Self Test: 1. b,c,e 2. b 3. b,c 4. c,d,e 5. c,d 6. b 7. a,d 8. a,c 9. c 10. b,c 11. d 12. c,e 13. a

Chapter 40

Concept Questions: 1. (a)nerve net, (b)cnidarians 2. nerve ring 3. (a)ladder-type, (b)cerebral ganglia 4. (a)ganglia, (b)ventral nerve cords 5. (a)central nervous system (CNS), (b)peripheral nervous system (PNS), (c)sympathetic and parasympathetic 6. (a)neural tube, (b)forebrain, midbrain, hindbrain 7. rhombencephalon 8. (a)cerebellum, (b)pons, (c)metencephalon, (d)medulla, (e)myelencephalon 9. brainstem 10. association 11. (a)superior colliculi, (b)inferior colliculi, (c)red nucleus 12. (a)prosencephalon, (b)thalamus, (c)cerebrum 13. (a)lateral ventricles, (b)third ventricle 14. (a)thalamus, (b)hypothalamus, (c)pituitary 15. corpus striatum 16. (a)hemispheres, (b)white matter, (c)cerebral cortex, (d)gyri, (e)sulci 17. (a)dura matter, arachnoid, pia matter, (b)choroid plexus 18. second lumbar 19. (a)white matter, (b)tracts 20. (a)ascending tracts, (b)descending tracts 21. (a)reflex action, (b)sensory, (c)association, (d)motor 22. (a)sensory, (b)motor, (c)association 23. (a)frontal lobes, (b)parietal, (c)central sulcus, (d)temporal, (e)occipital 24. brain stem and thalamus 25. cerebrum and diencephalon 26. (a)electroencephalogram (EEG), (b)delta, (c)alpha, (d)beta 27. (a)rapid eye movement (REM), (b)non-REM 28. (a)sensory, (b)short term, (c)long term 29. memory trace (engram) 30. cortex 31. (a)sense receptors, (b)sensory afferent neurons, (c)motor efferent neurons 32. (a)cranial nerves, (b)spinal nerves 33. (a)dorsal root, (b)ventral root 34. (a)sympathetic, (b)parasympathetic

Definitions: 1. hypothalamus 2. postganglionic 3. preganglionic 4. paravertebral 5. circumesophageal

Self Test: 1. a,c 2. d 3. a,c 4. a,b,d 5. a,c,d,e 6. a-e 7. b 8. b 9. b 10. e 11. b,d 12. d 13. b 14. a,d,e

Chapter 41

Concept Questions: 1. (a)receptor cells, (b)neuron 2. (a)touch, smell, taste, sight, hearing, (b)balance, pressure, pain, temperature, proprioception 3. (a)exteroceptors, (b)proprioceptors, (c)interoceptors 4. (a)mechanoreceptors, (b)electroreceptors, (c)thermoreceptors, (d)chemoreceptors 5. (a)electrical, (b)receptor 6. (a)graded, (b)depolarize, (c)action potential, (d)sensory 7. (a)greater, (b)receptor 8. more 9. brain 10. (a)number and identity, (b)frequency and total number 11. sensory adaptation 12. deformed 13. mechanoreceptors 14. (a)Pacinian corpuscles, (b)Meissner's corpuscles, Ruffini's end organs, Merkel's disks (c)pain 15. (a)statoliths, (b)hair cells 16. (a)canal, (b)receptor cells, (c)cupula 17. muscle spindles, Golgi tendon organs, joint receptors 18. (a)otoliths, (b)saccule and utricle 19. (a)angular acceleration, (b)endolymph, (c)crista 20. (a)cochlea, (b)mechanoreceptors 21. (a)tympanic membrane, (b)malleus, incus, stapes, (c)oval window 22. (a)basilar membrane, (b)organ of Corti, (c)cochlear nerve 23. (a)gustatation, (b)olfaction, (c)chemoreceptive 24. sweet, sour, salty, bitter 25. (a)olfactory epithelium, (b)50 26. pit vipers and boas 27. hypothalamus 28. rhodopsins 29. eyespots (ocelli) 30. (a)light, (b)vision, (c)lens 31. ommatidia 32. (a)sclera, (b)shape, (c)cornea 33. (a)lens, (b)retina, (c)rods, (d)rhodopsin 34. (a)cones, (b)fovea

Definitions: 1. proprioceptor 2. otolith 3. endolymph

Self Test: 1. c,d 2. b,d,e 3. b 4. b,d 5. a 6. c 7. c 8. d 9. c

Chapter 42

Concept Questions: 1. sponges, cnidarians, ctenophores, flatworms, nematodes   2. (a)circulatory, (b)diffusion   3. (a)blood, (b)heart, (c)vessels   4. (a)open circulatory, (b)hemocoel, (c)closed circulatory   5. gastrovascular cavity   6. (a)hemolymph, (b)interstitial   7. (a)arthropods, most mollusks (b)sinuses   8. (a)hemocyanin, (b)copper   9. (a)dorsal, (b)ventral, (c)five   10. plasma   11.(a)closed, (b)heart, (c)blood   12. (a)nutrients, oxygen, metabolic wastes, hormones, (b)maintenance of fluid balance, defense, (c)endotherms   13.(a)red blood cells, white blood cells, platelets, (b)plasma   14. (a)interstitial, (b)intracellular   15. (a)fibrinogen, (b)gamma globulins, (c)plasma proteins, (d)lipoproteins   16. (a)oxygen, (b)red bone marrow, (c)hemoglobin, (d)120   17. (a)neutrophils, (b)basophils and eosinophiles, (c)granular leukocytes   18. (a)lymphocytes, (b)monocytes, (c)macrophages   19. (a)thrombocytes, (b)hemostasis   20. (a)arteries, (b)veins   21. (a)tunica intima, (b)tunica media, (c)tunica adventitia   22. capillaries   23. (a)ventricles, (b)atria   24. (a)two, (b)sinus venosus, (c)conus arteriosus   25. (a)three, (b)sinus venosus, (c)conus arteriosus   26. (a)more, (b)higher, (c)endothermic   27. (a)pericardium, (b)pericardial cavity   28. (a)interventricular septum, (b)interatrial septum   29. (a)four, (b)atrioventricular, (c)tricuspid, (d)mitral valve, (e)semilunar valves   30. (a)sinoatrial (SA) node, (b)atrioventricular (AV) node, (c)atrioventricular (AV) bundle   31. (a)cardiac, (b)systole, (c)diastole   32. (a)lub, (b)AV valves, (c)dup, (d)semilunar valves, (e)diastole   33. electrocardiograph (ECG or EKG)   34. (a)cardiac output (CO), (b)heart rate, (c)stroke volume, (d)five liters/minute   35. (a)venous, (b)neural and hormonal   36. cardiac centers   37. (a)hypertension, (b)increase, (c)salt   38. diameter of arterioles   39. valves   40. (a)baroreceptors, (b)pressure changes, (c)vasomotor centers   41. (a)pulmonary circuit, (b)systemic circuit   42. (a)rich, (b)left   43. (a)aorta, (b)brain, (c)shoulder area, (d)legs   44. (a)superior vena cava, (b)inferior vena cava   45. (a)ischemic heart disease, (b)myocardial infarction   46. (a)interstitial fluid, (b)fat   47. (a)lymph, (b)interstitial fluids, (c)lymph nodes, (d)subclavian, (e)thoracic, (f)right lymphatic   48. edema

Definitions: 1. hemocoel   2. hemocyanin   3. erythrocyte   4. leukocyte   5. neutrophil   6. eosinophil   7. basophil   8. leukemia   9. vasoconstriction   10. vasodilatation   11. pericardium   12. semilunar   13. baroreceptor

Self Test: 1. a,b,d   2. c,d,e   3. a,c,e   4. a,c   5. b,d,e   6. b,c,d   7. c,e   8. a   9. b   10. b,c   11. a,d,e   12. a,b,c,e   13. c   14. c,e   15. d   16. c,e   17. d   18. d   19. a   20. d

Chapter 43

Concept Questions: 1. antigen   2. (a)nonspecific defense mechanisms, (b)specific defense mechanisms (immune responses)   3. immunology   4. antibodies   5. (a)phagocytosis, (b)inflammatory response   6. (a)skin (outer covering), (b)acid secretions, (c)mucous lining   7. (a)viral replication, (b)natural killer (NK) cells   8. histamine   9. redness, heat, edema, pain   10. (a)pyrogens, (b)hypothalamus, (c)interleukin-1 (IL-1)   11. (a)neutrophils and macrophages, (b)phagosome, (c)lysosomes   12. (a)stem, (b)bone marrow   13. (a)cytotoxic, (b)helper, (c)suppressor   14. antigen-presenting cells   15. thymosin   16. competent B cells   17. clonal selection   18. (a)macrophage, (b)antigen-MHC, (c)competent B cell, (d)helper T, (e)helper T, (f)competent B, (g)plasma cells   19. (a)immunoglobulins, (b)Ig, (c)antigenic determinants (epitopes), (d)binding sites   20. (a)IgG, IgM, IgA, IgD, IgE, (b)IgGs, (c)IgAs, (d)IgD, (e)IgE   21. opsonization   22. granules   23. lymphotoxins   24. (a)macrophage, (b)antigen-MHC, (c)competent T cell, (d)competent T cells, (e)cytotoxic T cells   25. (a)IgM, (b)IgG   26. immunization   27. passive immunity   28. theory of immune surveillance   29. blocking antibodies   30. (a)human leukocyte antigen (HLA), (b)T cells   31. (a)autograft, (b)allograft   32. (a)allergen, (b)IgE, (c)histamine   33. (a)acquired immune deficiency syndrome, (b)human immunodeficiency virus (HIV-1), (c)helper T

Definitions: 1. antibody 2. antihistamine 3. coelomocyte 4. phagosome
5. lymphocyte 6. leukocyte 7. monoclonal 8. autograft 9. autoimmune 10. lysozyme

Self Test: 1. a,c,d,e 2. b,c,d 3. b,c,e 4. a,b,c,e 5. c 6. a,d 7. c 8. a,e 9. a,c,d 10. b
11. a-d 12. b 13. b,c,e 14. b 15. a,d 16. a,c,d,e 17. a-e

Chapter 44

Concept Questions: 1. (a)respiration, (b)ventilation 2. (a)thin walls and large surface area, (b)moist, (c)blood vessels (or hemolymph) 3. body surface, tracheal tubes, gills, lungs 4. (a)surface-to-volume, (b)metabolic rate 5. (a)tracheal tubes, (b)spiracles, (c)tracheoles 6. (a)filaments, (b)countercurrent exchange system 7. (a)body surface, (b)body cavity, (c)book lungs 8. (a)density and viscosity, (b)more, (c)faster 9. (a)heme (iron-porphyrin), (b)globin, (c)hemocyanins, chlorocruorins, hemerythrins, (d)hemocyanins 10. (a)pharynx, (b)larynx, (c)trachea, (d)bronchi, (e)bronchioles, (f)alveoli 11. (a)epiglottis, (b)larynx 12. (a)three, (b)two 13.(a)inspiration, (b)expiration 14. (a)tidal volume, (b)500 ml, (c)vital capacity 15. (a)pressure (tension), (b)Dalton's law of partial pressure 16. (a)iron, (b)heme, (c)oxyhemoglobin (HbO_2) 17. (a)lower, (b)Bohr effect 18. (a)bicarbonate (HCO_3^-), (b)carbonic anhydrase 19. (a)medulla, (b)chemoreceptors 20. hypoxia 21. decompression sickness 22. chronic obstructive pulmonary diseases

Definitions: 1. hyperventilation 2. hypoxia

Self Test: 1. a,b,e 2. a 3. d 4. d 5. c 6. a,c 7. e 8. a,b,d 9. a,d 10. b,c,e 11. e
12. d 13. d 14. a,c,e 15. b,d,e 16. a-e 17. a,b,d,e 18. b,c,e 19. a,e 20. b,c,e

Chapter 45

Concept Questions: 1. heterotrophs (consumers) 2. (a)egestion, (b)elimination 3. herbivores 4. (a)carnivores, (b)shorter 5. omnivores 6. (a)mouth, (b)anus, (c)pharynx (throat), (d)esophagus, (e)small intestine 7. (a)submucosa, (b)mucosa, (c)adventitia, (d)muscularis 8. (a)incisors, (b)canines, (c)molars and premolars, (d)enamel, (e)dentin, (f)pulp cavity 9. amylase 10. (a)peristalsis, (b)bolus 11. (a)simple columnar epithelial, (b)perietal, (c)intrinsic factor, (d)chief, (e)pepsin 12. (a)duodenum, jejunum, ileum, (b)duodenum 13. (a)villi, (b)microvilli 14. (a)glycogen, (b)fatty acids and urea 15. (a)gallbladder, (b)emulsification 16. (a)proteases, (b)lipases, (c)amylases 17. (a)monosaccharides, (b)amino acids, (c)monoacylglycerols (fatty acids and glycerol) 18. (a)protein, (b)exopeptidases, (c)dipeptidases, (d)endopeptidases 19. (a)villi, (b)blood, (c)lymph 20. (a)cecum, (b)ascending colon, (c)transverse colon, (d)descending colon, (e)sigmoid colon, (f)rectum, (g)anus 21. (a)excretion, (b)elimination 22. cellular respiration 23. (a)Calories, (b)kilocalorie 24. fiber 25. (a)adipose, (b)glucose 26. (a)lipoproteins, (b)high-density lipoproteins (HDLs), (c)low-density lipoproteins (LDLs) 27. (a)liver, (b)deamination 28. (a)fat-soluble, (b)water-soluble, (c)B and C 29. sodium, chlorine, potassium, magnesium, calcium, sulfur, phosphorus 30. (a)increases, (b)decreases 31. (a)basal metabolic rate (BMR), (b)total metabolic rate

Definitions: 1. herbivore 2. carnivore 3. omnivore 4. anabolism 5. catabolism 6. submucosa 7. peritonitis 8. epiglottis 9. microvilli 10. chylomicron

Self Test: 1. c,d 2. a,b,c 3. a 4. b,e 5. d 6. b,c 7. a,e 8. d,e 9. b 10. b,e 11. a,b
12. a,b,c 13. c 14. b,c,e 15. a 16. a,b,e 17. a 18. a,c,d 19. d 20. e 21. a,b,d 22. c
23. d 24. b,c,d 25. a,e 26. c,d,e 27. a-e 28. c 29. a,d

Chapter 46

Concept Questions: 1. osmoregulation, homeostasis of body fluids 2. ammonia, urea, uric acid 3. (a)ammonia, (b)carbon dioxide 4. uric acid 5. (a)osmotic conformers, (b)osmotic regulators 6. (a)nephridial pores, (b)protonephridia, (c)metanephridia 7. (a)green glands, (b)crustaceans, (c)coelomic, (d)glandular 8. (a)hemocoel, (b)diffusion, active transport 9. filtration, reabsorption, secretion 10. skin, lungs or gills, digestive system 11. (a)ureters, (b)urethra 12. (a)Bowman's capsule, (b)renal tubule, (c)Bowman's capsule, (d)loop of Henle, (e)collecting duct 13. (a)filtration, (b)reabsorption 14. Bowman's capsule 15. renal tubules 16. urine 17. (a)cortical, (b)juxtamedullary 18. (a)descending, (b)ascending, (c)countercurrent mechanism 19. vasa recta 20. (a)antidiuretic hormone (ADH), (b)posterior pituitary, 21. hypothalamus 22. (a)adrenal cortex, (b)sodium

Definitions: 1. protonephridium 2. podocyte 3. juxtamedullary

Self Test: 1. a,c,e 2. b 3. a,b 4. a,d,e 5. c 6. c,d 7. b,d 8. d 9. c 10. e 11. c,e 12. c 13. a,c,d 14. b,e 15. b,d 16. a,d,e 17. e 18. b,c,d 19. a,c 20. c,d,e

Chapter 47

Concept Questions: 1. (a)glands and tissues, (b)nervous 2. target tissues (organs) 3. (a)neurohormones, (b)neurosecretory 4. steroids, proteins, derivatives of fatty acids or amino acids 5. (a)cholesterol, (b)lipid 6. (a)DNA protein, (b)mRNA 7. (a)cyclic AMP, (b)calcium ion 8. proteins (enzymes) 9. (a)cyclic AMP, (b)local hormones 10. double bonds 11. neurons 12. (a)BH or ecdysiotropin, (b)prothoracic, (c)ecdysone 13. releasing and release-inhibiting 14. (a)antidiuretic hormone (ADH), (b)oxytocin, (c)posterior pituitary 15. tropic 16. prolactin 17. (a)somatotropin, (b)anterior pituitary, (c)protein synthesis 18. (a)hypothalamus, (b)growth hormone-inhibiting hormone (GHIH or somatostatin), (c)growth hormone-releasing hormone (GHRH) 19. (a)gigantism, (b)acromegaly 20. (a)thyroxine, (b)iodine 21. (a)anterior pituitary, (b)thyroid-stimulating hormone (TSH) 22. cretinism 23. (a)bones, (b)kidney tubules, (c)thyroid, (d)calcitonin 24. (a)islets of Langerhans, (b)beta, (c)insulin, (d)alpha, (e)glucagon 25. (a)skeletal muscle and fat cells, (b)glycogenesis 26. (a)glycogenolysis, (b)gluconeogenesis 27. diabetes mellitus 28. epinephrine, norepinephrine 29. (a)androgens, mineralocorticoids, glucocorticoids, (b)androgen, (c)dehydroepiandrosterone (DHEA) 30. aldosterone 31. cortisol (hydrocortisone) 32. (a)corticotropin-releasing factor, (b) adrenocorticotropic hormone 33. thymosin 34. renin 35. atrial natriuretic factor 36. pineal

Definitions: 1. neurohormone 2. neurosecretory 3. hypersecretion 4. hyposecretion 5. hyperthyroidism 6. hypoglycemia 7. hyperglycemia 8. acromegaly

Self Test: 1. a,d 2. b,c 3. c,e 4. a,d 5. a,c,e 6. a,c,d 7. b 8. b 9. b,c,d 10. a,d,e

Chapter 48

Concept Questions: 1. budding 2. fragmentation 3. gamete 4. sexual and asexual 5. unfertilized egg 6. parasitic tapeworms 7. (a)scrotum, (b)seminiferous tubules, (c)interstitial cells, (d)testosterone 8. (a)spermatogonium, (b)primary spermatocyte 9. acrosome 10. inguinal canal 11. (a)epididymis, (b)vas deferens 12. (a)ejaculatory duct, (b)urethra 13. (a)400 million, (b)seminal vesicles and prostate gland 14. (a)shaft, (b)glans, (c)prepuce (foreskin), (d)circumcision 15. (a)cavernous bodies (corpora cavernosa), (b)blood 16. (a)FSH and LH, (b)testosterone 17. ovaries, uterine tubes, uterus, vagina, vulva, breasts 18. (a)oogenesis, (b)oogonia, (c)primary oocytes 19. (a)follicle, (b)FSH, (c)secondary oocyte 20. corpus luteum 21. uterine tube (Fallopian tube, oviduct) 22. (a)endometrium, (b)menstruation 23. cervix 24. labia minora, labia majora, clitoris, vaginal opening (introitus), mons pubis, (some also include hymen, vestibule)

25. (a)lactation, (b)prolactin, (c)anterior pituitary   26. gonadotropic and ovarian
27. ovulation   28. (a)estrogen (estradiol), (b)FSH and LH, (c)LH   29. estrogens and
progesterone   30. vasocongestion, increased muscle tension (myotonia)   31. excitement
phase, plateau phase, orgasm, resolution   32. sterility   33. progestin (synthetic
progesterone) and estrogen   34. sexually transmitted diseases (STDs)   35. (a)vasectomy,
(b)tubal ligation

Definitions: 1. parthenogenesis   2. spermatogenesis   3. acrosome   4. circumcision
5. oogenesis   6. endometrium   7. postovulatory   8. premenstrual   9. vasectomy
10. vasocongestion   11. intrauterine   12.contraception

Self Test: 1. a,e   2. c   3. c   4. a,b,c,e   5. a,c   6. a,d,   7. a,c   8. b,d   9. a–e   10. a,c,e
11. a,e   12. c   13. c,d,e   14. a   15. d   16. a,c,e   17. e   18. e   19. a,b,c   20. c   21. b   22. a,b
23. b,d   24. c,e

Chapter 49

Concept Questions: 1. (a)preformation, (b)epigenesis   2. (a)increases, (b)morphogenesis,
(c)cellular differentiation   3. (a)diploid, (b)sex   4. (a)zona pellucida, (b)acrosome,
(c)vitelline   5. bindin   6. (a)fertilization cone, (b)sperm   7. microtubules   8. (a)calcium,
(b)cortical reaction   9. cleavage   10. (a)number, (b)size, (c)same   11. (a)invertebrates and
simple chordates, (b)holoblastic, (c)radial, (d)spiral   12. (a)meroblastic, (b)blastodisc
13. (a)gastrulation, (b)germ layers, (c)endoderm, ectoderm, mesoderm   14. (a)endoderm,
(b)ectoderm, (c)mesoderm   15. archenteron   16. blastopore   17. primitive streak
18. (a)notochord, (b)neural groove, (c)neural tube, (d)central nervous system
19. (a)pharyngeal pouches, (b)branchial grooves   20. chorion, amnion, allantois, yolk sac
21. amnion   22. (a)allantois, (b)chorioallantoic membrane   23. embryonic chorion and
maternal   24. (a)umbilical cord, (b)yolk sac, allantois, umbilical arteries, umbilical vein
25. (a)zona pellucida, (b)morula, (c)blastocyst   26. (a)implantation, (b)seventh   27. (a)nine,
(b)not know   28. second and third   29. fetus   30. end of first trimester   31. (a)gestation
period, (b)280, (c)parturition   32. (a)cervix, (b)baby is delivered, (c)placenta is delivered
33. nutrition, vitamin and drug intake, cigarette smoking, exposure to disease-causing
organisms, etc.   34. (a)ultrasound, (b)amniocentesis   35. embryo, fetus, neonate, infant,
child, adolescent, young adult, middle age, old age

Definitions: 1. telolecithal   2. isolecithal   3. blastomere   4. blastocyst   5. blastopore
6. holoblastic   7. meroblastic   8. archenteron   9. trophoblast   10. trimester   11. neonate
12. teratogen

Self Test: 1. c   2. b,c,e   3. d   4. a–e   5. c   6. c,e   7. a   8. d   9. b   10. c,e   11. d   12. c
13. b,c   14. b,d

Chapter 50

Concept Questions: 1. signals (stimuli)   2. ethology   3. (a)diurnal, (b)nocturnal,
(c)crepuscular   4. biological rhythms   5. nervous, endocrine   6. instinct   7. sign stimulus
8. experience   9. (a)specific conditioned stimulus, (b)unconditioned   10. (a)positive
reinforcement, (b)negative reinforcement   11. behavioral bond   12. most important
13. (a)behavioral trigger, (b)orient and navigate   14. optimal foraging   15. foraging strategy
16. travel time, handling time, eating time   17. society   18. pheromones   19. testosterone
20. home range   21. courtship   22. cooperation, suppression of aggression, communication
23. mating, rearing young   24. Hymenoptera, termites   25. altruistic behavior

Definitions: 1. sociobiology

Self Test: 1. d   2. e   3. b   4. b   5. a,b   6. d   7. b   8. a   9. e   10. c

Chapter 51

Concept Questions: 1. (a)population, (b)community, (c)ecosystem   2. climate   3. permafrost   4. (a)rain forests, (b)deciduous forest   5. steppes   6. (a)Mediterranean, (b)sclerophyllous   7. allelopathy   8. (a)African, (b)South America and Australia   9. (a)80-180 inches (200-450 cm), (b)vegetation   10. light, minerals, oxygen   11. (a)plankton, (b)nekton, (c)benthos   12. (a)phytoplankton, (b)zooplankton   13. strength of the current   14. (a)littoral, (b)limnetic, (c)profundal, (d)littoral   15. (a)thermal stratification, (b)thermocline, (c)fall turnover, (d)spring turnover   16. (a)marshes, (b)swamps, (c)prairie potholes, (d)peat moss bogs   17. salt marshes   18. intertidal zone   19. (a)neritic province, (b)euphotic   20. (a)oceanic province, (b)abyss

Definitions: 1. phytoplankton   2. zooplankton   3. intertidal   4. thermocline   5. subtidal

Self Test: 1. a,d   2. e   3. d   4. b,e   5. b,d   6. b   7. e   8. b   9. b   10. c   11. d   12. a,d   13. b,d   14. a,c,e

Chapter 52

Concept Questions: 1. birth rate, death rate, sex ratios, age ranges   2. population density   3. (a),range, (b)random dispersion, (c)uniform dispersion, (d)clumped (aggregated) dispersion   4. arrival and departure rates   5. (a)change in number of individuals, (b)change in time, (c)natality, (d)mortality   6. (a)biotic potential (intrinsic rate of increase), (b)environmental resistance, (c)carrying capacity   7. (a)S-shaped, (b)J-shaped   8. high   9. disease   10. catastrophic events   11. (a)r selection, (b)K selection   12. zero population growth   13. (a)doubling time, (b)less   14. (a)replacement-level fertility, (b)lower, (c)higher   15. (a)total fertility rate, (b)higher   16. (a)preindustrial, (b)transitional, (c)industrial, (d)postindustrial

Definitions: no definitions in Chapter 52

Self Test: 1. g   2. h   3. d   4. f,i   5. b   6. b   7. b   8. a,e   9. a   10. d   11. c   12. a   13. b   14. b,c   15. a-e   16. a,d   17. c

Chapter 53

Concept Questions: 1. ecosystem   2. (a)autotrophs, (b)heterotrophs, (c)primary (or herbivores), (d)secondary (or carnivores), (e)omnivores, (f)saprophytes   3. detritivores (detritus feeders)   4. (a)facultative, (b)obligate   5. (a)fundamental, (b)realized   6. interspecific competition   7. law of tolerance   8. (a)geographical isolation, (b)edge effect   9. climax community   10. (a)primary succession, (b)pioneer, (c)mosses, (d)shrubs, (e)trees   11. secondary succession   12. facilitation   13. inhibition   14. tolerance   15. (a)organismic, (b)individualistic

Definitions: 1. carnivore   2. omnivore   3. autotroph   4. heterotroph   5. herbivore

Self Test: 1. e   2. d   3. e   4. a   5. b   6. a,d,e   7. a,c,e   8. a

Chapter 54

Concept Questions: 1. (a)ecosystem, (b)ecosphere   2. (a)biogeochemical cycles, (b)carbon, nitrogen, phosphorus, water   3. (a)photosynthesis, (b)respiration   4. (a)nitrogen fixation, (b)nitrification, (c)assimilation, (d)ammonification, (e)denitrification   5. (a)rocks, (b)soil   6. (a)runoff, (b)groundwater   7. (a)radiant, (b)chemical, (c)heat   8. (a)trophic level, (b)primary producers (autotrophs), (c)primary consumers (herbivores), (d)secondary consumers (carnivores)   9. food web   10. (a)numbers, (b)biomass, (c)energy   11. (a)gross primary productivity, (b)net primary productivity, (c)turnover   12. climate

13. (a)insolation, (b)distance between Earth and sun, (c)inclination   14. (a)equatorial, (b)poles   15. (a)high, (b)low, (c)Coriolis Effect   16. (a)currents, (b)gyres   17. latitude and inclination   18. (a)precipitation, (b)deserts   19. microclimate   20. (a)soil profile, (b)soil horizons, (c)O-horizon, (d)A-horizon (topsoil), (e)B-horizon, (f)C-horizon

Definitions: 1. biosphere   2. atmosphere   3. hydrosphere   4. lithosphere 5. microclimate

Self Test: 1. c   2. b,d,e   3. c,e   4. a-e   5. b   6. a,d   7. a,c   8. c   9. a   10. c   11. b   12. a,c 13. d   14. a   15. a,d,e

Chapter 55

Concept Questions: 1. (a)endangered, (b)threatened   2. habitat destruction   3. (a)nature (the wild), (b)human controlled settings   4. (a)Amazon, (b)Congo River basin, (c)slash-and-burn   5. (a)commercial logging, (b)cattle rangeland   6. (a)CO_2, CH_4, O_3, N_2O, chlorofluorocarbons (CFCs) (b)CO_2, O_3, N_2O (c)CH_4, (d)CFCs   7. greenhouse effect 8. (a)2-5, (b)50   9. ultraviolet radiation   10. (a)chlorine, fluorine, carbon (b)chlorine

Definitions: no definitions in Chapter 55